D0525468

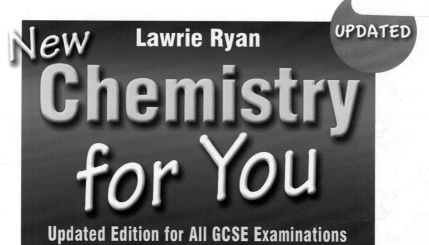

UPDATED

Lawrie Ryan

New Chemistry for You

Updated Edition for All GCSE Examinations

Nelson Thornes

Text © Lawrie Ryan 2006, 2011
Original illustrations © Nelson Thornes Ltd and Lawrie Ryan 2011

The right of Lawrie Ryan to be identified as author of this work has been asserted by him in accordance with the Copyright, Designs and Patents Act 1988.

All rights reserved. No part of this publication may be reproduced or transmitted in any form or by any means, electronic or mechanical, including photocopy, recording or any information storage and retrieval system, without permission in writing from the publisher or under licence from the Copyright Licensing Agency Limited, of Saffron House, 6-10 Kirby Street, London EC1N 8TS.

Any person who commits any unauthorised act in relation to this publication may be liable to criminal prosecution and civil claims for damages.

Published in 2011 by:
Nelson Thornes
Delta Place
27 Bath Road
Cheltenham GL53 7TH
United Kingdom

11 12 13 14 15 / 10 9 8 7 6 5 4 3 2

A catalogue record of this book is available from the British Library.

ISBN 978 1 4085 0921 0

Illustrations by Barking Dog, Peters & Zabransky, Jane Cope, IFA Design Ltd, Jordan Publishing Design, Harry Venning and Tony Wilkins
Page make-up by Tech-Set Ltd, Tyne & Wear with revisions by Fakenham Photosetting Ltd
Printed in China by 1010 International Ltd

Website:
The website at **www.chemistryforyou.co.uk** gives you details of exactly which pages in this book you need to study for your particular GCSE examination course.

Make sure you visit this website and save or print out the correct sections.
They will show you:
- which topics you need to learn for your particular examination, and
- which page numbers to read in this book.

'What is the use of a book,' thought Alice, 'without pictures or conversations?'
Lewis Caroll, *Alice in Wonderland*

'I think Mr C.S. Lewis is a very good writer. But he has one failing. There are no funny bits in his books.'
'You are right there,' Miss Honey said.
Roald Dahl, *Matilda*

Chemistry … is one of the broadest branches of science, if for no other reason than, when we think about it, everything is chemistry.
Luciano Caglioti, *The Two Faces of Chemistry*

Indulge your passion for science … but let your science be human, and such as may have a direct reference to action and society.
David Hume, *An Inquiry Concerning Human Understanding*

Introduction

Chemistry for You is designed to introduce you to the basic ideas of Chemistry. It will show you how these ideas help to explain the materials in our world, and how they can be changed. From your trainers to the space shuttle, chemists work to develop new materials.

Chemistry for You aims to be interesting and help you pass your exams, whether you are using it to study Core Science, Additional Science or Chemistry at GCSE. This Revised Edition covers all of the latest specifications from the Examination Boards. You can find more information at www.chemistryforyou.co.uk

The book is carefully laid out so that each new idea is introduced and developed on a single page or on two facing pages.

Throughout the book there are many experiments for you to do. A hazard sign means your teacher should give you further advice (for example to wear eye protection). Plenty of guidance is given on the results of these experiments, in case you don't actually do a particular practical or are studying at home. You will find the ideas introduced in Chapter 1 'How science works' useful throughout your GCSE. Refer back to it whenever you tackle an investigation.

Each new chemical word is printed in **bold** and important points are placed in a box. There is a summary of important ideas at the end of each chapter.

There are questions at the end of each chapter. They always start with an easy fill-in-the-missing-words question useful for revision and writing notes. At the end of each of the 7 main sections you will find plenty of further questions. These are taken from recent GCSE papers or have been written for the latest specifications by examiners.

There is an Extra Section at the end of the book. You will find advice on Key Skills, coursework, careers, revision and examination techniques in this section.

Throughout the book, cartoons and rhymes are used to help explain ideas. The 'Chemistry At Work' pages will also show you how Chemistry is useful to us in everyday life. You are encouraged to think about and discuss scientific issues that affect us all, both now and in the future.

However, above all I hope that you will find Chemistry easier to understand and that using the book will be fun!

I would like to thank my family (Judy, Nina and Alex) for their amazing support and understanding throughout this latest re-editioning of Chemistry for You.

Lawrie Ryan

Contents

Structure and Bonding

Earth Chemistry

Synthesis and Analysis

Extra sections

**Chemists make new materials that change
the world in many ways ...**

*From everyday materials, such as
cosmetics*

*... to the materials used in
the space shuttle*

chapter 1

How Science Works ⚖ EVIDENCE

▶ Why we need to know about how science works

Science has helped to shape the world we live in today.

Scientists have solved problems, using evidence, and have made modern technologies possible. This has brought great benefits for all of us but has also created problems. Pollution of our environment is one major concern – and scientists are also essential in dealing with this issue.

We all have to make decisions about how we lead our lives. Knowing more about 'how science works' will help you form your own opinions about important issues. These will include decisions about your diet, your transport, your work and your free time.

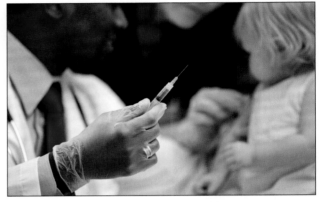

Some people have claimed that the MMR jab can cause autism in some babies. However, the evidence used to identify a link was only taken from a small sample of cases.

You will need to question evidence and to weigh up conflicting ideas. For example :
- Will you allow your children to have the MMR jab ?
- Will you buy a car ? If so, which one will be least harmful to our environment ?
- Will the latest food scare in the news affect what you eat ?
- Will you vote for a party who want to expand nuclear power ?

To answer such questions you should consider the evidence available.
Then you need to judge if the evidence is based on 'good science'. (You will find some people use evidence that is not scientific at all to answer scientific questions.)

> **Good scientists make sure their data provides evidence that is reliable and valid.**

Reliable evidence will be repeatable. If someone else collected the data again it would be the same (or similar). So you can *trust* reliable data. It is then valid evidence only if it also actually measures what you intended to find out about.
So scientists need reliable and valid evidence to draw firm conclusions.

> There are some questions that science really helps us to answer. These are usually questions that start 'How can we . . .?'
> But there are other questions based on beliefs and opinions that it can't answer. Society as a whole, which includes scientists, has to try and answer the 'Should we . . .?' type questions.

Example

You might want to find out about pollution in a river from a factory. You would need instruments that were sensitive enough to detect small quantities accurately to get reliable evidence. Then measuring the pollution downstream from the factory will only give you valid evidence if you also have data from the river before it passed the factory.

▶ Observing

Observing things is the link between the real world and science.
But observation is not as simple as you might think.
Surely two people looking at the same event will see the same thing, won't they? Sports fans will know that's not true!

But what about something like a freshly painted iron gate?
One person sees a nice colour that matches a house.
Another sees an iron barrier protected against rusting.
So what you see depends on what you know already.

> **You can use existing theories to explain ideas.**

This forms a **hypothesis**. You can then use your hypothesis to make a **prediction**. You can go on to test your prediction by collecting **evidence**. This may or may not support your prediction.

If an existing theory is not backed up by observations from your tests, first of all check the observations. If they prove to be correct, then the theory itself might need to be changed. This is one way in which science develops over time.

Observations are sometimes biased!

> **You need to observe things carefully to see if particular varibles are important, or not, in an investigation.**

Example

You are investigating how quickly different carbonates react with dilute acid.

On observing the reactions of carbonates in acid, you might decide that the type of test tube you use does not matter. It could be made of thick or thin glass.

However, you might see that the surface area (lumps or powder) of the carbonate *does* make a difference. So you will have to use the same surface area of carbonate, e.g. powder, in each test – but you can use different test tubes to do the tests in. Any slight differences this causes will probably be too small to affect your timings significantly.

Observations can also be the starting point for new investigations, experiments or surveys.

► Making measurements

Scientists must also think about the measurements they make when judging how strong their evidence is.

Are they reliable (are they repeatable) and valid (do they measure what you intended)?

There will usually be slight differences in measurements when you repeat them. This is because it is hard to measure the value of a variable under *exactly* the same conditions every time.

Example

You measure the volume of a gas given off in a reaction using a gas syringe.

You repeat the experiment but the temperature of the room might have changed. This will affect the volume of gas collected.

• If the temperature gets higher, how will this affect your volume reading?

dilute hydrochloric acid

syringe

marble chips

You should also consider the **accuracy** of your measuring instrument:

> **Accuracy tells us how near the *true reading* your measurement is.**

Example

An expensive pH meter will probably give readings nearer the true value than a cheaper model. The readings would also depend on you following the manufacturer's instructions on using and storing the pH probes. If you don't follow the instructions, the readings won't be reliable and valid.

If the pH of a solution is tested in different labs by different people, and they all get similar results, we call the pH *reproducible.*

The more expensive pH meter is also likely to have better **resolution**. It will respond to smaller changes in pH.

This may not matter in some experiments. But it could be significant, for example, in enzyme investigations.

The **precision** of a measuring instrument is also important:

> **A precise instrument will have small scale divisions. It will give the same reading again and again under exactly the same conditions.**

But precision and accuracy are not the same thing:

Example

A student weighs *exactly* 2.51 g of iron on two different balances. He repeats this 4 times on each balance.

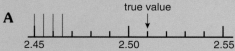

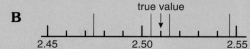

Balance A gives a precise set of repeat readings (the readings are grouped closely together) but they are not accurate.

Balance B produces an accurate mean value (the average of the 4 readings is 2.51 g) but the set of readings are not precise (the readings are spread out).

Errors

Then we also have **human error** in measurements.

Example

In the first cartoon the student is just careless. If he repeats his experiment, the results could be spread randomly, all over the place.

This is called **random error**.

In the second cartoon, his partner is more careful. However, she is not reading the volume in the burette correctly.

You should read to the bottom of the **meniscus** (curve in the surface of the solution). So her repeat readings will be consistently incorrect.

We call this **systematic error**.

- Will all her readings be too high or too low?
- Why won't it matter in this case?

In some investigations you should **repeat** your measurements.
This is especially important when measurements are tricky to make.
The number of repeats will depend on the accuracy you need.

Anomalous results

You should also look out for **anomalous** results.
These measurements do not match the general pattern found, or lie well outside the range of other repeat measurements.
If you find an anomalous measurement, try to work out why it happened.
If it was caused by poor measurement in the first place, you can just discard it. (See first cartoon above.)

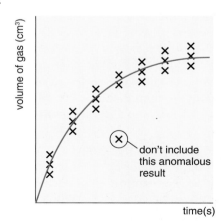

*Graph showing all repeat measurements plotted and a **line of best fit***

▶ Designing an investigation

The best way to answer some scientific questions is to design a fair test.

Example

You are looking at the reaction between metal carbonates and dilute hydrochloric acid. The reactions all produce carbon dioxide gas.

You might want to investigate this question:
'What might affect how quickly the reaction takes place?'
You can start by making a list of the variables that might affect the rate of the chemical reaction, e.g.:

- temperature,
- concentration of acid,
- surface area of carbonate,
- type of carbonate,
- type of acid.

You can select any one of these variables to investigate. Imagine you choose to vary *the concentration of the acid* you use.
This will be your **independent variable**.

But how will you judge how quickly the reaction takes place?
There are different ways to judge this.
Imagine you decide to measure *the time it takes to give off a certain volume of carbon dioxide gas*.
The time it takes is called the **dependent variable**.

To make this a **fair test,** we should only vary one thing at a time (in this case it will be the concentration of the acid).

So we should try to keep all the other variables that might matter constant. These are the **control variables**. They will include temperature, surface area and the type of acid and carbonate used.

Choosing values of variables

You would have to carry out a few **trial experiments** to decide on how much carbon dioxide to collect each time. These might show that for reasonable amounts of acid and carbonate, 20 cm³ would give sensible times – not too long and not too short.

You could also choose your **range** of concentrations to use. You could systematically dilute down the acid provided using water, whilst keeping the total volume the same.

Five different concentrations would be needed to draw a decent line graph of the results. These should be well spaced out. If there are any interesting parts on your graph, you can choose to test more concentrations. These might be more repeats to test possible anomalous results. Or they might be grouped around an area where the graph looks like it changes, e.g. where the slope gets much steeper or levels off.

It's important to get your range right! Always quote your maximum and minimum values when asked for the range of your data (and don't forget the unit).

▶ Different types of variable

We can classify variables into the following types:

- **Categoric variables**: These are variables that we can describe using words.

  Examples are: 'type of acid' e.g. hydrochloric acid, or 'type of carbonate' e.g. calcium carbonate.

- **Ordered variables**: These are a type of categoric variable that we can put into an order.

  An example is 'surface area' of calcium carbonate, if you use 'large lumps / medium lumps / small lumps' as a way of describing the values of the variable.

- **Continuous variables**: These are variables that we can describe by any number, as their values are measurements e.g. $17.5\,cm^3$. Examples are: 'temperature' or 'concentration'. This type is the 'most powerful' to investigate. That's because the relationship between the variables can be described by a mathematical equation. You can use the equation or the line graph produced (see page 12) to make the most accurate predictions to test.

- **Discrete variables**: These are variables that can only have whole number values.

  An example is: 'the number of marble chips'.

'Not-so-fair' tests!

In some investigations it is easy to change one variable at a time, and keep all the others constant, to make a fair test.

However, in other investigations controlling all the variables that matter is difficult (or even impossible). An example is an investigation into the effects of fertiliser on plant growth.

If this investigation takes place **'in the field'**, there are factors such as the weather conditions that you can't control. But you should make sure all the plants tested experience the same weather.

You would also use a large sample of the same types of plant to test. It's no good picking just one plant to test with the fertiliser and one without it. There could be differences between plants which might affect the result.

Using a large **sample size** and choosing your sample carefully will help to overcome the problems of setting up a fair test involving living things.

Evaluating your data

Remember that the data you collect must be **reliable** (see page 6). You should also design your investigation so that the data will answer the question posed. Consider the investigation on the previous page:

Is it **valid** to judge the rate by timing how long it takes to collect $20\,cm^3$ of gas? After all, as soon as the reaction starts the concentration of acid will start to decrease!

▶ Presenting your data

Tables

As you carry out an investigation, you can record your data in a results table. Scientists arrange their tables so that the independent variable (the one you change deliberately, step by step) goes in the first column.

The dependent variable (used to judge the effect of varying the independent variable) goes in the second column.
For example, if you were doing an investigation like the one on page 10, but varied temperature, the table would have these headings:

Independent variable: Temperature (°C)	Dependent variable: Time to collect 20 cm³ of gas (s)

If you decide to repeat your tests at each temperature to get more reliable data, you can split the second column up:

Temperature (°C)	Time to collect 20 cm³ of gas (s)			
	First test	Second test	Third test	Mean (average)

Graphs

Then you can show the relationship (link) between the two variables by drawing a graph.

The independent variable goes along the horizontal axis.
The dependent variable goes up the vertical axis.
Look at the axes opposite:

The graph you draw depends on the type of independent variable you investigated. (See page 11.)

If it is a continuous variable, you can display your results on a **line graph**.
That's because the independent variable is measured in numbers.
You might choose to do the tests at 20, 30, 40, 50 and 60 °C.
However, you could equally have chosen any points in between.
That's why we can join the points with a continuous line.

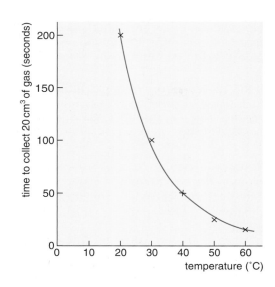

Look at the graph opposite:
In this case, the independent variable is 'Type of metal'.

If your independent variable is a categoric variable – you have to display your results on a **bar chart**.
That's because there are no values between each bar.
(Remember that categoric variables are usually described by words.)

Of course, the dependent variable (up the side) has to be continuous (measured), otherwise we can't draw any type of graph!

Spotting patterns in data

Once you have your graph to refer to, you can see what the relationship is between the two variables.

Examples you will find in chemistry include:

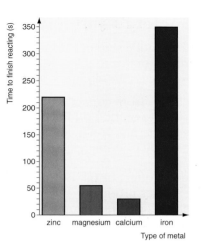

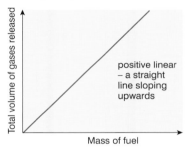

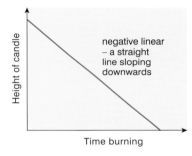

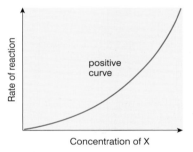

In this case, the variables are directly proportional because the straight line passes through the origin (0, 0) on the graph.

The relationship is only shown between the range of values in your investigation. The link may or may not hold true outside this range. This will be a **limitation** of your investigation.
You can extend the line on your graph to make predictions, but you cannot be sure until you test them by experiment.

▶ Evaluating your whole investigation

You should consider the reliability and validity of the measurements you make and the method you used.
You can also improve reliability and validity by:

- Looking up data from secondary sources, e.g. on the Internet.

- Checking your results by using an alternative method. For example, in the investigation on page 10, you might measure the rate by monitoring the loss in mass as the reaction progressed. You could then plot initial rate against different concentrations.

- Seeing if other people following your method get the same results, i.e. are your results reproducible?

Try to use lines of best fit to give 'average' lines. You can also plot all your data to show the range within sets of repeat measurements. This indicates the precision and reliability of your data. The larger the spread of repeat readings the less precise your data is. Also the less reliable as other people are less likely to get exactly the same results as you.

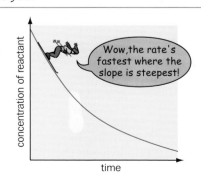

▶ Society's influence on evidence

Some scientific studies are of interest to society in general – not just to the scientific community. For example, investigations into:

- pollution from cars and industry,
- the safety of food additives,
- the properties of new materials,
- the purity of drinking water.

Bottled water contains

New wonder drug can cure cancer

Pollution will make life intolerable by 2020

These will affect us all. But how do we hear about the evidence these studies produce? Often it is through the media – TV, radio, the Internet and newspapers.
So how reliable are the reports we get? Can we trust them?

Tabloid newspapers are likely to 'play up' scare stories. Sensational headlines will sell their newspapers. So often scientists will find their research reported badly in the media.
Dangers can be exaggerated. The very small chances of harmful effects may not be mentioned.

On the other hand, successful breakthroughs on a small scale may be reported as findings that will change the world – now!
These make great headlines but are often unrealistic.

There are lots of factors that influence the way evidence is presented to us. If it is going to be unpopular with the public, evidence might be 'watered down' or hidden in publications.
This might also happen if the research results show a problem that will be expensive to put right.

The reputation of the scientist doing the research can also influence people's response to research findings. Famous, eminent scientists are likely to get their ideas listened to more favourably than less well known scientists.

When we look at evidence for a certain viewpoint we should also consider any bias. For example, do we know who paid for the research? If a food company produce data that a colouring is safe, should we trust that as much as data from a government 'watch-dog'? Or what about an independent consumer group? Are they unbiased?

These protestors believe that genetically modified crops are a danger to the environment

The ideas we have looked at in this chapter should help you to consider evidence critically. You might want to know about the sensitivity and accuracy of instruments used in tests.
The sample size might be important. How scientists chose sites to test and how many times they repeated their tests could also affect the trust you have in their evidence.

▷ Chemistry at work : Accepting new ideas

In 1989 two electrochemists claimed to have made a discovery that would solve the world's energy crisis! Their findings suggested that they had managed to reproduce the nuclear reactions that produce energy on the Sun. The two scientists, Pons and Fleischmann of Utah University in the USA, said that they had achieved 'cold fusion' in their lab.

Pons and Fleischmann held a press conference to tell the world about their great discovery. However, doubts soon began to surface about their claims. The press conference had been set up before their findings had been published in any scientific journal. They hadn't even presented their work to colleagues at a conference. This is not normal practice in the world of science.

Physicists have built machines that can generate temperatures of over 50 000 000 °C to make nuclear fusion happen on Earth, but they still can't produce energy economically. So imagine their surprise when a couple of chemists said that they had managed to achieve fusion using only simple lab apparatus.

Scientists usually submit new findings to a journal to share with other interested scientists. But before they get published, fellow scientists check to make sure that proper scientific methods have been used. So this had not been done.

Once the details were out, scientists all over the world tried to reproduce their results. Fleischmann and Pons claimed to have got a lot more energy out from their experiment than they put in the first place. But other scientists failed to get the same effect when they set up the experiment. They showed that the original evidence was **unreliable**. It could not be reproduced, so it could not be trusted.

Naturally, other scientists were sceptical. They put down the original results to experimental error. So cold fusion was never accepted into the main body of scientific knowledge.

However, a few people are still convinced in the technology that could revolutionise the world. Of course, the rewards for inventing a clean, cheap way to produce energy would be incredible. So money is still invested in funding further research. Conferences are held and favourable findings are shared by the faithful.

And so the search for the elusive 'fusion in a test tube' goes on. One day it could make somebody rich beyond their wildest dreams. It could also save our environment from the pollution caused by most of our current sources of energy.

But the lucky person who solves the problem must make sure their data is reliable. Fellow scientists will then need to check their findings before 'going public' and publishing results.

Summary

- People need to understand 'how science works' to get involved in scientific issues that will affect their lives.
- Science can help us with 'Can we . . . ?' type questions but not with 'Should we . . . ?' type questions.
- Scientists can test the links between variables. Sometimes they will carry out fair tests. To set these up, you change the variable under investigation (the **independent variable**) whilst keeping other variables constant (the **control variables**). To judge the effect of varying the independent variable, you measure the **dependent variable**.
- Scientists should make sure that their evidence is reliable (i.e. you can trust it) and it is valid (i.e. it actually answers the question it is supposed to).
- Science is also influenced by a variety of social factors, e.g. funding, vested interests, reputation of researchers, political pressure.

▶ Questions

1. Copy and complete:
   a) The a . . . . of a measurement tells us how near the true value it is.
   b) The s . . . . of an instrument tells us its ability to measure small changes in a variable.
   c) We can improve the r . . . . of data by checking against other people's data.
   d) We can describe our data as p . . . . if repeat readings are grouped very closely together.
   e) The . . . . variable is the one we choose to investigate and vary in an investigation.
   f) In an investigation we judge the effect of varying the variable in part d) by measuring the . . . . variable.
   g) In a fair test we must keep all the . . . . variables constant.
   h) If the design of an investigation allows you to answer the original question, we can describe the data collected as v . . . . .

2. A class was given four solutions to test for acidity.
   Pete decided to use universal indictor paper to dip in each solution. Then he matched the colour of the indicator paper to the colours on the pH scale.
   Dean used a pH sensor and data logger to measure the pH of the solutions.
   Here are their results:

Solution	Pete's pH values	Dean's pH values
A	3	3.6
B	6	5.3
C	2	1.9
D	4	3.5

   a) i) Whose results are likely to be the more accurate?
       ii) Explain your answer.
   b) Did Pete and Dean get the same order of acidity? Write down each student's order.
   c) Sara used a different pH sensor to measure the pH of the same solutions.
      Her results for A to D respectively were 3.8, 5.5, 2.1 and 3.7. Like Dean's, these numbers were read off the computer screen.
      Are the differences between Dean's and Sarah's results likely to be through random error? Explain your answer. What might have caused the differences?

3. A student heated samples of metal carbonates to see which decomposed most easily.

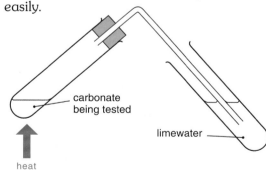

   carbonate being tested

   limewater

   heat

   She used equal masses of magnesium carbonate powder, a lump of calcium carbonate, zinc carbonate powder and copper carbonate powder.
   When the carbonate decomposed it gave off carbon dioxide gas.

a) When she heated copper carbonate it turned from a pale green colour to black. Magnesium carbonate also decomposed, but there was no colour change in the reaction.

How did the student know there had been a reaction?

(*HINT*: Look at the diagram at the start of the question.)

b) i) How could she judge how quickly each carbonate decomposed?
    ii) What measuring equipment would she need to make her judgements?
    iii) Do you think her judgements will be precise? Explain your answer.
    iv) What could the student do to improve the reliability of her data collected?

c) i) Give one way mentioned in the question in which the student tried to carry out a fair test.
    ii) Give one variable mentioned in the question that the student failed to control. Say how this would affect her results.

d) i) Name the independent variable investigated.
    ii) Is this a categoric, ordered, discrete or continuous variable?

e) i) Name the dependent variable in her investigation.
    ii) Is this a categoric, ordered, discrete or continuous variable?

f) What type of graph should the student use to display her results?

4. The residents of a village on the banks of a river are worried about pollution from a new factory upstream.
Mercury, a toxic heavy metal, is involved in the manufacturing process in the factory.
Some people in the village enjoy fishing. They believe that there are less fish to catch since the factory opened a month ago.
The managers in the factory promised to carry out some tests.

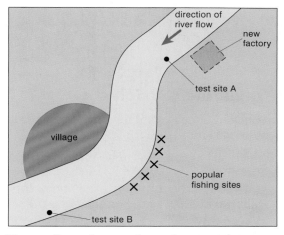

They collected water from the sites shown above. Then they analysed the water in the lab.
Two months later they reassure the residents that mercury levels are perfectly safe.

> TECHNICAL INFORMATION
> Mercury levels found in a water sample taken at 8.00 a.m. on 5th May at each site:
> Site A = 0.000 004 g / litre
> Site B = 0.000 002 g / litre
> Accuracy of method used to analyse water samples + or − 50%.
> Safe level of mercury in drinking water is 0.000 002 g / litre

a) Some people in the village want the County Council or Environmental Agency to carry out their own tests for mercury. Why?
b) When questioned, the factory owners said that the machine used to analyse the water sample was 20 years old. Why did this worry the concerned villagers?
c) Give two ways in which the factory could have made sure the data collected was more reliable and valid?
d) i) How could the factory argue that levels of mercury were safe?
    ii) How could the villagers argue that they were not convinced even if the data was accurate?

Further questions on page 63.

chapter 2

Basic ideas

▶ Invisible particles

The ancient Greeks were the first to suggest that everything is made up from particles.

Scientists still believe this now. Yet these particles are too small to be seen.
Even if you use the most powerful microscopes, you still can't really 'see' them. So why are we so sure that they exist?

Nowadays, scientists have developed special probes that can measure very small forces.
These probes can be used to detect particles, and even position them where you want. (See page 336.)

Around 400BC Democritus put forward his ideas about particles – but most people didn't believe him

Diffusion

If somebody spills some perfume in your classroom, you soon know about it. The smell quickly spreads through the room. However, you don't **see** any perfume in the air. Yet your nose tells you that it really is there.

The perfume gives off invisible particles of itself. These particles mix with the gas particles in the air. When particles mix like this it is called **diffusion.**

Diffusion happens by itself. You don't need to mix or stir the substances. The particles move from an area of **high** concentration to an area of **low** concentration. You can see for yourself in the next experiment:

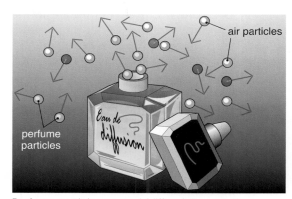

Perfume particles spread (diffuse) through the air particles

Experiment 2.1   Diffusion through a liquid

Use tweezers to pick up a few potassium manganate(VII) crystals.
Gently place them at the bottom of a beaker of water.
Observe the beaker for a few minutes.
● What do you see happen?

Leave your beaker until next lesson.
Draw a diagram to show your results.
● Explain what you think happens to the purple particles.
  Use the words **particles** and **diffuse** in your answer.

Investigation 2.2   Does temperature affect the rate of diffusion?

Plan an investigation to see how temperature affects the rate of diffusion.
Let your teacher check your plan before you start any practical work.

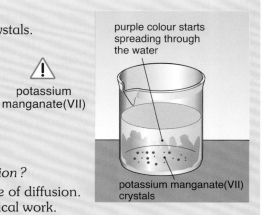

potassium manganate(VII)

purple colour starts spreading through the water

potassium manganate(VII) crystals

▶ Particles in solids, liquids and gases

Our model of solids, liquids and gases describes:
- the distance between particles, and
- the movement of the particles.

Look at the diagrams and poems below that describe the three **states of matter**:

Solid

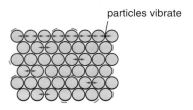

particles vibrate

Solid

Squashed together, the particles are tight,
Some think they're still – but that's not right.
Even though you can't see them shaking,
Believe it or not – they are vibrating !

Liquid

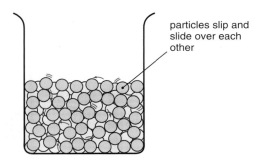

particles slip and slide over each other

Liquid

Particles mingling and inter-twined,
There's not much room in here to find !
The forces of attraction are still quite strong,
But particles are free to move slowly along.

Gas

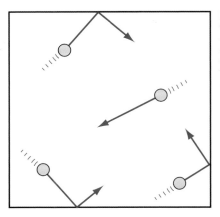

particles move very quickly in all directions; as the particles bash against the walls of the container, they exert a force that causes pressure

Gas

Particles whizzing everywhere,
Moving randomly, they just don't care !
Speeding left and speeding right,
With space in between, they're amazingly light.

▶ Atoms

Scientists believe that everything is made up from tiny particles.

But what are these particles like?

This is a difficult question because nobody has *ever* seen one. They are too small.
So scientists have had to think up ideas which explain the things they *can* see.

Around 1805 John Dalton put forward his ideas.
He thought that the smallest particles were like tiny, hard snooker balls. They could not be split.
He called these **atoms**.

Dalton suggested that there were just a few dozen different types of atom. He made up a list from substances which could not be broken down at that time.

He explained the millions of different substances on Earth quite simply. The atoms could join together in different combinations to make new substances.

Many of Dalton's ideas are still useful today.
Can you think of any things he got wrong?

There are 92 different types of atom found naturally on Earth. Your body is made up from just 26 of these!

John Dalton (1766–1844) was born in Cumbria. For most of his life he taught in Manchester. His ideas were based on experiments, unlike the Greek philosophers who first thought up the idea of particles.

All these words in this dictionary... from just 26 different letters!

All those substances in this person... from just 26 types of atom!

Symbols

Each atom has its own name and chemical symbol.
Most symbols are taken from the English names.
Some are based on their Latin names.

Look at the table opposite:
● Which atoms have symbols based on their Latin names?
● Why is helium's symbol He, and not H?

Notice that you always use a capital letter for the first letter of a symbol.
If the symbol has a second letter, it is always a small letter. Examples include Cu, Fe and Ne.

Atom	Symbol
Hydrogen	H
Helium	He
Nitrogen	N
Neon	Ne
Oxygen	O
Chlorine	Cl
Iron	Fe
Copper	Cu
Lead	Pb

Can you see where the word 'plumber' came from? Water pipes used to be made of lead. Why don't we use it now?

▶ Molecules

Atoms are nature's building blocks.
When atoms are joined together by
chemical bonds they make **molecules**.

A hydrogen molecule

> **Molecules are groups of 2 or
> more atoms bonded together.**

Atoms are nature's building blocks

Chemical formula

Chemists use a kind of short-hand
to describe molecules.
It's called the **chemical formula**.

Can you guess the most well-known
chemical formula in the world?
It's H_2O – a molecule of water.
Lots of people know the formula,
but many have no idea what it means.

The formula tells you which atoms
are in the molecule by looking at
the symbols. It also tells you how many
of each atom there are by small numbers
after each symbol.
Notice that if there is no number after
a symbol, it means there is
just one of those atoms in the molecule.

So, look at the molecule of water opposite:
It is made from 2 hydrogen atoms and
just 1 oxygen atom.

This table shows some common molecules.

A formula helps chemists who can't spell!

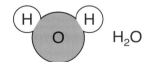

H_2O

A water molecule

nitrogen	N_2	
ammonia	NH_3	
hydrogen chloride	HCl	
chlorine	Cl_2	

21

▶ Elements

Some substances can't be broken down into anything simpler.
This is because all the atoms in them are the same type.
These substances are the chemical **elements**.

> **Elements contain only one type of atom.**

You might remember from page 20, that there are
92 different types of atom found naturally on Earth.
About 20 others have been made artificially by
scientists.

Elements just can't be split into simpler substances !

- So how many natural elements could there be ?

Look at these examples of elements :

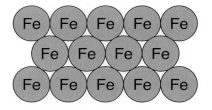

*Iron is an element
(it is a solid, made from only Fe atoms)*

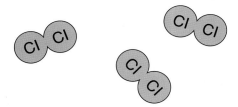

*Chlorine is an element
(it is a gas, made from Cl_2 molecules)*

Can you see how we can have *molecules* of elements ?

▶ Compounds

There are only 92 elements in nature,
and therefore 92 different types of atom.
However, there are millions of different substances.
So as you can imagine, most substances must
be made from more than one element. They must contain
more than one type of atom !

These substances are called **compounds**.

> **Compounds contain 2 or more types of atom (chemically bonded together).**

Many of us use gas to cook with or to heat our homes.
This gas is made up mainly of a compound called methane.
Its formula is CH_4.
It is made from 1 carbon atom and
4 hydrogen atoms bonded together.
So there are 2 elements in CH_4 – carbon and hydrogen.

Look back to the table on page 21.
- Which substances are elements ?
- Which substances are compounds ?

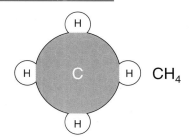

A molecule of methane (a compound)

Experiment 2.3   Making a compound

In this experiment you can look at the difference
between a compound and the elements it is made from.

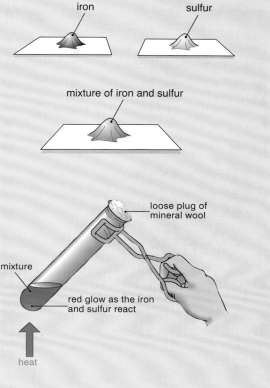

A. Collect a spatula of iron filings on a piece of paper.
- Describe what is looks like.

Collect a spatula of sulfur powder.
- Describe what it looks like.

B. Mix the iron and sulfur together with a spatula.
This is called a **mixture**.

Use a hand-lens to look closely at your mixture.
- What does the mixture look like?
- How could you separate the iron and sulfur?
Can you think of 2 ways?
If you have time, try out your ideas.

C. Put your mixture into a small test-tube.
Heat it strongly in a ***fume-cupboard***.

⚠ Take care to keep your flame away from
the mouth of the tube. If sulfur catches fire
toxic sulfur dioxide gas is given off.
Stop heating once the reaction starts.
- What do you see happen in your test-tube?

D. Give your test-tube to your teacher when it has cooled down.
Your teacher can wrap the tube in a cloth
and crack it carefully to get your new compound out.
- What are the differences between the mixture
of iron and sulfur and the new compound you have made?

The compound made from the elements, iron and sulfur,
is called iron sulfide.
The reaction can be shown by a **word equation** :

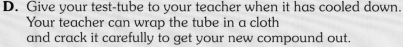

iron + sulfur ⟶ iron sulfide
reactants ⟶ product

The things that we start with before the reaction are called **reactants**.
The substances made in reactions are called **products**.

Notice how sulf**ur** changes to sulf**ide** in the compound.
- If a compound's name ends in **-ide** it usually shows that an element is
bonded to one other element. For example, magnesium oxide is MgO or
calcium chloride is $CaCl_2$. (Hydroxides are an exception, e.g. $NaOH$ is
sodium hydroxide.)
- If a compound's name ends in **-ate**, then oxygen is present in the
compound. For example, copper sulfate is $CuSO_4$ or silver nitrate is $AgNO_3$.

▶ Chemical reactions

Do you remember from Experiment 2.3, how iron reacts with sulfur? It forms iron sulfide.

This change was shown by a word equation. However, a **chemical** (or **symbol**) **equation** gives you more detail. It shows you the formula of each substance in the reaction:

iron + sulfur ⟶ iron sulfide   **Word equation**

Fe + S ⟶ FeS   **Symbol equation**

This tells us that 1 atom of iron reacts with 1 atom of sulfur. They make iron sulfide, which contains 1 iron atom for each sulfur atom.

The atoms have been chemically bonded together. The new compound, iron sulfide, is not like iron or sulfur at all. It has totally different properties. A new substance has been made.

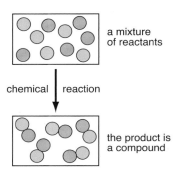

a mixture of reactants

chemical | reaction

the product is a compound

It is also difficult to get the iron and sulfur back from the iron sulfide. This is typical of a **chemical change** or **reaction**.

Another example of a chemical change is the reaction between sodium and chlorine:

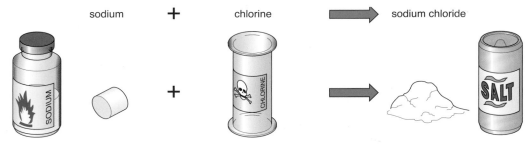

sodium + chlorine ⟶ sodium chloride

Sodium is a highly reactive, dangerous metal. Chlorine is a toxic gas.

Yet when they react together, they make something as harmless as sodium chloride.

You know sodium chloride better as common salt. So luckily its properties are very different from sodium or chlorine!

Sodium and chlorine were pretty wild when they were single. Now they've 'tied the bond', they're totally different!

Classifying changes

Chemical changes are very different from
physical changes, such as melting.

- List some chemical changes that take place in everyday life.
  Which of these changes are useful and which are not?

Chemical change	Physical change
new substance(s) made	no new substances made
not easily reversed	easily reversed

*New products are formed when
we cook foods.
Is baking a cake a physical
change or a chemical change?*

In the next experiment you can try to classify
some changes for yourself:

Experiment 2.4   Physical and chemical changes

Heat each substance in a test tube, gently at first.
If there is no sign of change, you can heat it
more strongly.

Observe carefully what happens as you are heating.
Record this and what is left after heating
in a table as shown below:

⚠
copper carbonate
copper sulfate

Substance tested	Observations	
	During heating	After heating
copper carbonate		
salol		
zinc oxide		
copper foil		
paraffin wax		
copper sulfate		
sand		

Your teacher might show you some other changes.
- Which changes do you think are physical changes?
- Do any substances show no signs of change?
- Which are chemical changes? Explain how you decided.

▶ Balancing equations

You are used to working with equations
in maths lessons.
You know that each side of an equation
must balance. They are equal.

It is the same in chemical equations.
The number of atoms on either side of
the equation must be equal. Remember that
no new atoms can be made or destroyed in
a chemical reaction.

A balanced equation. The same number and type
of atoms are on both sides of the equation.

mass of reactants = mass of products

▲

Let's look at an example.
Do you know the test for hydrogen gas?
A lighted splint 'pops'.

The hydrogen (H_2) reacts with oxygen molecules (O_2)
in the air.
They make the compound called water (H_2O).

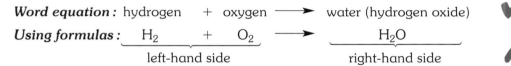

Word equation : hydrogen + oxygen ⟶ water (hydrogen oxide)

Using formulas : $\underline{H_2 \quad + \quad O_2}$ ⟶ $\underline{\quad H_2O \quad}$

left-hand side   right-hand side

But this equation is *not balanced.*

Counting atoms

If you count the atoms on either side of the equation,
they are not equal.

On the left-hand side (reactants):   we have two H atoms
         and two O's (from the O_2 molecule).
On the right-hand side (products): we've got 2 H's (that's good),
         but only one O atom.

This can't be correct because you know that
an oxygen atom cannot just disappear in
the reaction.

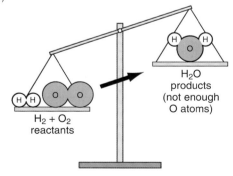

H_2O
products
(not enough
O atoms)

$H_2 + O_2$
reactants

Not balanced

Get the balance right!

So how do you go about balancing the equation?

Well, you need 1 more O atom
on the right-hand side.
Unfortunately, you can't simply change the
formula of water from H_2O to H_2O_2!

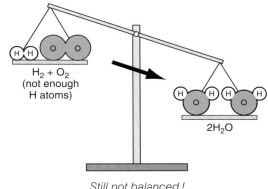

$H_2 + O_2$
(not enough
H atoms)

2H_2O

Still not balanced!

But you can change the number of H_2O's
made in the reaction.
For each O_2 molecule, you can make **2** H_2O's
(because you start with 2 oxygen atoms).
Let's try that:

$$H_2 + O_2 \longrightarrow 2H_2O$$

This has now solved the oxygen problem.
You have 2 on each side.
Unfortunately, in doing this, you now
need 2 extra H's on the left-hand side.

But that's easy to put right:

$$2H_2 + O_2 \longrightarrow 2H_2O$$

This is the **balanced equation**.

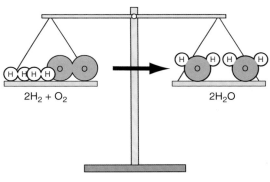

$2H_2 + O_2$

$2H_2O$

Balanced at last!

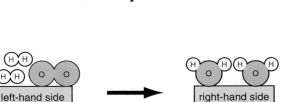

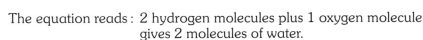

left-hand side
4 H's
2 O's

right-hand side
4 H's
2 O's

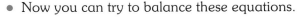

The equation reads: 2 hydrogen molecules plus 1 oxygen molecule
gives 2 molecules of water.

Remember you should ***never change a formula*** when
balancing an equation. You can only put numbers
in front of a formula.

- Now you can try to balance these equations.

  i) $H_2 + Cl_2 \longrightarrow HCl$

  ii) $Na + Cl_2 \longrightarrow NaCl$

  iii) $Li + O_2 \longrightarrow Li_2O$

▶ State symbols

Sometimes we find it useful to add even more information to our equations. We can show the **state** of reactants and products.
Remember that there are 3 states of matter – solid, liquid and gas. It is also useful to know if any substances in the equation are dissolved in water.

Chemists use a short-hand called **state symbols** to show these things:

A balanced equation,
Now what shall we do?
I know, we'll put in state symbols too.

		state symbol
solid	=	(s)
liquid	=	(l)
gas	=	(g)
solution (dissolved in water)	=	(aq)

(aq) comes from the Latin word for water, aqua.

States of matter
There are three,
We'll show them by an s, l and g.

Look at this equation with all 4 state symbols:

sodium + water $\longrightarrow$ sodium hydroxide + hydrogen
$2\,Na(s) + 2\,H_2O(l) \longrightarrow 2\,NaOH(aq) + H_2(g)$

It tells us that solid sodium reacts with liquid water.
They form a solution of sodium hydroxide and hydrogen gas.

Dissolved in water?
Now this one is new.
I know!
It's shown by the letters aq!

State symbols are also useful when substances in equations are not in their normal state at room temperature. For example,

$Mg(s) + H_2O(g) \longrightarrow MgO(s) + H_2(g)$

Which substance in the equation is not in its usual state?
The state symbols tell us that the magnesium has reacted with steam. Notice that the H_2O in the equation is a gas, not a liquid.

Summary

- **Diffusion** is the movement of particles of one substance through another.
  The particles move from an area of higher concentration to one of low concentration.
- All substances are made from tiny particles called **atoms**.
- Groups of 2 or more atoms chemically bonded together are called **molecules**.
- Some substances contain only one type of atom. These are **elements**.
  Elements can't be broken down into simpler substances.
- If a substance contains more than one type of atom, it is a **compound**.
  The different elements in a compound can't be separated easily (they are chemically bonded together), unlike the substances in a mixture.
- Chemical changes make new substances, but physical changes don't.
- We can show chemical changes (called reactions) by equations.
  We use either words or chemical symbols (and formulas).
- Symbol equations must be balanced. New atoms are not created or destroyed in a chemical reaction.
- State symbols – (s), (l), (g) or (aq) – give us extra information about a reaction.

▶ Questions

1. Copy and complete :
   a) The smallest part of an element is called an . . . . . All the . . . . in an element are the same. Elements can't be broken down into . . . . substances.
   b) Atoms joined, or bonded, together chemically are called . . . . . If a substance is made from more than one type of atom, it is called a . . . . .
   c) In a . . . . change, new substances are formed. However, no new substances are made in a . . . . change.

2. The density of a substance is worked out using this formula :

$$\text{density} = \frac{\text{mass}}{\text{volume}}$$

   a) An $8\,cm^3$ block of iron has a mass of $62.88\,g$. Work out the density of iron in (g/cm^3).

   Look at the densities in the table below :

Element	Density (g/cm³)
oxygen	0.001 33 (at room temp. and pressure)
copper	8.92
zinc	7.14
nitrogen	0.001 17 (at room temp. and pressure)

   b) What is the physical state of each element in the table at 25 °C ?
   c) Explain the big differences in the densities of the elements shown. Mention particles in your answer.
   d) Why is the temperature and pressure important when giving the density of oxygen and nitrogen ?

3. Explain each of the statements below in terms of particles :
   a) You can smell a fish and chip shop from across the road.
   b) Sugar dissolves faster in hot water than cold water.
   c) Condensation forms on the inside of your windows in winter.

4. Look at this candle burning :

   Which label (A or B) shows a chemical change and which shows a physical change ? Explain your answer.

5. Look at the boxes below :

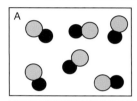
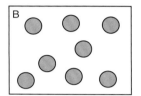
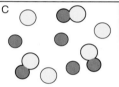
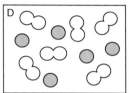

   Which box contains :
   a) one element
   b) a mixture of elements
   c) a pure compound
   d) a mixture of elements and a compound ?
      What might be happening in this box ?

6. Some rescue flares contain aluminium (Al) powder. It reacts with oxygen (O_2) in the air to form aluminium oxide (Al_2O_3) – a white solid.
   a) Is this a physical or a chemical change ?
   b) Write a word equation for the change.
   c) Now write a balanced symbol equation. Include state symbols.

7. Balance these equations :
   a) $H_2 + F_2 \longrightarrow HF$
   b) $CaCO_3 + HCl \longrightarrow CaCl_2 + CO_2 + H_2O$
   c) $CH_4 + O_2 \longrightarrow CO_2 + H_2O$
   d) $NaNO_3 \longrightarrow NaNO_2 + O_2$

Further questions on page 64.

chapter 3

ATOMIC STRUCTURE

▶ History of the atom

The word atom comes from a Greek word meaning 'something that can't be split'.
This fits in nicely with Dalton's ideas about atoms. However, as you probably already know, atoms can be split!

In the late 1800s and early 1900s, scientists had to think up new pictures of atoms to explain new observations.

For example, in 1897 J.J. Thomson put forward his 'plum pudding' theory. He thought atoms were balls of positive charge with tiny negative particles stuck inside. The negative particles were called **electrons**. He said they were like the currants in a bun or Christmas pudding.

This model explained Thomson's experiments with electricity very well. However, later experiments using radioactive particles needed a new picture. By 1915, scientists, like Ernest Rutherford and Niels Bohr, had developed a model that is still useful today.

400 BC   Democritus suggests that all things are made of particles. (See page 18.)

1805   John Dalton's atomic theory. Atoms of the same element are all alike. They combine to make compounds. (See Chapter 2.)

1897   J.J. Thomson finds the electron. (See page 31.)

1909   Ernest Rutherford discovers the proton. (See page 31.)

1911   Ernest Rutherford discovers the nucleus. (See below.)

1913   Niels Bohr suggests that electrons are found in shells around the nucleus. (See page 32.)

1932   James Chadwick proves that neutrons exist. (See page 31.)

▶ Inside the atom

There are 3 types of particle inside an atom. These are **protons**, **neutrons** and **electrons**.

The protons and neutrons are found squashed together in the centre of the atom.

The centre, called the **nucleus**, is incredibly small and dense.

The tiny electrons whizz around this nucleus.

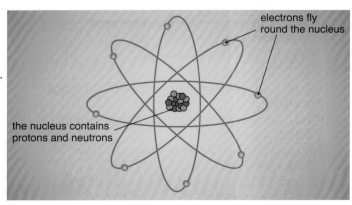

electrons fly round the nucleus

the nucleus contains protons and neutrons

This table shows the charge and mass of the sub-atomic particles in an atom.

Sub-atomic particle	Charge	Mass (in atomic mass units)
proton	1+	1
neutron	0	1
electron	1−	0 (almost)

Things to notice:

- protons and neutrons have the same mass.

- the electrons are so light that you can ignore their mass.
  In fact it takes almost 2000 electrons to weigh the same as a proton or a neutron.

- remember the charge on each particle like this:
  protons are **p**ositive,
  neutrons are **neut**ral,
  (so electrons must be the negative ones).

Protons

*I'm Penny the **Proton** and I'm pretty large,
I'm considered a plus, with my **positive charge**.
My friends and I, **in the nucleus** we huddle,
It's nice and cosy **with neutrons** to cuddle!*

Neutrons

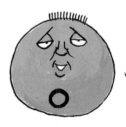

*I'm Ned the **Neutron** and I'm pretty **heavy**,
I'm fat and lazy and take things steady.
You could call me 'cheap' – I've **no charge** at all,
I really am just a dense, **neutral** ball!*

Electrons

*I'm Elvis the **Electron** and I'm pretty quick,
I fly round the nucleus at a fair old lick!
The protons and I, we tend to attract,
I'm **negative** you see and that's a fact!*

▶ Electron shells

The electrons in the atom are arranged
in **shells** around the nucleus. The shells
are sometimes called **energy levels**.

The **first shell** (nearest to the nucleus) can hold just **2 electrons**.
The **second shell** can hold up to **8 electrons**.
The **third shell** can also hold **8 electrons**.

Electrons go round the nucleus in shells !

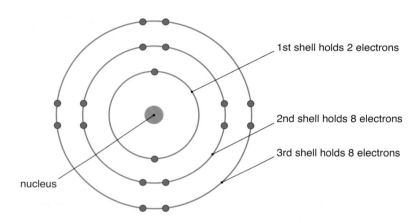

1st shell holds 2 electrons

2nd shell holds 8 electrons

3rd shell holds 8 electrons

nucleus

Filling the shells

The electrons start filling up the shells from
the inner shell outwards. They like to get
as close to the nucleus as possible.

You can see this from the examples below :

helium
(has 2 electrons)

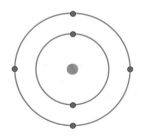

carbon
(has 6 electrons)

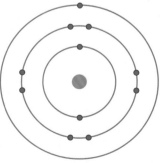

sodium
(has 11 electrons)

▷ Describing atoms

You don't need to draw pictures of
atoms every time. You can use a short-hand
called the **electronic structure** (sometimes called the **electronic configuration**).
This just shows the number of electrons in each shell,
working outwards from the nucleus.

So the examples drawn on the previous page can be shown as:

helium   2   carbon   2, 4   sodium   2, 8, 1

- Can you write down the electronic structures of:
  a) nitrogen (with 7 electrons)   b) neon (with 10 electrons)
  c) aluminium (with 13 electrons)   d) calcium (with 20 electrons – the last 2 electrons start filling
                                         the 4th shell)?

Atomic number (or proton number)

Each element has its own **atomic number**,
sometimes called the proton number.
It tells us how many protons there are in
one atom of that element.
(All the atoms of a particular element have
the same number of protons.)

Atomic number = the number of protons (which is the same as the number of electrons)

Atoms are neutral. The positive charges are cancelled out
by an equal number of negative charges.

So the number of protons ($+$) in an atom
must always equal the number of electrons ($-$).

Mass number

You already know that protons and
neutrons are the heavy particles which
make up an atom's mass.
The mass number tells us how many
of these protons and neutrons there
are in an atom.

You have 13 protons and 14 neutrons. Your mass number is 27. Try Weight-watchers Al.

SPEAK YOUR MASS

Mass number = the number of protons + the number of neutrons

(N.B. Mass number is sometimes called nucleon number.)
You can work out the number of
neutrons from this, as long as you know
the atomic number as well.

Number of neutrons = mass number − number of protons
                   = mass number − atomic number

▶ Using atomic numbers and mass numbers

We can use the atomic number and mass number
to build up a complete picture of an atom.

Let's look at an example :

Example
Lithium's atomic number is 3. Its mass number is 7.
a) How many protons, neutrons and electrons are in a lithium atom ?
b) Draw a diagram of a lithium atom.

a) The atomic number $= 3$, so there are 3 protons and 3 electrons.
   The mass number $= 7$
   Therefore, $7 =$ number of protons $+$ number of neutrons.
   We know there are 3 protons, so $7 = 3 +$ the number of neutrons.
   It follows that the number of neutrons must be 4.
   Or you might just remember :

   number of neutrons $=$ mass number $-$ atomic number
   $\qquad\qquad\qquad = \qquad\quad 7 \qquad\quad - \qquad\quad 3$
   $\qquad\qquad\qquad = \qquad\quad 4$

Answer : **3 protons, 3 electrons and 4 neutrons**.

b)

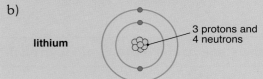

lithium

3 protons and
4 neutrons

More short-hand !

The atomic number and mass number can be
shown using chemical short-hand.
The information about lithium in the example above
can be shown as :

$$^{7}_{3}\text{Li}$$

The top number is the mass number.
The bottom number is the atomic number.

| mass number $\longrightarrow$ | $^{27}_{13}\text{Al}$ |
| atomic number $\longrightarrow$ | |

● Work out the number of protons, neutrons and electrons in
   $^{16}_{8}\text{O}$, and $^{31}_{15}\text{P}$.   Now show how the electrons are arranged.

▷ Isotopes

There are often stories about nuclear power and atomic weapons on television. So you might have heard the phrase 'radioactive isotopes'. But what is an **isotope**?

These Easter eggs look the same, but one is heavier. Why? (See below.)

Some elements are made up of atoms with different masses. The number of protons in atoms of the same element is always the same. So isotopes must have **different numbers of neutrons.**

We find that some isotopes are radioactive but many are non-radioactive.

Let's look at this example:

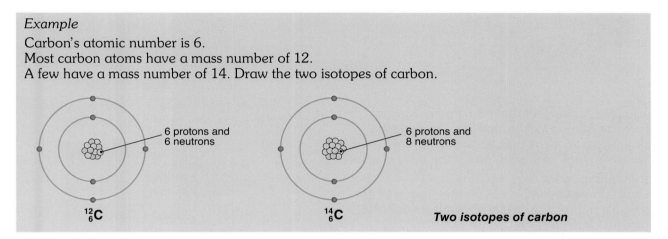

Example

Carbon's atomic number is 6.
Most carbon atoms have a mass number of 12.
A few have a mass number of 14. Draw the two isotopes of carbon.

6 protons and 6 neutrons

$^{12}_{6}C$

6 protons and 8 neutrons

$^{14}_{6}C$

Two isotopes of carbon

You can see that the isotopes are exactly the same, apart from their numbers of neutrons.

Isotopes of an element have the *same chemical reactions*. This is because their electrons are arranged in exactly the same way.
(Chemical reactions only involve electrons. The protons and neutrons in the nucleus take no part in chemical reactions.)

Isotopes are a bit like Easter eggs which have the same chocolate shell, but different numbers of sweets inside!

Isotopes are sometimes written showing just their mass numbers. In the example above the isotopes can be written as carbon-12 and carbon-14.

> **Isotopes are atoms with the same number of protons, but different numbers of neutrons.**

Radioactive isotopes have many uses. For example, they are used in hospitals to kill cancer cells.
They are also used in industry to find leaks in underground pipes without having to dig the pipes up.
Uranium-235 is the energy source in a nuclear power station.

▷ Relative atomic mass (R.A.M.)

Each type of atom has a different mass.
However, the mass of an atom is too small to measure.
So we compare their masses on a scale which gives
the lightest element of all, hydrogen, a mass of 1.
This gives us a number called the **Relative Atomic Mass (R.A.M.)**,
sometimes given the symbol A_r.

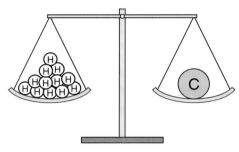

Each element has a R.A.M. !

Let's look more closely at the hydrogen atom :
Hydrogen's atomic number is 1. Its mass number is also 1.
This means that a hydrogen atom has 1 proton, 1 electron,
but no neutrons.
You can see why it is so light !

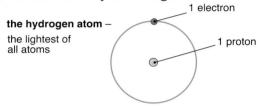

the hydrogen atom –

the lightest of
all atoms

1 electron

1 proton

You might think that there is no point having
Relative Atomic Masses and mass numbers.
Both seem to tell us how heavy atoms are.
However, look at the R.A.M.s in the table opposite :

Do you notice anything strange ?
Could you ever get a mass number that was not
a whole number ? You can't have fractions of
protons or neutrons in atoms.

In fact, the Relative Atomic Mass of an element
takes into account the element's *different isotopes*.
Some elements are found as mixtures of their isotopes.
The R.A.M. is their average mass, taking into account
the *different proportions* of each isotope in the natural mixture.

A carbon atom is 12 times as heavy as a hydrogen atom. Its R.A.M. is 12.

Element	R.A.M. (A_r)
chlorine	35.5
copper	63.5
iron	55.8

Example

On the last page, you worked out the difference between
chlorine-35 and chlorine-37. (Chlorine-37 has 2 more
neutrons than chlorine-35.)
Any sample of chlorine gas is a mixture of the 2 isotopes –
75 % is chlorine-35 and 25 % is chlorine-37.

So if you have 100 chlorine atoms, 75 will be chlorine-35 and
25 will be chlorine-37.
We can work out the total mass of the 100 atoms
(relative to a hydrogen atom) :
$(75 \times 35) + (25 \times 37) = 3550$
Therefore, the average mass (or Relative Atomic Mass) of chlorine
$= 3550 \div 100 = \mathbf{35.5}$

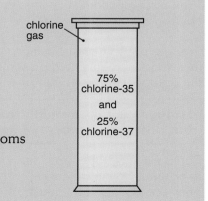

chlorine gas

75%
chlorine-35
and
25%
chlorine-37

▷ Relative Formula Mass

Once you have a list of relative atomic masses,
you can work out the relative mass of any molecule.
You just need to know the chemical formula.

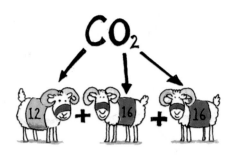

Let's look at carbon dioxide as an example :

The formula of carbon dioxide is CO_2.
It's molecules are made up of 1 carbon atom and 2 oxygen atoms.
The R.A.M. of carbon is 12.
The R.A.M. of oxygen is 16.

If we add up the R.A.M.s as in the formula,
we get the **relative formula mass (or molecular mass)**,
sometimes given the symbol M_r.
So for CO_2 we have :

Add up the R.A.M.s to get the Relative Formula Mass

$$
\begin{aligned}
1 \text{ carbon} &= 1 \times 12 = \quad 12 \\
2 \text{ oxygen} &= 2 \times 16 = \underline{+32} \\
&\qquad\qquad\qquad 44
\end{aligned}
$$

Therefore the relative formula mass of carbon dioxide is 44.

How many times heavier than a hydrogen atom is
a molecule of carbon dioxide ?

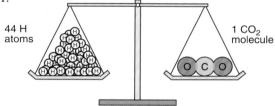

44 H atoms   1 CO_2 molecule

These masses are useful to chemists when
working out the masses of reactants and products
in chemical reactions. (See page 350.)

Now let's try a harder example :

CO_2 is 44 times as heavy as one H atom

Example

What is the relative formula mass of aluminium sulfate, $Al_2(SO_4)_3$?
(R.A.M.s : Al = 27, S = 32, O = 16)

Sometimes a formula has a number outside a pair of brackets.
This tells us that all the atoms inside the brackets
are multiplied by the number outside.
So in aluminium sulfate, we have :
2 aluminiums, 3 sulfurs and 12 oxygens.

Now let's add up their R.A.M.s to get the answer :

$$
\begin{aligned}
2\,Al &= 2 \times 27 = \quad 54 \\
3\,S &= 3 \times 32 = \quad 96 \\
12\,O &= 12 \times 16 = \underline{+\,192} \\
&\qquad\qquad\qquad\quad 342
\end{aligned}
$$

Therefore the relative formula mass of aluminium sulfate is **342**.

*Formula masses help us to do calculations
(See pages 237 and 350.)*

Summary

- Atoms contain **protons**, **neutrons** and **electrons**.
- Protons are positively charged. Electrons are negatively charged.
  Neutrons have no charge – they are neutral.
- Protons and neutrons are the heavy particles in an atom.
  They each have a mass of 1 atomic mass unit, and are found in the **nucleus** (centre) of an atom.
  We can ignore the tiny mass of the electrons.
- The electrons orbit the nucleus in **shells** (or **energy levels**).
  The 1st shell can hold 2 electrons.
  The 2nd shell can hold 8 electrons, as can the 3rd shell.
- The **atomic number** = the number of protons (which equals the number of electrons).
- The **mass number** = the number of protons + the number of neutrons.
- Isotopes are atoms with the same number of protons, but different numbers of neutrons.

▷ Questions

1. Copy and complete :
There are 3 types of particle found inside
atoms : a) . . . . , b) . . . . and c) . . . . .
This table shows their mass and charge :

Sub-atomic particle	Charge	Mass (atomic mass units)
proton	. . . .	. . . .
. . . .	0	1
. . . .	. . . .	0 (almost)

The protons and neutrons are found in the
. . . . of the atom, called the nucleus.
The . . . . zoom around the nucleus in shells.
The 1st shell, which is . . . . the nucleus, can
hold . . . . electrons, whereas the 2nd and 3rd
shells can hold . . . . electrons.

2. a) What is the *atomic number* of an atom ?
b) What is the *mass number* of an atom ?
c) What is the atomic number of the atom below?
What is its mass number ?

$$^{40}_{18}Ar$$

3. Give the numbers of protons, electrons and
neutrons in the atoms below :
a) $^{14}_{7}N$ b) $^{20}_{10}Ne$ c) $^{19}_{9}F$
d) $^{39}_{19}K$ e) $^{60}_{27}Co$ f) $^{235}_{92}U$

4. Draw fully labelled diagrams of the atoms below :
a) $^{4}_{2}He$ b) $^{9}_{4}Be$ c) $^{27}_{13}Al$ d) $^{40}_{20}Ca$

5. Copy this table and fill in the gaps.
Use Table 1 on pages 390 and 391 to help you.

Element	Atomic number	Electronic structure
lithium	. . . .	2, 1
. . . .	14	. . . .
. . . .	19	. . . .

Do you have to look up the atomic number of
lithium, given the information in the table
above? Explain your answer.

6. Hydrogen (atomic number 1) has 3 isotopes.
They can be shown as $^{1}_{1}H$, $^{2}_{1}H$ and $^{3}_{1}H$.
a) What are *isotopes* ?
b) What is the difference between the 3 isotopes?
c) Hydrogen reacts with chlorine in sunlight,
forming hydrogen chloride :
$$H_2 + Cl_2 \longrightarrow 2\,HCl$$
Would you expect the same reaction for
each isotope of hydrogen ? Why ?
d) Chlorine exists naturally as 2 isotopes.
75 % is $^{35}_{17}Cl$, and 25 % is $^{37}_{17}Cl$.

Show why the relative atomic mass of
chlorine is 35.5.
e) The element chlorine is a gas. Its formula is
Cl_2. How many different masses of the Cl_2
molecule would you expect to find in a
sample of chlorine gas? Explain your answer.

7. Work out the relative formula mass of :
a) H_2O b) C_2H_5OH c) Na_2SO_4
(R.A.M.s : H = 1, O = 16, C = 12, Na = 23,
S = 32)

Further questions on page 65.

The Periodic Table

Sorting out the elements

Scientists like to find patterns.
Around 200 years ago, scientists were discovering lots of new elements. However, they struggled to find any links between the different elements.

Look at the picture opposite :

At the time, some substances, which were thought to be elements, were in fact compounds.
Other elements had not yet been discovered.
No wonder finding a pattern was tricky !

How would you like to do a jig-saw with no picture to work from ? Some pieces are missing and others don't belong in this jigsaw ! This was the state of chemistry at the start of the 1800s.

Finding the pattern

Real progress was made around 1865 by John Newlands.
He put the elements in order of their atomic mass.
He found that every eighth element was similar.

Unfortunately, his pattern only worked for the first 15 elements known at that time. After that, he could see no links between the rest of the elements. Other scientists made fun of his ideas. They suggested that he could have done better by sorting the elements into alphabetical order !

In 1869 the problem was solved by a Russian chemist called Dmitri Mendeleev.
He also tried putting the elements in order of their atomic mass.

He made a table of elements.
New rows were started so that elements which were alike could line up together in columns.
He wanted a table of regular (periodic) patterns.

However, Dmitri was not afraid to take risks.
When the pattern began to go wrong, he would leave a gap in his table. He claimed that these gaps were for elements that had not yet been discovered.
He even changed the order round when similar elements didn't line up.

As you might expect, people doubted his 'Periodic Table'.
However, he used his table to predict the properties of elements which could fill the gaps.
In 1886, the element germanium was discovered.
The new element matched Dmitri's predictions.
Finally other scientists accepted his ideas.

Dmitri was a chemistry teacher in Russia !

▷ The Periodic Table

Although Mendeleev's table was accepted, there was one thing that he could not explain.
Why did he sometimes need to change the order of atomic masses to make the pattern carry on?
The answer lies inside the atoms. The atoms of elements in the Periodic Table are not arranged in order of mass.
It is their number of protons (atomic number) which really matters.

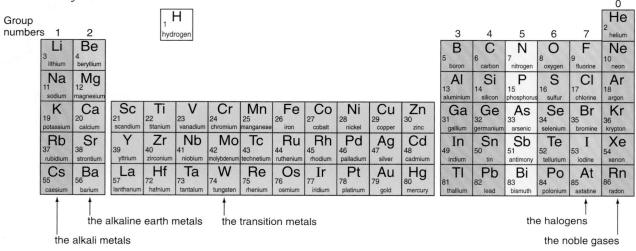

| the alkaline earth metals | the transition metals | the halogens |

the alkali metals | the noble gases

Each colour shows a chemical 'family' of elements with similar properties. You can see the full Periodic Table on pages 392–3.

Groups

There are 8 groups in the Periodic Table.
A group is a **vertical column**.
All the elements in a group have similar properties.
They are a 'chemical family'.

Look at the Periodic Table above:
Some groups have special 'family' names.
Can you find Group 7? What is the group called?
Have you met any elements from this 'family' before?
Other groups are just known by their group number.
Notice that the transition metals form a block on their own.

Groups are families of elements. The members of the family are similar but not exactly the same.

Periods

Periods are the **rows across** the Periodic Table.
You read the table like a page. Start at the top, and work your way down, reading from left to right.
So there are 2 elements in the 1st period, H and He.
The 2nd period has 8 elements, starting with Li.
● Which is the last element in the 2nd period?
● Count how many elements are in the 3rd period.

> *Example*
> **C** is in Group 4.
> It is in the 2nd period.
> **Al** is in Group 3.
> It is in the 3rd period.

► Metals, non-metals and the Periodic Table

As you know from the previous page, the Periodic Table has 8 groups of elements. Groups 1 and 2 are all metals, whereas Groups 7 and 0 contain only non-metals.

However, the elements in the middle groups start with non-metals at the top, but finish with metals at the bottom.

For example, look at Group 4 :

Silicon is a semi-metal

C	carbon	non-metal
Si	silicon	silicon and germanium are called **semi-metals** or **metalloids**. They are on the borderline between metals and non-metals
Ge	germanium	
Sn	tin	metal
Pb	lead	metal

Silicon is the most well-known semi-metal. It behaves like a metal in some ways, but like a non-metal in others. For example, it is shiny like a metal, but brittle like a non-metal.

Tin oxide reacts like a metal oxide, but also reacts like a non-metal oxide! It is an **amphoteric** oxide (behaves like an acid and a base).

This shows us that science is not always 'black and white'. The semi-metals are an example of a 'grey' area. However, we can draw an imaginary line in the Periodic Table to divide the metals and non-metals. You can think of the line as a staircase. (See the bold line below.) Below stairs you find metals, above stairs you find non-metals.

Silicon is used in the micro-electronic industry. It is a 'semi-conductor'. The silicon chip has made it possible to make circuits incredibly small.

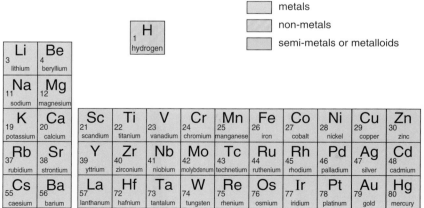

- metals
- non-metals
- semi-metals or metalloids

▶ Group 1 – The alkali metals

The elements in this first group don't have many uses
as the metals themselves. They are too reactive.
However, you will certainly use some of their compounds every day.
You can read about these on pages 44, 45, 105 and 254.

Li	lithium
Na	sodium
K	potassium
Rb	rubidium
Cs	caesium

You will find the elements in Group 1
an exciting bunch!
They are metals, but they have some very
unusual properties.
You can see for yourself below:

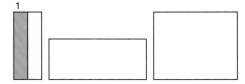

Demonstration 4.1   Looking at sodium

Your teacher will show you a piece of sodium.
What is it stored under? Why?
Does the sodium look like a metal?

⚠ sodium

*Sodium is stored under
oil. This keeps it away
from oxygen and
moisture in the air.*

SODIUM

shiny surface

Your teacher will cut a piece of sodium with a knife.
• How soft is it?
• What does it look like inside?
  Is this more like a metal now?

Your teacher will warm a small piece of sodium
gently on a combustion spoon.
• How easily does sodium melt?

sodium melts
at 98°C

*Sodium can be cut with a knife.
It's like cutting a piece of cheese.*

Now you can see why sodium, and the other
alkali metals, are unusual metals. They have
low melting points and are *very soft*.
For metals, they also have very *low densities*.
You will see in the next demonstrations that
lithium, sodium and potassium float on water!

Help!

*The alkali metals
can be cut with a
knife*

Look at the table below:

Alkali metal	Atomic number	Melting point (°C)	Boiling point (°C)
lithium	3	180	1360
sodium	11	98	900
potassium	19	63	777
rubidium	37	39	705
caesium	55		669

• What patterns can you see going down the group?
• Try to predict the melting point of caesium.

Typical metals have much higher melting points.
For example, iron melts at 1540°C.

Reactions of the alkali metals

The alkali metals are the most reactive group of metals in the Periodic Table. They are too dangerous for you to use in experiments. However, your teacher can show you some reactions of lithium, sodium and potassium.

Demonstration 4.2   Lithium with water

Your teacher will drop a small piece of lithium into a trough of water. What do you see happen? The gas given off can be collected as shown:
Test the gas with a lighted splint.
- Which gas is given off?
Now add a little universal indicator solution to the trough.
- Is the solution left acidic or alkaline?

lithium

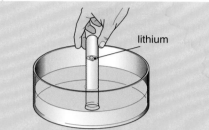
lithium

Demonstration 4.3   Sodium and potassium with water

Your teacher can now try sodium and potassium with water.

You can see that sodium gets hot enough to melt itself as it reacts. Potassium gets so hot that it lights the hydrogen gas given off. It burns with a lilac flame.

- What happens to the reactivity of the alkali metals going down the group?

sodium
potassium

sodium reacting with water   potassium reacting with water

The Periodic Table is very useful. Its groups make chemistry easier! You only have to learn the reactions of one element in a group. The others are usually similar. For example,

lithium + water $\longrightarrow$ lithium hydroxide + hydrogen
(**alkaline solution**)

$$2\,Li(s)\ +\ 2\,H_2O(l)\ \longrightarrow\ 2\,LiOH(aq)\ +\ H_2(g)$$

Knowing this, we know the equations for the other alkali metals:

sodium + water $\longrightarrow$ sodium hydroxide + hydrogen

$$2\,Na(s)\ +\ 2\,H_2O(l)\ \longrightarrow\ 2\,NaOH(aq)\ +\ H_2(g)$$

- Write a word and symbol equation for potassium reacting with water.

> **The alkali metals get more reactive going down the group.**

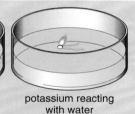

Group 1

| Li |
| Na |
| K |
| Rb |
| Cs |

least reactive

most reactive

This pattern is explained on page 61

Other reactions

The alkali metals react well with non-metals. For example,

lithium + oxygen $\longrightarrow$ lithium oxide

$$4\,Li(s)\ +\ O_2(g)\ \longrightarrow\ 2\,Li_2O(s)$$

sodium + chlorine $\longrightarrow$ sodium chloride

$$2\,Na(s)\ +\ Cl_2(g)\ \longrightarrow\ 2\,NaCl(s)$$

> **Compounds formed by Group 1 metals are usually white solids (that dissolve in water to form colourless solutions).**

▷ Chemistry at work : Group 1 metals and their compounds

Lithium

Lithium gets its name from the Greek word for 'stone'.
As the metal itself, we use it to strengthen other metals,
such as magnesium, in alloys. (See page 271.)
It is also used in batteries.
Lithium batteries are powerful and lightweight.
In fact you might be nearer than you think to a
lithium battery now!
They are suitable for calculators, watches and cameras.
They are also used in heart pace-makers.

Lithium batteries are used in heart pace-makers

Lithium compounds

Lithium compounds have been used more and more
since the Second World War.
Look at the diagram below.
It shows many of their uses today :

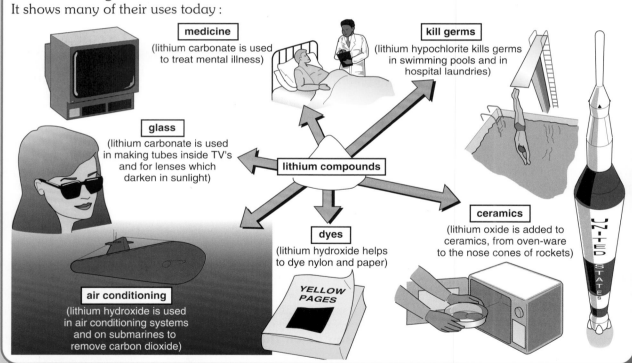

medicine
(lithium carbonate is used to treat mental illness)

kill germs
(lithium hypochlorite kills germs in swimming pools and in hospital laundries)

glass
(lithium carbonate is used in making tubes inside TV's and for lenses which darken in sunlight)

lithium compounds

ceramics
(lithium oxide is added to ceramics, from oven-ware to the nose cones of rockets)

dyes
(lithium hydroxide helps to dye nylon and paper)

air conditioning
(lithium hydroxide is used in air conditioning systems and on submarines to remove carbon dioxide)

Potassium compounds

Potassium is essential for the healthy growth of plants.
Potassium nitrate is used as a fertiliser. It is also used
in making explosives and fireworks. (See page 197.)

▷ Chemistry at work : Group 1 metals and their compounds

Sodium

The element sodium is used in street lamps
and in nuclear reactors.
You can read about the uses of sodium on page 106.

Sodium compounds

The two most important sodium compounds are
sodium chloride and sodium hydroxide. Their uses
are covered in Chapter 8.

Food additives

Do you eat cereal for your breakfast? If you do,
have you ever read the box?
You will almost certainly find some sodium compounds
in the list of ingredients, even if it's only salt.
Common salt, sodium chloride ($NaCl$), was the first
food additive. You can read about it on page 102.

Nowadays, we use many other sodium compounds
as additives. Sodium sulfate(IV), sodium nitrite and
sodium nitrate are all preservatives. They stop
bacteria growing. Other sodium compounds
improve the texture of foods.

But perhaps the most well-known (or infamous!) additive is
mono-sodium glutamate. It brings out the flavour in foods, and
is used a lot in Chinese restaurants. However, in some people
it can cause dizziness, headaches, nausea and is dangerous
to asthmatics. Not surprisingly, it is banned from baby foods!

*Fast food contains additives. You can read
about E numbers on page 180.*

Sodium carbonate

The materials needed to make sodium carbonate are
brine (sodium chloride solution), limestone and
ammonia. We make about 26 million tonnes each year!
Look at the diagram showing its uses :

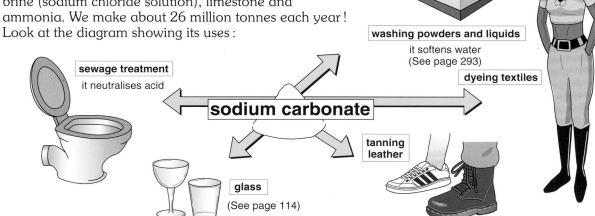

sewage treatment
it neutralises acid

washing powders and liquids
it softens water
(See page 293)

dyeing textiles

sodium carbonate

glass
(See page 114)

**tanning
leather**

▷ The transition metals

The transition metals lie in between Group 2 and Group 3.
They have many important uses. (See page 48.)
Some well-known transition metals are iron, copper,
chromium, nickel and gold.

Metals in the whole block have similar
properties. As you know, this is
strange for the Periodic Table. Usually
families of elements line up in
the columns we call groups.

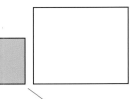

		Ti	V	Cr	Mn	Fe	Co	Ni	Cu
									Ag
				W				Pt	Au

Physical properties

> **The transition metals are 'typical' metals.**

They are hard, dense and shiny. They are good conductors
of heat and electricity. They are malleable and ductile.
The transition metals also have high melting points and
boiling points, except for mercury.

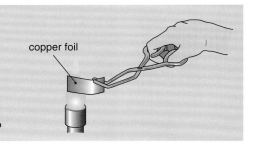

Why is copper used for water pipes?

Experiment 4.4   Magnetic metals
Try touching different transition metals with a magnet.
- Which ones are attracted to it?

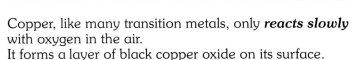

> **Iron, cobalt and nickel are the magnetic metals.**

Chemical properties

Experiment 4.5   Heating copper
Hold a strip of copper foil in some tongs.
Heat it strongly for a few minutes in a Bunsen flame.

- Does the copper burst into flames, like magnesium does?
- What does the copper look like after it has cooled down?
- Which gas in the air has it slowly reacted with?
- What do you think the black coating on its surface is called?

copper foil

Copper, like many transition metals, only **reacts slowly**
with oxygen in the air.
It forms a layer of black copper oxide on its surface.

copper + oxygen ⟶ copper oxide
$2\,Cu(s)$ + $O_2(g)$ ⟶ $2\,CuO(s)$

Q. *What does an alloy of
nickel and titanium wear in bed?*

A. *a Ni-Ti!*

46

Experiment 4.6   Transition metals with water

Put a transition metal in the apparatus as shown :
You can repeat this using iron, copper, nickel and any other
samples you have.
- Are there any signs of a reaction?

Leave them in the water for a week.

- Have any transition metals reacted with the water?
  If they have, how can you tell that there has been a reaction?

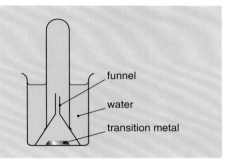

The transition metals react very slowly, if at all, with water.
You can read about iron rusting on page 76.

> **The transition metals are *less reactive* than Group 1 metals.**

Transition metal compounds

Look at the compounds of transition metals :
What do they have in common?
Look at the photo of the weathered copper
roof on page 203. The copper compound
formed on it is green.

copper(II) chloride   nickel chloride   iron(III) chloride

cobalt(II) chloride   manganese(II) chloride

> **Transition metals form *coloured* compounds.**

On the other hand, the compounds of Group 1 metals
are usually white.

Which formula?

Most transition metals form compounds which can
have more than one formula. For example,
on the last page you saw copper oxide, CuO, formed.
However, there is another form of copper oxide
whose formula is Cu_2O. The compounds are different colours.
Roman numbers in the names tell us which compound
we mean.
Copper(II) oxide is CuO. Copper(I) oxide is Cu_2O.
Other examples are iron(II) oxide, FeO, and
iron(III) oxide, Fe_2O_3.
You can find out how to work out formulas on page 251.

*The presence of chromium (Cr^{3+})
makes this emerald green*

Catalysts

We use catalysts in industry to speed up reactions.
(See page 212.) Look at the table opposite :

Manufacture of ...	Catalyst
ammonia	iron
sulfuric acid	vanadium(V) oxide
nitric acid	platinum / rhodium
margarine	nickel

> **The transition metals and their compounds are important catalysts.**

▶ Chemistry at work : Transition metals and their compounds

The transition metals and their compounds are widely
used in industry and in everyday life.
Iron is the most commonly used metal, mostly as steel.
You can read about its uses on page 78.

Here are some uses of the transition metals around the home :

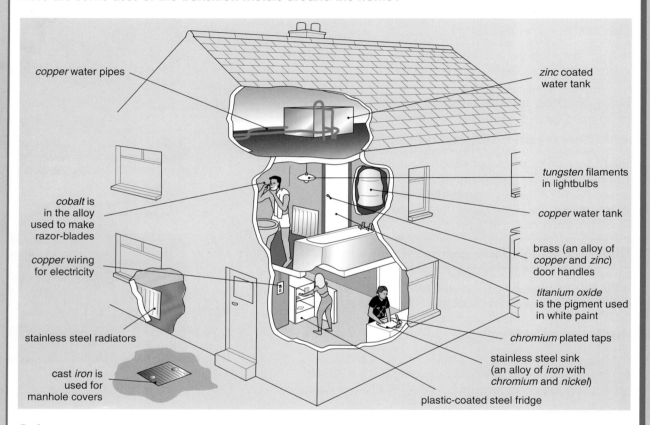

copper water pipes

zinc coated
water tank

tungsten filaments
in lightbulbs

copper water tank

cobalt is
in the alloy
used to make
razor-blades

copper wiring
for electricity

brass (an alloy of
copper and *zinc*)
door handles

titanium oxide
is the pigment used
in white paint

stainless steel radiators

chromium plated taps

stainless steel sink
(an alloy of *iron* with
chromium and *nickel*)

cast *iron* is
used for
manhole covers

plastic-coated steel fridge

Other uses

A radioactive
isotope of **cobalt**
is used to treat
patients with
cancer.

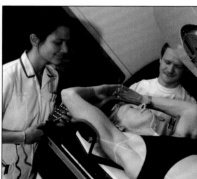

Platinum is used in catalytic converters,
fitted to car exhausts. It cuts down the
amount of pollution from cars. (See page 148.)

► Chemistry at work : Transition metals and their compounds

Gold

People often think it's just jewellers that use gold. (See page 273.) It is used for jewellery because of its attractive appearance and its low reactivity.

However, gold is also used in a great variety of industries – and that's despite its high price.
Here are some of the other users of gold:

Industry	Use of gold
computer, telecommunication and home appliance industries	gold-coated electrical connections
satellite manufacturers	gold-plated shields and reflectors to protect equipment from solar radiation
the latest laser technology in hospitals	in lasers to perform eye operations and kill cancerous cells. The lasers use gold reflectors to concentrate light energy
motor industry	gold-coated contacts in sensors that activate air bags

Gold jewellery

Many of these are 'new technologies', at the cutting edge of industry.

- What properties of gold have resulted in its various uses?
  Link each use in the table to gold's useful properties.

Silver

Gold, like silver, has been used to make jewellery for thousands of years.

- Why is silver one of the 'jewellery metals'?
- Which metal, gold or silver, is more likely to tarnish?
- Name another expensive metal used by jewellers.

Gold helps to save lives. How?

Silver also shares many of gold's other useful properties.

Did you know that TVs, dishwashers, telephones and computers all contain silver in contacts and switches? A typical washing machine will contain more than 15 silver contacts!

However, the biggest use of silver (44%) is in the silver compounds in photographic film. (See page 254.)

The latest developments use nano-scale, tiny particles of silver. (See pages 336–7.) These have very effective anti-bacterial, anti-viral and anti-fungicide properties. They are used in sprays to clean operating theatres.

- Why are materials with nanoparticles of silver used to line fridges?

Silver is the most reflective of all metals. We use it to plate the surface of cutlery and in mirrors.

Summary

- The **Periodic Table** arranges the elements in order of atomic number.
- Elements with similar properties line up in vertical columns. These columns are called **groups**.
- There are 8 groups in the Periodic Table.
- A row across the Periodic Table is called a **period**.
- The elements can be divided into **metals** and **non-metals** (with a few semi-metals or metalloids in between).
- **Group 1 – The alkali metals**
    - are soft metals, stored under oil
    - have low densities and low melting points
    - quickly tarnish in air, but are shiny when freshly cut
    - are the most reactive group of metals
    - get more reactive going down the group.
- **Transition metals**
    - are hard and dense
    - have high melting points
    - are less reactive than Group 1 metals
    - form coloured compounds
    - can form compounds with more than one formula, for example, iron(II) oxide, FeO, and iron(III) oxide, Fe_2O_3
    - are important catalysts in industry.

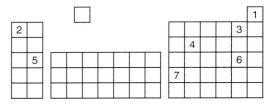

▶ **Questions**

1. Copy and complete :
   a) The elements line up in order of atomic
      .... in the Periodic Table. There are ....
      groups. The elements in each group have
      .... properties. Groups form v_ _ _ _ _ _ _
      columns, whereas periods are
      h_ _ _ _ _ _ _ _ _ rows. Most of the elements
      can be divided into 2 sets – the .... and
      the non-metals.
   b) Group 1 elements are called the ....
      metals. They have .... melting points and
      .... densities. They are the most ....
      group of metals, and are stored under
      ..... Lithium is the .... reactive member
      of the group. They all react with cold
      water, giving off .... gas and leaving an
      .... solution.
   c) The .... metals are typical metals. They
      are hard, have .... melting points and
      .... densities. They are .... reactive than
      the metals from Group 1. They form ....
      compounds, which can often have more
      than one .....

2. Look at the cartoons in the Summary box
   above.
   a) Which alkali metal is used in street lights?
      (See page 106.)
   b) List some uses of named transition metals.
      (See page 48.)

3. The numbers in this Periodic Table represent
   elements.

   a) Which 2 elements are in the same group?
      Give the name and number of this group.
   b) Which elements are in the 2nd period?
   c) Which elements are metals?
   d) Which group is element 7 in? Which
      period is it in?
   e) Which element is a semi-metal (or
      metalloid)?

4. Use Table 1 on pages 390 and 391 to help you answer this question.
   a) Record the melting points and atomic numbers of the Group 1 metals in a table.
   b) Plot a graph like the one shown below :

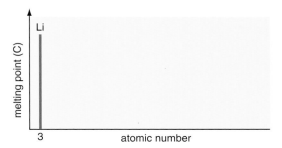

   c) What pattern do you see going down Group 1?
   d) Look up the melting points and atomic numbers of the Group 2 metals and put them in a table.
   e) Now do another graph, showing the melting points of the Group 2 metals.
   f) Compare your graphs for Groups 1 and 2. How are they alike? How do they differ?

5. a) Describe what you **see** when sodium reacts with water.
   b) Write a word equation for the reaction in a).
   c) When the reaction in a) has finished what would you add to the solution left to test its pH?
   d) What is the pH of the solution left?
   e) Which substance makes the solution an alkali?
   f) We can collect the gas given off when lithium reacts with water. How can you test that the gas is hydrogen?
   g) Why would it be impossible to collect the gas given off when potassium reacts with water?
   h) Predict what would happen when rubidium, which is under potassium in Group 1, reacts with water.
   Include a word and symbol equation in your answer.
   i) State the pattern in the reactivity of the Group 1 metals going down the group.

6. The alkali metals, and particularly their compounds, are very useful.
   Draw a spider diagram to show their uses. You could make a poster from your research. You will have to refer to information in Chapter 8 to find out more about sodium.

7. a) Where in the Periodic Table do you find the transition metals?
   b) Name 5 common transition metals.
   c) Make a list of the properties of a typical transition metal.
   d) Compare the reactions of sodium (Group 1) and copper (a transition metal) with air. What can you say in general about the reactivity of the transition metals compared to metals from Group 1?
   e) Look at these two compounds :

 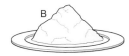

   Which one is the transition metal compound, and which is the alkali metal compound? How can you tell the difference?

8. Look at this table of data about 4 metals, labelled A, B, C and D.

Metal	Melting point (°C)
A	63
B	1494
C	1060
D	98

   Use letters A, B, C or D to answer a) and b).
   a) Which metals are probably transition metals?
   b) Which metals do you think are from Group 1?
   c) What would the data quoting the density of each metal show?

Further questions on pages 66 to 68.

chapter 5

We have looked at some groups of metals. Now let's move across to the other side of the Periodic Table and look more closely at some non-metal elements.

▶ Group 7 – The halogens

F	fluorine
Cl	chlorine
Br	bromine
I	iodine
At	astatine

Halogen molecule	Colour	State (at 25 °C)
F_2	pale yellow	gas
Cl_2	yellow / green	gas
Br_2	orange / brown	liquid
I_2	grey / black (violet vapour)	solid

The Group 7 elements are called the **halogens**.
Look at the table above :
Notice that all the halogens' atoms 'go round in pairs'.
They form **diatomic** molecules. In other words, 'two-atom' molecules, like F_2 and Cl_2.

You can also see some patterns in the table going down the group.
● Do they get darker or lighter in colour?

Look at their states :
● What is the pattern?
● Do their melting points and boiling points get higher or lower as we go down the group?
● Using the table, make some predictions about astatine (the element, at the bottom of Group 7, below iodine).

Now let's look at some of the reactions of the halogens :

The halogens form 'diatomic' molecules. (They are atomic twins !)

Q. Why can iodine molecules see so well?

A. Because they have two I's !

Experiment 5.1   The halogens with water

In this experiment you will use halogens dissolved in water.
Use a dropper to put a few drops of solution on to some universal indicator paper as shown :
● What do you see happen?
● Which halogen solution is most acidic?
● Which is the strongest bleach?

solutions of chlorine and bromine

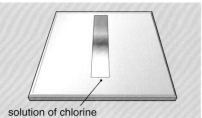

solution of chlorine

Chlorine is used as a bleach. (See page 54.)
It reacts with water :

chlorine + water ⟶ hydrochloric acid + chloric(I) acid (bleach)
$Cl_2(g)$ + $H_2O(l)$ ⟶ $HCl(aq)$ + $HOCl(aq)$

Bromine is less acidic than chlorine.
It is also a weaker bleach. Iodine is even weaker.

Reactivity of the halogens

Demonstration 5.2   Chlorine with iron

Your teacher will show you this experiment in a fume-cupboard. The chlorine can be made by dropping hydrochloric acid on to sodium chlorate(I). Heat the iron wool strongly as the chlorine passes over it.

● What do you see happen?

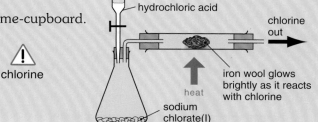

The iron and chlorine react vigorously.
Red/brown iron(III) chloride is made in the reaction.

iron + chlorine $\longrightarrow$ iron(III) chloride
$2\,Fe(s) + 3\,Cl_2(g) \longrightarrow 2\,FeCl_3(s)$

Notice how chlor**ine** changes to chlor**ide** in its compounds. In general, we say that the halogens form **halides**. Bromine forms bromides. Iodine forms iodides. Bromine does not react with iron as quickly as chlorine. The reaction with iodine is even slower. We find that:

> **The halogens get *less* reactive going down the group.**

● Which is the most reactive of all the halogens?

Halogens in competition

We can put the halogens into competition with each other to see which is more reactive.

The halogens also react with other non-metals. For example:

$H_2(g) + Cl_2(g) \longrightarrow 2\,HCl(g)$

The compounds formed are made of small molecules. They have low melting/boiling points.

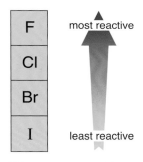

This pattern is explained on page 61

Experiment 5.3   Displacement from solution

Add chlorine solution to a solution of potassium bromide.
● What do you see happen?
  Do the same with a solution of potassium iodide.
● What happens?
  Now try mixing solutions of bromine and potassium iodide.
● What happens?
  If you have time, mix solutions of bromine, then iodine, with potassium chloride.

chlorine solution
bromine solution

chlorine solution

potassium bromide turns yellow as bromine is displaced

Chlorine is more reactive than iodine or bromine.
It can 'push' (**displace**) bromide and iodide out of solution.

chlorine + potassium bromide $\longrightarrow$ potassium chloride + bromine
$Cl_2(aq) + 2\,KBr(aq) \longrightarrow 2\,KCl(aq) + Br_2(aq)$

● Bromine is more reactive than iodine. Can it displace iodide from solution?
  Write a word and symbol equation for the reaction of bromine with potassium iodide in solution.

▷ Chemistry at work : Halogens and their compounds

Fluorine

Fluorine is the most reactive of all the
non-metal elements. It even attacks glass!
However, some of its *compounds* are useful.
Remember that the properties of elements are completely
different from those of the compounds they form.

Teflon is a fluorine compound. It is the non-stick lining on pans. (See page 155.)

Most toothpastes contain fluoride to prevent tooth decay. Some places have fluoride added to their water supplies. (See page 254.)

Chlorine

Chlorine and its compounds have many uses :

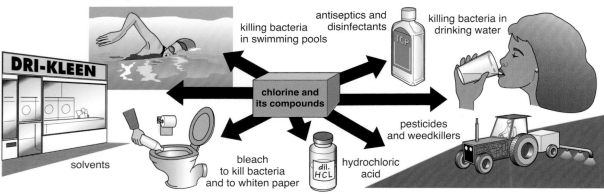

killing bacteria in swimming pools

antiseptics and disinfectants

killing bacteria in drinking water

DRI-KLEEN

solvents

bleach to kill bacteria and to whiten paper

dil. HCL

hydrochloric acid

pesticides and weedkillers

chlorine and its compounds

The most common chlorine compound is common salt,
sodium chloride. You can see some products made from it on page 105.

Bromine

Bromine is used in pesticides. It is also used to make
medicines. Silver bromide is used in photographic film.
(See page 254.)

Iodine

Iodine is an antiseptic. It is dissolved in alcohol,
and put on to cuts.

Iodine can be used before operations

▷ Halogens – working for or against us?

The halogens and their compounds have had both good and bad effects on the world.

Chlorine has had a bad reputation since it was used as the first chemical weapon in the First World War. Its compounds are also used in biological nerve gases.

However, chlorine's use in killing bacteria in **drinking water** has saved millions of lives around the world. It has greatly reduced the number of cases of cholera. However, about 25 000 people still die around the world every day from diseases spread in water.

There are still many people who have to drink untreated water

Yet even this use is now being questioned. Scientists are carefully measuring the poisonous compounds that the chlorine makes after it is added to our water supply.

These poisons were first noticed in discharges from paper mills. Chlorine is used to make paper white. Paper manufacturers now use less chlorine gas, and are looking for new ways to bleach paper.

Halogen-based **pesticides** have reduced deaths caused by diseases carried by insects, such as malaria. They also reduce the loss of crops in storage. On the other hand, these compounds can affect other wildlife in the food chain.

Halogen compounds are also damaging the **ozone layer**. This layer in the upper atmosphere protects us from harmful ultra-violet rays from the Sun. (See page 306.) **CFC**s (chloro-fluoro-carbons) were once hailed as new wonder compounds. They are very unreactive and were thought to be completely harmless to living things. They found uses in aerosols and as coolants in fridges.

However, scientists in the 1980s discovered a hole in our ozone layer – and it was growing. They predicted that more cases of skin cancer and eye cataracts would soon be observed. This has led to a search for different compounds to do the jobs of the CFCs.

CFCs are now being phased out, as are other halogen compounds used in fire extinguishers, flame-proofing and solvents. However, because they are so long-lasting, CFCs will remain in the atmosphere for many years to come.

These aerosol cans contain no CFCs

► Group 0 – The noble gases

He	helium
Ne	neon
Ar	argon
Kr	krypton
Xe	xenon
Rn	radon

Q. *What do you call a chemical element with no chemistry?*
A. *A noble gas!*
The most striking thing about this family of elements is their lack of reactivity.
They used to be called the inert gases.
Inert means 'having no reactions'.
However, in 1962 their first compound was made.
They were then re-named the **noble gases**.

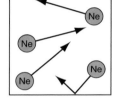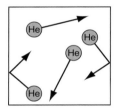

The noble gases prefer to be alone

As you know, most of the gases we have met so far are made up of molecules. Gases like oxygen, O_2; hydrogen, H_2; and chlorine, Cl_2. However, the noble gases are so unreactive that they are found as the atoms themselves. They are called **monatomic** gases ('one-atom' gases).

The noble gases are 'monatomic'

> **The noble gases are very unreactive.**

Why are the noble gases so unreactive?

Chemical reactions involve the electrons in the outer shells of atoms. Most atoms have a tendency to have their outer shell completely full of electrons. With **full outer shells**, atoms are **stable**.

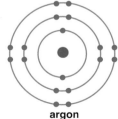

Atoms can give, take or share electrons with other atoms to get full outer shells. (See Chapters 19 and 20.)
So let's look at the atoms of the first 3 noble gases:
(Remember that the 1st shell can hold 2 electrons, the 2nd and 3rd shells can each hold 8 electrons.)

helium	neon	argon

What do you notice about their outer shells?

> **The noble gases are stable because their outer shells are full.**

Q. *Why are the noble gases so lonely?*

A. Because their friends argon!

56

Physical properties of the noble gases

Boiling points

Do you think that it takes a lot of energy to boil a noble gas?
What evidence have you got for your answer?
By the time we get to room temperature the noble gases
have already boiled at much lower temperatures.
In general, ***the forces between molecules get stronger
as the molecules get larger.***
So what do you think happens to the boiling points
(or boiling temperatures) of the noble gases
as we go down the group? Do you expect a pattern?
Look at the graph opposite:
So we can say that:

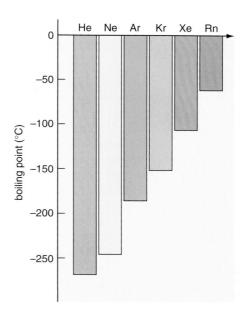

> **The boiling points (or boiling temperatures) of the noble gases increase going down the group.**

*Liquid helium boils at
−269 °C. That's only 4 °C
above the lowest
temperature you could
ever have in theory (called
absolute zero).*

Density

How would you expect the density of the noble gases to
change as we go down the group? Can you think why?
Look at the graph opposite:

*Helium has a very low
density*

> **The density of the noble gases increases going down the group.**

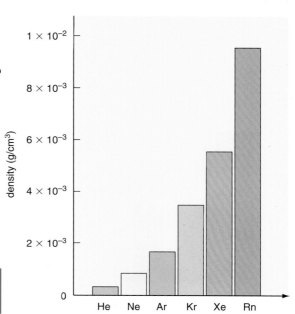

Helium

If we cool down metals to very low temperatures, they lose their electrical resistance. They become perfect conductors, or super-conductors. (See page 269.) Liquid helium is used to cool metals down.
Remember the graph on the previous page :
Helium boils at −269 °C! It is used to cool down the coils in body scanners in hospitals.
The coils can then make the very strong magnetic fields that scanners need.
These machines have become very important.
Doctors can now see a complete picture of the organs inside your body. This is very useful in finding cancers and helps doctors to judge what effect their treatment is having on tumours.

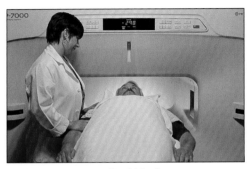
Body scanners use liquid helium

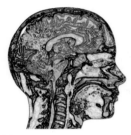 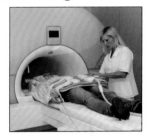

Have you ever seen an airship? Perhaps at a sporting event? Cameras on board provide those fantastic views looking down on the arena and crowds.
As you know, helium is much less dense than air.
It is used to fill airships to provide lift.
However, the first large airships built in the 1920's used an even lighter gas – hydrogen.

Aerial view from an airship

- Can you think of any dangers using this gas?
  Look at the photo below:
  Helium is much safer? Why?

Hydrogen is not safe to use in airships

Helium is used nowadays

▷ Chemistry at work : Noble gases

Neon

Late in the 1800's scientists were investigating gases and electricity.
They found that a gas at low pressure can be made to glow if we apply a high voltage to it. These discharge tubes helped scientists, such as J.J.Thomson, find out more about electrons and atoms.
Nowadays, we use the noble gases to make many of the brightly coloured lights you see in cities.
Neon is the gas used for red signs.

Argon

Light bulbs are filled with argon gas.
Why do you think that air is replaced by argon?
Argon won't react with the tungsten metal filament inside the bulb, even when it is white-hot.

Its lack of reactivity also helps in welding.
It acts as a shield around the hot metal in the weld.
● What might happen otherwise?
Argon is used in the extraction of titanium metal.
We also pump the gas into the molten mixture to stir it up when making steel.

Krypton

Krypton gas is in the lasers that doctors use to operate on eyes.

The laser opposite is being used to repair the retina at the back of the eye:

Radon

There is hardly any radon gas in the air.
But in some parts of the country people have had to check the levels of this gas in their homes.
In rocks, such as granite, there are small amounts of naturally occurring radium. This radioacive element breaks down into radon gas which is itself radioactive.
So the gas can build up and become a danger to health. Some houses have pumps installed to make sure the air inside is changed regularly.
Ironically, radon is used to help treat cancer.
● Research and write a brief report on 'The discovery of the noble gases'.

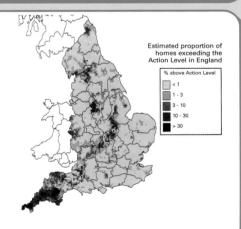

Estimated proportion of
homes exceeding the
Action Level in England

% above Action Level
< 1
1 - 3
3 - 10
10 - 30
> 30

► Atomic structure and the Periodic Table

We already know that the elements in the Periodic Table
are arranged in order of **atomic number**.
The atoms of each element have one more proton (and electron)
than the one before it.

The table below shows the way electrons are arranged
in the first 20 elements.(You don't need to know any more than this.)

Atomic number	Element	Electronic structure	Atomic number	Element	Electronic structure
1	H	1	11	Na	2,8,1
2	He	2	12	Mg	2,8,2
3	Li	2,1	13	Al	2,8,3
4	Be	2,2	14	Si	2,8,4
5	B	2,3	15	P	2,8,5
6	C	2,4	16	S	2,8,6
7	N	2,5	17	Cl	2,8,7
8	O	2,6	18	Ar	2,8,8
9	F	2,7	19	K	2,8,8,1
10	Ne	2,8	20	Ca	2,8,8,2

Can you see any patterns?

Look at these atoms from the first and last groups in the Periodic Table:

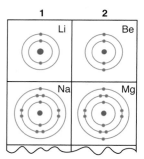

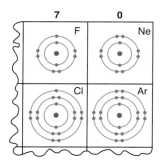

Can you see a link between the group number and the number of electrons in the outer shell?

> **Atoms of elements in the same group in the Periodic Table
> have the same number of electrons in their outer shell.**

For example, N – 2,5 and P – 2,8,5 are both in Group 5.

It is the electrons in the outer shell that get transferred
or shared in chemical reactions. This explains why elements
in the same group have **similar chemical reactions**.

But what about the transition metals? They are placed in the
4th period. That's because after the first 2 electrons go in the
4th shell, the next 10 electrons go into the inner 3rd shell.
You can think of the 3rd shell as holding 8 electrons, but
with space for another 10 in reserve!

*Although iodine has 53 electrons, you know
it has 7 electrons in its outer shell because
it is in Group 7*

Explaining patterns in the Periodic Table

Why are the alkali metals (from Group 1) so reactive?

Like all atoms, they want to react to get a full outer shell.
As elements, they have just 1 electron in their outer shell.
So when they react, they want to lose that 1 electron to
leave a full outer shell.

Group 1 elements are so reactive because it is easy for them
to get rid of just 1 electron.

- Why do you think that sodium, from Group 1, is more reactive
  than its neighbour from Group 2, magnesium?

Why do the metals get more reactive going down Group 1?

The outer electron gets easier to remove as you go down
the group.

Remember that electrons are negatively charged. They are
attracted to the positive protons in the nucleus. As you go down
the group, the atoms get bigger. Therefore the outer electron
gets further away from the attractive force of the nucleus.
This makes it easier for an electron to escape from
a bigger atom (even though the positive charge on
the nucleus is greater).

Why do the halogens get less reactive going down Group 7?

An atom of a halogen has 7 electrons in its outer shell.
It needs to gain 1 electron to fill up its outer shell.

A fluorine atom is a lot smaller than an iodine atom.
Therefore, an electron entering the outer shell of
a fluorine atom is nearer to the attractive force
of the nucleus. The electron is attracted more strongly.

Charged particles

Remember that when the alkali metals react,
they lose the electron from their outer shell.

Once the electron is lost, the atom is no longer neutral.
It becomes a charged particle called an **ion**.
It has one more proton than electrons, so its charge is $1+$.

All the elements in a group form ions with the same charge
(for example, Li^+, Na^+ and K^+).

On the other hand, the halogens gain an electron.

- How many protons and electrons does fluorine have
  after it has gained an electron?

The halogens all form $1-$ ions (F^-, Cl^-, Br^- and I^-).

Sodium is more reactive than lithium

The shielding effect
As well as distance from the
nucleus, another factor affects the
attraction of a nucleus for the
electrons in the outer shell. The
inner shells of electrons **shield** the
outer electrons from the pull of the
nucleus.
Does this fit with the patterns in
reactivity for Group 1 and Group 7?

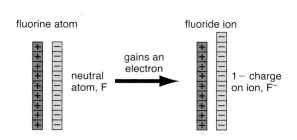

Summary

The elements of Group 7 are called the **halogens**.
They exist as diatomic molecules (F_2, Cl_2, Br_2, and I_2).
They are a reactive group of non-metals.
They get less reactive going down the group.

The elements in the last group (numbered 0) are called the **noble gases**.
They exist as single atoms.
They are all very unreactive.
Their atoms are very stable because their outer shells are full of electrons.

All the atoms of elements in the same group of the Periodic Table have the same number of electrons in their outer shell.

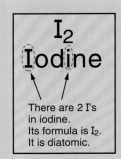

I_2
Iodine

There are 2 I's in iodine.
Its formula is I_2.
It is diatomic.

▷ Questions

1. Copy and complete :
   The elements in Group . . . . are called the halogens.
   Fluorine and . . . . are gases, ... is a liquid and iodine is a . . . . . They all exist as . . . . molecules, e.g. Cl_2 and Br_2.
   They get . . . . reactive as you go down the group. Therefore, ... is the most reactive halogen.

   The elements in Group 0 are called the . . . . gases. They are the most . . . . group in the Periodic Table, having almost no chemistry at all. This is because the outer shell of their atoms is . . . . of electrons.

   The number of the group an element is in usually tells us how many electrons are in the . . . . shell of its atoms. For example, iodine is in Group 7; therefore it has . . . . electrons in its outer shell.

2. Copy and complete these sentences about the uses of the noble gases :
   a) Helium is used to fill airships and balloons because . . . . .
   b) Liquid helium is used in body scanners because . . . . .
   c) Neon is used in advertising signs because . . . . .
   d) Argon is used inside light bulbs because . . . . .

3. Draw a picture in your book, or make a poster, showing the uses of the halogens and their compounds.

4. a) Draw a table showing the colour and physical state of fluorine, chlorine, bromine and iodine.
   b) Which is the **least** reactive of the halogens in a) ?
   c) You want to collect a gas-jar of chlorine gas. Copy and complete the diagram below. (Hint : chlorine is more dense than air.)

   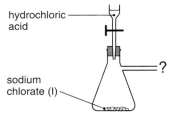

   hydrochloric acid

   sodium chlorate (I)

   ?

   d) Write a word and symbol equation for the reaction between chlorine and iron.

5. Sodium (Na) reacts with chlorine (Cl_2), making sodium chloride (NaCl).
   a) Write a word equation for this reaction.
   b) Now write a symbol equation. (Don't forget to balance your equation.)
   c) Write word and symbol equations for the reaction between sodium and fluorine.
   d) How would you expect the reaction with sodium to differ if you used fluorine instead of chlorine. Explain your answer.

Further questions on page 69.

▷ How science works

1. A student is asked to study the rate at which magnesium reacts with hydrochloric acid solutions of different concentrations.
The time taken for a piece of magnesium ribbon to completely disappear in the acid is measured. The experiment is repeated using different concentrations of acid.
a) What is the independent variable? [1]
b) What is the dependent variable? [1]
c) Suggest two control variables. [2]

2. A student planned an investigation to compare the temperature rises which occur when different metals are added to hydrochloric acid.
a) Suggest **three** variables which must be controlled so that the investigation is a fair test. [3]
b) Suggest **two** ways in which the student could make the temperature measurements as accurate as possible. [2]

3. A class of students were given the task of measuring the volume of hydrogen gas produced when 24 g of magnesium reacts with an excess of hydrochloric acid. The apparatus shown in the diagram was used.

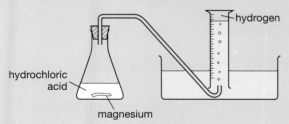

- They weighed a small piece of magnesium ribbon.
- The magnesium was dropped into the acid in the conical flask.
- The bung was quickly placed in the top of the flask.
- The hydrogen produced was collected in the measuring cylinder.
- The volume of hydrogen that would be produced by 24 g of magnesium was then calculated from their results.

The students all had slightly different results.
Suggest **three** possible reasons for the students not getting exactly the same result. [3]

4. Tom works as a scientist with a company that makes rubber products.
He carries out tests on two samples of rubber.
Sample **A** has been heated with chemicals. This is called vulcanised rubber.
Sample **B** has not been treated. This is unvulcanised rubber.
He tests the samples for the following:
- the resistance to breaking when it is stretched (tensile strength)
  The higher the value, the less likely it is to break.
- hardness
  The higher the value, the harder the rubber is.

Tom does each test five times.
The table shows Tom's results.

	Tensile strength in MPa					Hardness				
Sample A (vulcanised rubber)	33	34	33	34	36	43	38	45	65	40
Sample B (unvulcanised rubber)	3	4	2	5	3	22	29	24	28	25

a) i) Work out the best estimate for the tensile strength of sample A by calculating an average value.
You **must** show how you work out your answer. The unit will be MPa. [2]
ii) Explain why Tom repeated each test and took an average. [2]
iii) There are small differences in the repeat measurements.
Suggest **one** reason why. [1]
b) Look at Tom's results for the hardness of sample A.
Tom thinks that the best estimate for the hardness of sample A lies in the range 38 to 45.
Explain why. [2]
c) The discovery of the process of vulcanising rubber was important in the manufacture of car tyres.
How do Tom's results confirm this? [2] (OCR)

▶ Basic ideas

5. In the boxes below different atoms are represented by ● and ○

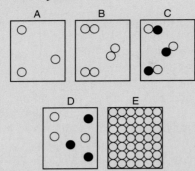

Match the letter on the box to the following descriptions:
a) a mixture of gases
b) a solid
c) a gaseous compound
d) oxygen (O_2)
e) a pure gas made up of single atoms.
[5] (WJEC)

6. The list below contains three elements and three compounds.

aluminium	copper
ammonia	sulfur
carbon dioxide	water

Say which are elements and which are compounds. *[3]* (AQA)

7. Choose words from this list to complete the sentences below.

hundreds	tens
millions	thirty
ninety	twenty

There are about . . . . naturally occurring elements.
From these elements . . . . of different substances are made. *[2]* (AQA)

8. Iron is a metal and sulfur is a non-metal. Explain why it is easier to separate iron from a mixture of iron and sulfur than from a compound of iron and sulfur. *[2]* (AQA)

9. The fertiliser a farmer uses is based on ammonium nitrate (NH_4NO_3). Is this substance an element or a compound? Explain your answer. *[2]* (AQA)

10. Chemists often give names to compounds, which end in '-ide' or '-ate'. What are the correct names for these compounds?
a) HCl   b) CuS   c) MgO
d) $CuSO_4$ e) $MgCO_3$ f) KNO_3 *[6]*

11. Bicarbonate of soda is a compound used in cooking.
The chemical formula of bicarbonate of soda is:

$NaHCO_3$

Name all the elements present in bicarbonate of soda. *[2]*

12. When calcium carbonate is heated it decomposes to form calcium oxide and the gas called carbon dioxide.

$$CaCO_3(s) \rightarrow CaO(s) + CO_2(g)$$

Choose one of the following phrases to complete this sentence.

more than	**the same as**	**less than**

The total mass of the products formed will be . . . . the total mass of the reactant. *[1]*

13. Each molecule of a compound contains two carbon atoms, six hydrogen atoms and one oxygen atom.
Write the chemical formula of this compound. *[2]*

14. a) Balance these chemical equations.
   i) $H_2 + O_2 \longrightarrow H_2O$ *[1]*
   ii) $Al + O_2 \longrightarrow Al_2O_3$ *[1]*
b) Briefly explain why an unbalanced chemical equation cannot fully describe a reaction. *[2]* (AQA)

15. The reaction between sodium and water may be represented by the following equation:

$$2Na(s) + 2H_2O(l) \longrightarrow 2NaOH(aq) + H_2(g)$$

Write down the meaning of **each** of the state symbols in brackets above.
[2] (WJEC)

▷ Atomic structure

16.

Element	Atomic number	Mass number
Oxygen	8	16
Chlorine	17	35
Gallium	31	70
Zinc	30	65
Tungsten	74	184

Use the table above, where necessary, to help you to answer these questions.
a) How many protons would you expect to find in an atom of oxygen ?
b) Where in the atom would the protons be found ?
c) How many neutrons would you expect to find in an atom of zinc ?
d) Which element in the table above has atoms which contain the same number of protons as neutrons ?   *[4]*   (EDEXCEL)

17. Use the information given about a sodium atom to complete the table.

mass number $^{23}_{11}\text{Na}$
proton (atomic) number

In one atom of sodium	the number of protons is	a)
	the number of neutrons is	b)
	the number of electrons is	c)

[3]   (AQA)

18. a) A diagram of the nucleus of an atom is shown below.

8p 8n

i) Draw a diagram to show the electronic arrangement of this atom.   *[1]*
ii) What is the mass number of this atom ?   *[1]*
b) Copy and complete the table shown below.

Name	Relative charge	Relative mass
proton	i)	1
ii)	zero (0)	iii)
electron	negative (-ve)	$\frac{1}{1840}$

[3]   (AQA)

19. Which two particles **W, X, Y** and **Z** represent different isotopes of the same element ?

Particles	Number in one particle		
	Protons	Neutrons	Electrons
W	1	1	1
X	1	2	1
Y	9	10	9
Z	9	10	10

[1]   (AQA)

20. Boron is a non-metallic element. The table shows information about two types of boron atom.

atom	symbol	protons	electrons	neutrons
Boron–10	$^{10}_{5}\text{B}$			5
Boron–11	$^{11}_{5}\text{B}$			

a) Copy and complete the table.   *[3]*
b) The relative atomic mass of boron measured precisely is 10.82. What does this tell you about the relative amounts of the two different boron atoms in a sample of boron ? (No calculation required.)   *[1]*
c) What is the electron arrangement in a boron atom ?   *[1]*   (OCR)

21. Look at the table.

Substance	Formula	Relative Formula Mass (M_r)
Decane	$C_{10}H_{22}$	142
Ethene	C_2H_4	
Octane	C_8H_{18}	

The relative atomic mass (A_r) of hydrogen is 1 and the relative atomic mass of carbon is 12. Work out the relative formula mass (M_r) of ethene and octane.   *[2]*   (OCR)

22. a) The formula for ammonia is NH_3. What does the formula tell you about each molecule of ammonia ?   *[3]*
b) Ammonia is used to make nitric acid (HNO_3). Calculate the formula mass (M_r) for nitric acid. (Show your working.)
(H = 1, N = 14, O = 16)   *[3]*   (AQA)

▷ **The Periodic Table**

23. Part of the Periodic Table which Mendeleev published in 1869 is shown below.

	Group 1	Group 2	Group 3	Group 4	Group 5	Group 6	Group 7
Period 1	H						
Period 2	Li	Be	B	C	N	O	F
Period 3	Na	Mg	Al	Si	P	S	Cl
Period 4	K Cu	Ca Zn	* *	Ti *	V As	Cr Se	Mn Br
Period 5	Rb Ag	Sr Cd	Y In	Zr Sn	Nb Sb	Mo Te	* I

Use page 392 to help you answer this question.

a) i) Give the symbols of **two** elements in Group 1 of Mendeleev's Periodic Table which are **not** found in Group 1 of the modern Periodic Table. *[1]*

    ii) Name these two elements. *[2]*

b) Which group of elements in the modern Periodic Table is missing on Mendeleev's table? *[1]*

c) Mendeleev left several gaps on his Periodic Table. These gaps are shown as asterisks (*) on the table above. Suggest why Mendeleev left these gaps. *[1]*

d) What is the missing word in the following sentence.
In the **modern** Periodic Table the elements are arranged in the order of their . . . . numbers. *[1]*

e) Mendeleev placed lithium, sodium and potassium in Group 1 of his Periodic Table. This was because they have similar properties.

Some properties of elements are given in the following table.

Four of them are properties of lithium, sodium and potassium. One of these properties has been ticked for you. Write down the other three properties.

Property	
They react with water to give alkaline solutions.	
They are gases.	
They are non-metals.	
They form an ion with a 1+ charge.	
They react with water and give off hydrogen.	✓
They form an ion with a 1− charge.	
They are metals.	
They react with water to give acidic solutions.	

[3]

24. Below is the outline of a Periodic Table. The letters represent elements but are **not** their chemical symbols.

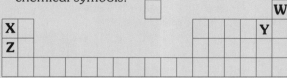

Which two elements are in the same period?
A **W** and **X**
B **W** and **Z**
C **X** and **Y**
D **Y** and **Z**   *[1]*   (AQA)

25. This is an outline of part of the Periodic Table.

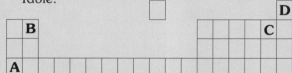

The four letters show the positions of four elements. Which letter represents an element with one electron in the outermost shell of each atom? *[1]*   (AQA)

26. The element potassium is in the same group of the Periodic Table as sodium. Potassium reacts with chlorine to make potassium chloride which is sometimes used instead of common salt in cooking.
a) Predict the formula of potassium chloride. *[1]*
b) By reference to the electronic structures of potassium and sodium, explain why the reaction of potassium with chlorine is similar to the reaction of sodium with chlorine. *[1]*   (AQA)

27. Sodium reacts violently with cold water.
a) Complete the word equation for the reaction.

sodium + water ⟶ sodium hydroxide + ... [1]

b) Describe in detail what you would **see** when a small piece of sodium is placed in a dish of cold water. [4]
c) At the end of the reaction, the solution in the dish contains sodium hydroxide. If universal indicator is added to this solution it turns purple.
  i)   What colour is universal indicator in water? [1]
  ii) Explain why the indicator goes purple in sodium hydroxide solution. [2]
  iii) Suggest a likely pH value for sodium hydroxide solution. [1]
d) Name a metal which reacts:
  i)   more violently than sodium with water,
  ii) less violently than sodium with water. [2]   (EDEXCEL)

28. a) The diagram shows the electronic structure of a particular element.

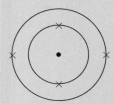

In a similar way, show the electronic structure of another element from the same group in the Periodic Table and name the element you select. [2]
b) The element lithium gives a moderate reaction with cold water, releasing hydrogen and forming a solution of lithium hydroxide.
Describe how sodium is:
  i)   similar to lithium and
  ii) how it is different from lithium, in its chemical reaction with cold water.
Explain any similarity or difference in terms of their atomic structure. [5]   (AQA)

29. The table below shows some properties of the Group 1 metals.

Element	Reaction with water	Melting point (°C)	Density (g/cm³)
Lithium	Floats, forms hydrogen and an alkaline solution. Burns with a crimson flame when lit with a splint.	180	0.53
Sodium	Floats, forms hydrogen and an alkaline solution. Burns with a yellow flame when lit with a splint.	98	0.97
Potassium	Floats, forms hydrogen and an alkaline solution. Burns with a lilac flame. Explosive reaction.	64	0.86
Rubidium	i)	39	1.53
Caesium	Explodes on contact with water. Forms hydrogen and an alkaline solution.	ii)	1.88

a) Use the table to predict:
  i) rubidium's reaction with water, [2]
  ii) caesium's melting point. [1]
b) In terms of electronic structure, what do the Group 1 metals have in common? [1]
     (EDEXCEL)

30. Which of the following represents the electronic structure of three elements in the same group of the Periodic Table?

A 2.1   2.8.1   2.8.8.1
B 2.7   2.8   2.8.1
C 2.8   2.8.1   2.8.2
D 2.8.2   2.8.3   2.8.4   [1]   (AQA)

31. **G**, **H**, **J** and **K** are four elements. These are their electronic structures:

G 2.1   **H** 2.6
J 2.8.1   **K** 2.8.8

Which two elements have similar chemical properties?
A **G** and **H**
B **G** and **J**
C **H** and **J**
D **J** and **K**   [1]   (AQA)

32. Use the Periodic Table on page 392 to help you to answer this question. In the diagram that follows, **e** represents an electron, **n** represents a neutron and **p** represents a proton.

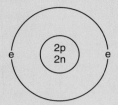

a) Identify the element, the atom of which is drawn above. [1]
b) Name another element that is in the same group. [1]
c) What do the outer electron shells of the other elements in this group have in common with each other? [1] (AQA)

33. Look at the Periodic Table on page 392.
a) Describe and explain the pattern in the electronic structure of the first 18 elements. [3]
b) Explain why the charge on a sodium ion is +1 but the charge on a chloride ion is −1. [2] (EDEXCEL)

34. Here is some information about the elements vanadium, manganese and cobalt.

	Vanadium	Manganese	Cobalt
Symbol	V	Mn	Co
Atomic number	23	25	27
Density (g/cm³)	6.0	7.2	8.9
Melting point (°C)	1130	1240	1492
Electrical conductivity	good	good	good
Common oxides	V_2O_3 V_2O_5	MnO MnO_2 Mn_2O_7	CoO Co_2O_3

a) Choose **three** properties from the table that show why these three elements are examples of *transition metals*. [3]
b) Give **two** more properties of transition metals or their compounds **not** mentioned in the table above. [2] (AQA)

35. Iron, nickel and platinum can act as catalysts. Several industrial reactions use catalysts to lower the cost of the manufactured products.
a) What type of metals are able to act as catalysts? [1]
b) Name an industrial process that uses a catalyst and name the catalyst. [2] (AQA)

36. Lithium, sodium, potassium and rubidium appear in this order in Group 1 of the Periodic Table. This table contains data about three of these elements.

Element	Melting point in °C	Boiling point in °C
Lithium	180	1330
Sodium	98	890
Potassium	—	—
Rubidium	39	688

Which of **A, B, C** or **D** is most likely to be potassium?

	Melting point in °C	Boiling point in °C
A	34	700
B	64	774
C	83	1007
D	120	817

[1] (AQA)

37. The table below lists some of the properties of the element silicon.

Property	Silicon
Appearance	black solid
Melting point	1410°C
Electrical conductivity	medium
Reaction with dilute acid	no reaction
Type of oxide	acidic

It is difficult to classify silicon. Give **two** pieces of evidence, from the table, for silicon being a metal and **two** pieces of evidence for it being a non-metal. [4] (EDEXCEL)

▷ **Groups of non-metals**

38. a) Fluorine is the first member of a group called the halogens.
Give the **names** of **two** other halogens in this group. *[2]*
b) Give **two** properties which the vapours of the halogens have in common.
[2]   (WJEC)

39. The table shows observations made when each of the halogens chlorine, bromine and iodine are added to solutions of sodium halides.
a) What are the missing entries labelled i)–iii) in the table?

	Sodium halide solution		
	sodium chloride	sodium bromide	sodium iodide
Chlorine	no reaction	ii)	brown solution formed
Bromine	no reaction	no reaction	iii)
Iodine	i)	no reaction	no reaction

[3]

b) Fluorine is the element at the top of Group 7.
Predict what will happen when fluorine is added to a solution of sodium chloride. *[1]*

40. The table shows the name and symbol of four elements.

Name	Symbol
Fluorine	F
Chlorine	Cl
Bromine	Br
Iodine	I

These elements are placed in the same group of the Periodic Table. Explain why.
[3]   (EDEXCEL)

41. The table below shows the elements of Group 7 of the Periodic Table.

Symbol	Name	Atomic number	Melting point (°C)
F			
Cl			
Br			
I			
	Astatine	85	–

a) Copy the table above. Use Table 1 on page 390 and the Periodic Table on page 392 to help you complete the table. *[2]*
b) The element astatine does not exist naturally on Earth. Draw a suitable graph to enable you to estimate its melting point. *[5]*
c) Use your knowledge of the properties of the elements in Group 7 to predict:
  i) the number of electrons in the outer shell (energy level) of an astatine atom,
  ii) the formula of a molecule of astatine,
  iii) the electrical charge on an astatide ion. *[3]*
d) From a knowledge of electronic structure, explain why scientists predict that astatine is the least reactive element in Group 7.
[3]   (AQA)

42. Look at the table below.
It shows some of the noble gases and their atomic numbers.

Noble gas	Atomic number
Helium	2
Neon	10
Argon	18
Krypton	36

a) Write down one use for argon. *[1]*
b) Write down the name of one **other** noble gas from the table. Write down one use of this gas. *[1]*
c) Which of the noble gases in the table has the highest density? *[1]*
d) What is the meaning of the term atomic number? *[1]*   (EDEXCEL)

chapter 6

Extraction of Metals

Chemistry is important in all our lives.
We rely on chemical reactions to make most
of the materials that we all take for granted.
Materials like the metals used to make bicycles,
cars, ships, trains and aeroplanes.

We will look at the chemistry involved in getting
these metals from their raw materials, **ores**, in
this chapter. Ores are rocks from which it is
economical to extract metals that they contain.
With new techniques, we can now extract metals
from rock once thought of as waste. (See page 97.)

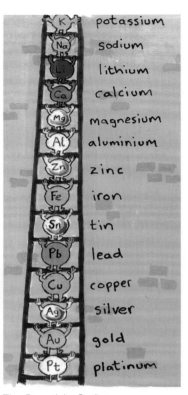

The Reactivity Series

The metals in ores are chemically bonded to
other elements. So how can we extract the metals?
To answer this we must understand the **Reactivity Series** of metals.

The Reactivity Series is like a 'league table' for metals.
The most reactive metals are at the top of the league.
The least reactive ones are at the bottom.
We can start putting the metals in order by looking at
their reactions with the oxygen in air, water and dilute acid.

Order of reactivity	Reaction when heated in air	Reaction with water	Reaction with dilute acid
potassium sodium lithium calcium magnesium aluminium zinc iron	burn brightly, forming oxide	fizz, giving off hydrogen; alkaline solutions (hydroxides) are formed	explode
		react with steam, giving off hydrogen; the metal oxide is formed (but no immediate reaction with cold water)	fizz, giving off hydrogen
tin lead copper	oxide layer forms without burning	only a slight reaction with steam	react slowly with warm acid
silver gold platinum	no reaction	no reaction, even with steam	no reaction

▶ Displacement reactions

We have now seen how to use the reactions of metals
to get an order of reactivity.
We can also judge reactivity by putting the metals
into competition with each other.

In the next two experiments the metals
will 'fight' each other to 'win their prize' – oxygen.
The more reactive metal will win the competition.

Experiment 6.1   Metals in competition – iron v copper

Mix a spatula of iron filings and copper oxide
in a test tube. Heat the mixture strongly.
- Is there a reaction? Look for a red glow
  spreading through the mixture.

When the tube has cooled, empty it into a dish.
- Can you see any pink copper metal left?

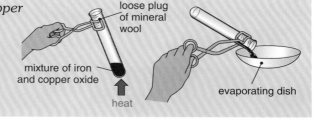

Copper starts off with the oxygen in copper oxide.
However, iron is more reactive, so it takes the oxygen
away from copper. You see a reaction take place,
and copper is left by itself.
We say that iron has **displaced** the copper.

copper oxide + iron ⟶ iron oxide + copper

This is a **displacement reaction**.

It shows us that iron is more reactive than copper.
Why wouldn't you expect to see a reaction between
iron oxide and copper?

iron oxide + copper ⟶✗ no reaction

You can now try some other displacement reactions.

Experiment 6.2   Displacement reactions

Try heating the mixtures of metals and oxides
shown in the table:
Look for any signs of a reaction.
✓ = a reaction   ✗ = no reaction
(Be careful when looking for signs of a reaction.
Zinc oxide turns yellow when you heat it by itself.
It turns white again when it cools down.)

Metal \ Metal oxide	zinc oxide	iron oxide	copper oxide
zinc	✗		
iron		✗	✓
copper			✗

- Why can you put some ✗'s in the table before you do the experiment?
- Write word equations for the reactions you have ticked.

We find that:

> **a more reactive metal can displace a less reactive metal from its compounds.**

▶ Metals of low reactivity

You can now start to understand how we get metals from their ores.

Most metals are found naturally in rocks called ores. They are in compounds, chemically bonded to other elements. However, the unreactive metals at the bottom of the Reactivity Series can be found as the elements themselves. We say that they are found **native**.

We can find copper, silver, gold, and platinum as the metals in nature. (Copper and silver compounds are also mined as ores.)
Look at the photo below :

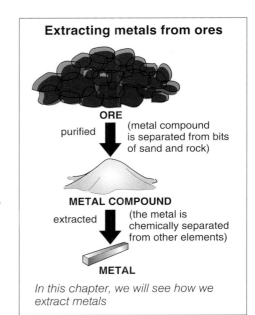

Extracting metals from ores

ORE
purified (metal compound is separated from bits of sand and rock)

METAL COMPOUND
extracted (the metal is chemically separated from other elements)

METAL

In this chapter, we will see how we extract metals

Gold is found as the metal itself

● Why can gold be found native ?

Gold is very expensive because it is difficult to find. For example, there is plenty of gold in the sea. Unfortunately, it is spread all round the world. This makes it too costly to extract from sea-water.

These people are 'panning' for gold in Brazil

Roasting ores

Many metal ores contain oxides or sulfides of the metal. Copper is found in an ore called chalcocite. This has copper(I) sulfide in it.
We can get the copper from this ore just by heating it in air.
Look at the equation below :

copper(I) sulfide + oxygen $\xrightarrow{\text{heat}}$ copper + sulfur dioxide
$$Cu_2S(s) \ + \ O_2(g) \longrightarrow 2\,Cu(s) + SO_2(g)$$

Care must be taken to stop sulfur dioxide gas escaping into the atmosphere. This gas causes acid rain. (See page 148.)
However, the gas can be piped to a nearby chemical plant that makes sulfuric acid. (See page 135.)

The copper in this ore is chemically bonded to sulfur and iron. We must heat it to extract the copper.

► Metals of medium reactivity

The metal above copper in the Reactivity Series is lead.
Lead can be extracted from an ore that contains
lead sulfide. Look at the experiment below:

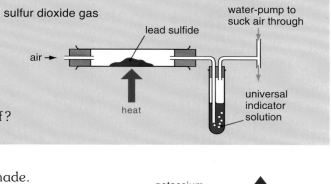

Demonstration 6.3   Roasting lead sulfide
This should be done in a fume-cupboard.

Turn on the water-pump to suck air
over the lead sulfide.
Now heat the lead sulfide.

⚠ sulfur dioxide gas

lead sulfide

water-pump to
suck air through

air →

heat

universal
indicator
solution

- What happens to the indicator solution?
- What does this tell you about the gas given off?
- Can you see any lead metal forming?

When we roast copper(I) sulfide, we get copper.
However, when we roast lead sulfide, no lead is made.
Instead, we get lead oxide (plus sulfur dioxide):

lead sulfide + oxygen $\longrightarrow$ lead oxide + sulfur dioxide
$2\,PbS(s)$ + $3\,O_2(g)$ $\longrightarrow$ $2\,PbO(s)$ + $2\,SO_2(g)$

The sulfur dioxide can be used to make sulfuric acid.
(See page 135.)
There is a link between a metal's place in the Reactivity Series
and how *easy* it is to extract:

> **The more reactive a metal is, the more difficult it
> is to extract from its ore.**

Reactive metals are eager to join chemically with non-metals.
Once they have formed compounds, they don't want
to change back into the metal again!
Less reactive metals are easier to persuade!
Their compounds are easier to break down.

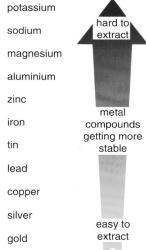

potassium
sodium
magnesium
aluminium
zinc
iron
tin
lead
copper
silver
gold

hard to
extract

metal
compounds
getting more
stable

easy to
extract

After our experiment above, we are left with lead oxide.
How can we get the lead from the lead oxide?
This is where **carbon** comes into the story.

The whole story is on the next page!

Lead is left by itself as carbon flies off with the oxygen!

► Extracting metals with carbon

Carbon and the Reactivity Series

Carbon is a non-metal. However, we can put it into our league table of metals. It slots in between aluminium and zinc. This means that carbon can displace any metal below aluminium in the Reactivity Series.

We get carbon from coal. Coal is cheap and there's plenty of it at present. Therefore, extracting metals with carbon is not too expensive.

You can try to extract lead and copper from their oxides in the next experiment:

potassium		carbon cannot be
sodium		used to extract the
magnesium		more reactive
aluminium		metals
CARBON		
zinc		
iron		these metals can
tin		be extracted
lead		using carbon
copper		

Experiment 6.4   Extracting lead and copper

Mix a spatula of carbon powder with a spatula of lead oxide.
Set up the apparatus as shown:
Heat it gently at first, then more strongly.
Look for signs of a reaction in the test-tube.
• Can you see any silvery beads left after the reaction? This is lead metal.

Try the same experiment, using carbon and copper oxide.
• Can you see any copper metal (a pinkish brown powder)?

lead oxide
copper oxide

loose plug of mineral wool (to stop mixture shooting out)

mixture of carbon and lead oxide

heat

Carbon is more reactive than lead.
Therefore, it can displace lead from lead oxide.

lead oxide + carbon ⟶ lead + carbon dioxide
$2\,PbO(s)$ + $C(s)$ ⟶ $2\,Pb(s)$ + $CO_2(g)$

Reduction and oxidation

Look at the equation above.
The lead oxide loses its oxygen. We say that lead oxide is **reduced**.

The carbon gains oxygen. We say that carbon is **oxidised**.

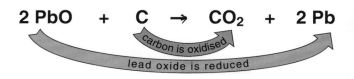

$$2\,PbO \quad + \quad C \quad \rightarrow \quad CO_2 \quad + \quad 2\,Pb$$

carbon is oxidised
lead oxide is reduced

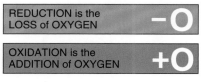

REDUCTION is the LOSS of OXYGEN   −O

OXIDATION is the ADDITION of OXYGEN   +O

Reduction and oxidation are chemical opposites. (For more details see page 99.)

▶ Extraction of iron

Carbon is important in the extraction of iron.
We use a giant **blast furnace** to get the iron
from its ore. The raw materials are fed into the top
of the furnace. The raw materials are:

- **iron ore** (mainly haematite, iron(III) oxide),
- **coke** (a cheap form of carbon, made from coal),
- **limestone** (to get rid of sandy waste).

Look at the diagram:
Can you see why it is called a **blast** furnace?
It's a bit like a huge barbecue. The temperature
gets above 1500 °C.

Reactions in the blast furnace

1. The coke (carbon) reacts with oxygen in the
   hot air to make carbon dioxide.

$$C(s) \ + \ O_2(g) \longrightarrow CO_2(g)$$

2. This carbon dioxide reacts with more hot coke
   to make **carbon monoxide** gas.

$$CO_2(g) \ + \ C(s) \longrightarrow 2\,CO(g)$$

3. The carbon monoxide then **reduces**
   the iron oxide to iron.

$$Fe_2O_3(s) \ + \ 3\,CO(g) \longrightarrow 2\,Fe(l) \ + \ 3\,CO_2(g)$$

4. Limestone breaks down into calcium oxide (a base).
   This gets rid of the sandy bits (acidic impurities) in
   the iron ore. They form a liquid **slag**.

Notice that the iron oxide is reduced to iron
by carbon monoxide gas in step 3.

$$Fe_2O_3 \ + \ 3\,CO \longrightarrow 2\,Fe \ + \ 3\,CO_2$$

We say that carbon monoxide is the **reducing agent**.

> **Reducing agents take away oxygen.**

At the high temperatures in the furnace,
the iron formed is molten (a liquid). It sinks
to the bottom of the furnace.
The impure iron is run off into moulds.

The molten slag floats on top of the iron.
The slag is tapped off, cooled and
used for making roads.

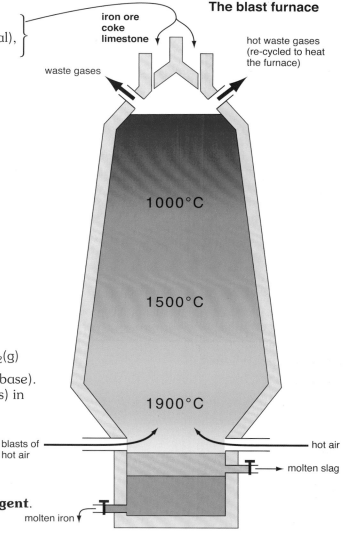

The blast furnace

iron ore
coke
limestone

waste gases

hot waste gases
(re-cycled to heat
the furnace)

1000°C

1500°C

1900°C

blasts of
hot air

hot air

molten slag

molten iron

▶ Iron into steel

The iron from the blast furnace is impure.
It contains 3 % to 4 % carbon, plus some other non-metals.
The impure iron is very brittle. It smashes easily.
Most of it gets turned into steel which is much tougher.

Steel is made up mainly of iron. It has a tiny amount
of carbon left in it. Other metals can also be added.
There are 2 main steps in turning iron into steel :

1. Removing most of the carbon
This is done by blowing oxygen into molten iron
from the blast furnace. Carbon burns
and escapes as carbon dioxide gas.

2. Adding other metals
We can also add other metals to give the steel
special properties. For example, we make
stainless steel by adding chromium
and nickel. This type of steel does not rust.
It is used to make cutlery.

You can find out more uses of steel on page 78.

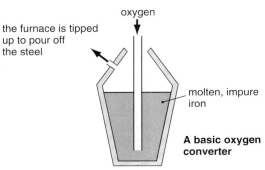

oxygen

the furnace is tipped
up to pour off
the steel

molten, impure
iron

**A basic oxygen
converter**

*Iron is turned into steel. The oxygen burns off the
carbon and other non-metal impurities.*

*Stainless steel
does not rust*

Changing a rusty exhaust

▶ Corrosion of iron

The rusting of iron costs us millions of pounds every year.
Rust forms on the surface of iron (or steel).
Unfortunately, it is a soft, crumbly substance.
It soon flakes off, then more iron rusts.
Let's find out what causes iron to rust :

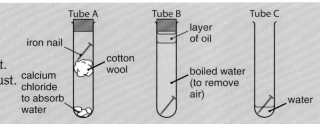

Experiment 6.5   What causes iron to rust?

Set up the test-tubes as shown :
Tube A tests to see if air alone will make iron rust.
Tube B tests to see if water alone will make iron rust.
Tube C tests to see if air and water will make iron rust.
Leave the tubes for a week.
• What do you see in each test tube?

Tube A — iron nail, cotton wool, calcium chloride to absorb water
Tube B — layer of oil, boiled water (to remove air)
Tube C — water

Rust is a form of iron(III) oxide. It has water
loosely bonded to it.
It is called **hydrated iron(III) oxide**.

> **Both air (oxygen) and water
> are needed for iron to rust.**

*In order for iron to rust
Both air and water's a must.
Air alone won't do
Without water there too,
So protect it, or get a brown crust!*

We can show the reaction as :

iron + oxygen + water ⟶ hydrated iron(III) oxide

Experiment 6.6 Comparing rates of corrosion

Set up the test tubes as shown:
Leave the tubes for a week.

- Which tube has most rust in it?

Test some other metals, including aluminium,
in the same way.

- Compare their corrosion with iron's.

water | salt water (sodium chloride solution) | weakly acidic solution (acid rain)

Salt speeds up the rusting of iron.
- Car owners who live near the sea
  should wash their cars regularly. Why?
- Why is salt a problem for car owners in winter?
  What can they do to reduce the problem?
Acid rain also speeds up corrosion of metals.
- Why does aluminium resist corrosion? (See page 89).

Preventing rust

We know that air and water are needed for iron to rust.
Therefore, if we can keep these away from iron, it can't rust.
We can do this by coating the iron or steel with :
1. paint   2. oil or grease   3. plastic
4. a less reactive metal, or   5. a more reactive metal.

Let's look at how good these methods are :

How has the fence around these tennis courts been protected?

Experiment 6.7 Stop the rust

Set up the test tubes as shown :
The rust indicator turns blue at the first signs of rust.
Look at your test tubes every few minutes.

- Which method is best at preventing rust?
- Explain why we set up a control experiment.

water with rust indicator

greased nail | painted nail | nail wrapped in zinc | nail in copper | nail in tin | nail in magnesium | control experiment

Most methods for preventing rust rely on
keeping the iron away from air and water.
The iron will rust if there is even a tiny gap in the coating.
Then the rust soon spreads under the coating.

However, this does not happen if you use
a more reactive metal. Even if the coating is scratched,
the iron does not rust.
Zinc is often used to protect iron.
We say that the iron is **galvanised**.

The zinc is more reactive than the iron. Therefore,
any water or oxygen reacts with the zinc rather than the iron.
This is called **sacrificial protection**.
The zinc sacrifices itself to protect the iron. (See page 79.)

*If the tin coating is broken, the can will rust
– harmful bacteria will then get to the food.
Why can't we use zinc to coat food cans?*

► Chemistry at work : Uses of steel

Steel is used more than any other metal.
It is very important in the building industry.
It is used for girders and for the rods inside
reinforced concrete.
You will have seen the steel tubes, called scaffold,
used when buildings are made or repaired.

Workers erecting steel scaffold

Different types of steel

As you know from page 76, steel is made mainly
from iron. It has a small amount of carbon in it.
The amount of carbon affects its properties.
Look at this table :

Carbon steels

Type of steel	Amount of carbon	Hardness	Uses
mild steel	0.2%	can be easily shaped	car bodies, wires, pipes, bicycles
medium steel	0.3% to 0.6%	hard	girders, springs
high-carbon steel	0.6% to 1.5%	very hard	drills, hammers, other tools

- What happens to the hardness of the steel as we
  increase the amount of carbon ?
- Why do you think that mild steel is used to make cars ?

Why don't we use mild steel for rail tracks ?

Alloy steels

You can change the properties of steel by adding other metals.
For example, **tungsten** in steel keeps it hard *and* tough, even
when it gets very hot. It is used to make high-speed cutting tools.

Mixtures of metals are called **alloys**. Steels with less
than 5 % other metals are called low alloy steels.
Those with up to 25 % other metals, such as stainless
steel, are high alloy steels.
You can read more about alloys in Chapter 21.

*How can steel cutting tools be
used to shape steel objects ?*

► Chemistry at work : Preventing rust

Stainless steel

Unfortunately, iron and steel rust. (See 'Cars for scrap', page 91.)
However, we have found some very effective ways to fight against rust.

You have just read about alloy steels.
If **chromium** and **nickel** are added to steel, you get **stainless steel**, a steel which does not rust.
Stainless steel is expensive, but is used for small items, such as knives and forks.

● Why don't we make cars from stainless steel ?

Stainless steel is used to make a surgeon's instruments. It is important that they do not rust.

Sacrificial protection

You have seen on page 77 that iron can be coated in zinc (galvanised). This works even if the coating gets scratched.

● How does this method work ?

Magnesium can be used instead of zinc.
It is used in harsh conditions.

Look at the photos opposite :

● Why are these things likely to rust very quickly without the protection of magnesium ?

This pier is built on steel legs

Magnesium bars are bolted onto this ship's hull

Electrical protection

SHOCKING RUST !

Concrete is the most widely used building material.
It is reinforced by steel rods which make it much stronger. However, some concrete buildings and bridges are showing signs of weakness. The steel inside the concrete is rusting away!

Rust takes up much more space than steel so the concrete cracks !
Scientists have found a way to stop this.
A small electric current passing through the steel will protect it. Many new structures will be protected this way.

► The highly reactive metals

The highly reactive metals are the most difficult
to extract from their ores. These metals are found
in very stable compounds.

Reduction by carbon won't work. You know
carbon's place in the Reactivity Series. It lies
just under aluminium. Therefore carbon can't displace
the highly reactive metals from their ores.

However, we do have a way to get these metals.
It is called **electrolysis**. We will look at this
in more detail in the next chapter. You will see
how aluminium is extracted by electrolysis.
Once you have separated the metal compound from the ore,
the next 2 steps are :

1. *melt* it, then
2. *pass electricity* through it.

Both steps use up a lot of energy.
Therefore highly reactive metals are expensive to extract.

*Reactive metals form stable compounds. This
makes them difficult to extract !*

*It takes a lot of energy to split up compounds of
reactive metals, like potassium – and that costs a
lot of money, making these metals expensive.*

Summary

The way we extract a metal from its ore depends on
its place in the Reactivity Series.
The more reactive a metal is, the harder it is to extract.

Look at the list of metals below :

Order of reactivity	Method of extraction
potassium (K) sodium (Na) lithium (Li) calcium (Ca) magnesium (Mg) aluminium (Al)	**Electrolysis** The metal compound is : 1. melted, then 2. has electricity passed through it.
zinc (Zn) iron (Fe) tin (Sn) lead (Pb)	**Reduction** by carbon For example, $ZnO + C \longrightarrow Zn + CO$ Carbon monoxide is formed when we extract zinc. (If the ore is a sulfide, it is roasted first to get the oxide.)
copper (Cu) silver (Ag) gold (Au) platinum (Pt)	These metals can be found uncombined, as the metal itself. We say that they are found **native**. (Copper and silver are often found as ores but they are easy to extract by roasting the ore.)

difficult
to
extract

easy to
extract

► Questions

1. Copy and complete:
   The method used to extract a metal from its ore depends on its position in the .... *Reactivity series* :
   a) Metals of low reactivity can be found . . . . , as the metal itself.
   b) Metals of medium reactivity must be extracted from their ores by heating with . . . . . The metal oxide has its oxygen . . . . by the carbon. We say that it has been . . . . by the carbon.
   Iron is extracted in a . . . . furnace. Carbon, in the form of . . . . , is mixed with iron ore and . . . . . The main reducing agent in the furnace is . . . . gas.
   c) Highly reactive metals are extracted by . . . . . There are 2 main steps:
   1. the metal compound is . . . . , then
   2. . . . . is passed through it.

2. This table shows when some metals were discovered:

Metal	Known since:
potassium	1807
sodium	1807
zinc	before 1500 in India and China
copper	ancient civilisations
gold	ancient civilisations

   a) What pattern can you see between a metal's place in the Reactivity Series and its discovery?
   b) Can you explain this pattern?

3. Lead is found in the ore galena. The lead is in a compound called lead sulfide, PbS.
   To extract the lead, we first roast the ore. It is then reduced by carbon.
   a) Write word and symbol equations to show what happens when lead sulfide is roasted in air.
   b) i) What useful product can be made from the waste gas made when lead sulfide is heated?
      ii) Why is it important to stop this gas escaping into the atmosphere?
   c) Explain what happens in the reduction with carbon. Include a word equation and a symbol equation in your answer.

4. Look at this table which shows how much some metals cost:

Metal	Cost ($ per tonne)
lithium	61 500
magnesium	2250
aluminium	1650
zinc	980

   a) Can you see a pattern between the position of these metals in the Reactivity Series and their cost?
   b) Can you explain this pattern?
   c) Gold costs $12 300 000 per tonne. Platinum costs about the same as gold.
      i) Is your pattern from a) true for the whole Reactivity Series?
      ii) Can you explain the high cost of gold and platinum?

5. a) Copy this diagram of a blast furnace and fill in the blanks:

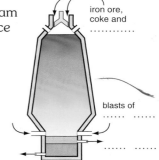

   b) Which substance is the main reducing agent in a blast furnace?
   c) Some iron(III) oxide is reduced by carbon in the furnace.
      i) Write a word equation for this reaction.
      ii) Write a symbol equation for this reaction.
   d) What is the main impurity in the iron we get from a blast furnace?

6. a) How is iron turned into steel?
   b) List some uses of steel.
   c) Name the metals added to iron when making stainless steel.

7. a) How could you show in an experiment that *both* **air (oxygen)** *and* **water** are needed for iron to rust?
   b) List 5 ways to prevent iron rusting.
   c) *Explain* which method is best to protect i) a bicycle chain, and ii) a dustbin.

Further questions on page 119 to 121.

81

chapter 7

Electrolysis

We met the word **electrolysis** in the last chapter.
You've probably seen it in adverts for beauty shops.
It can be used to remove unwanted hair.
More importantly for us, it is the way that we extract
highly reactive metals from their compounds.

> **Electrolysis is the break-down of a substance
> by electricity.**

Now you can find out more about electrolysis.

▶ Electrolytes

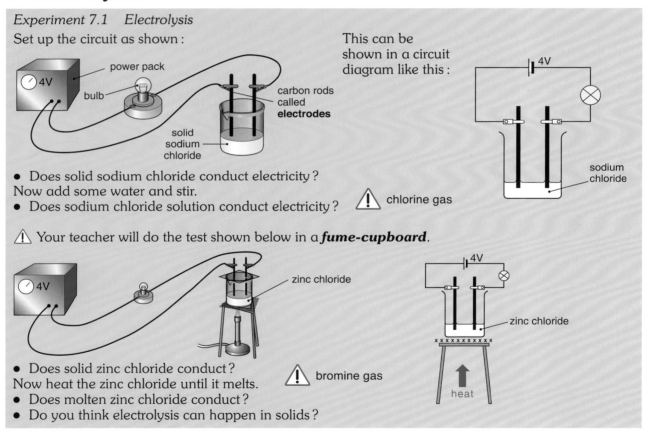

Experiment 7.1   Electrolysis

Set up the circuit as shown :

power pack
bulb
carbon rods called **electrodes**
solid sodium chloride

This can be shown in a circuit diagram like this :

4V
sodium chloride

• Does solid sodium chloride conduct electricity?
Now add some water and stir.
• Does sodium chloride solution conduct electricity? ⚠ chlorine gas

⚠ Your teacher will do the test shown below in a **fume-cupboard**.

zinc chloride

4V
zinc chloride
heat

• Does solid zinc chloride conduct?
Now heat the zinc chloride until it melts. ⚠ bromine gas
• Does molten zinc chloride conduct?
• Do you think electrolysis can happen in solids?

We find that electrolysis only happens in liquids.
Liquids which can be electrolysed are called **electrolytes**.

> **Electrolytes are liquids (molten compounds or compounds in
> solution) which are broken down when they conduct electricity.**

Experiment 7.2   Which solutions are electrolytes?
The substances in the table below have been dissolved in water.
You can test which conduct electricity.

Set up the circuit as shown :
Do not breathe in any gases given off.

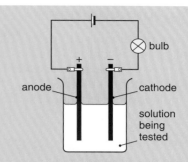

Make sure you don't let the electrodes touch each other.
Rinse your beaker and electrodes well after each test.
Put your results in a table like this :

Solution	Does it conduct ?	What do you see at each electrode ?	
		+ (anode)	− (cathode)
potassium chloride (KCl)			
sucrose ($C_{12}H_{22}O_{11}$)			
ethanol (C_2H_5OH)			
sodium iodide (NaI)			
zinc bromide ($ZnBr_2$)			
calcium hydroxide ($Ca(OH)_2$)			
dilute sulfuric acid (H_2SO_4)			
glucose ($C_6H_{12}O_6$)			
copper sulfate ($CuSO_4$)			

• Which solutions conduct electricity ?
These solutions are the **electrolytes**.
• Did you see the reactions at the electrodes ?
Electrolytes usually contain compounds of metals and non-metals.
• Which is the exception in the table above ?

When electrolytes conduct, we see reactions at the electrodes.
The reactions at the electrodes are explained later in the chapter.

Why do electrolytes conduct ?

Most electrolytes contain metals and non-metals.
These compounds are all made up of *ions*.
We have met ions before on page 61.
Remember that ions are *charged particles*.

We saw in Experiment 7.1 that solid compounds
don't conduct electricity.
Solids like sodium chloride contain ions,
but they are fixed in position.
Their ions can't move to the electrodes.
However, when they are melted or dissolved in water,
the *ions become free to move* around.

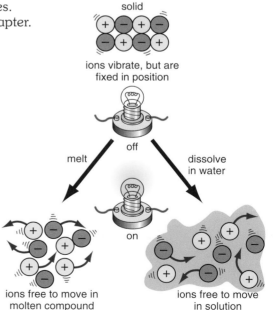

| Electrolytes contain ions. |
| When molten or dissolved in water, the *ions are free to move*. |

▶ Moving charges

We know that liquids containing ions conduct electricity.
Yet solids which contain ions don't conduct.
We think it must be the **movement** of ions in liquids
that is important in electrolysis.
Let's look at some evidence for the theory of
the moving ions:

In order to make your lamp bright,
You need an elec-tro-lyte.
The ions must flow,
Through liquids they go,
In solids they're just packed too tight!

Experiment 7.3   Ions on the move

We can use coloured ions to see which way
ions move during electrolysis.

a) Potassium manganate(VII)

Set up the apparatus as shown:
Use tweezers to place the potassium manganate(VII)
in the middle of the damp filter paper.
Leave the power pack on for 20 minutes.

The potassium ions are **colourless**.
The manganate(VII) ions are **purple**.

potassium manganate(VII) starts in the middle

damp filter paper on a glass slide

- What happens on the filter paper?
- Which electrode do the manganate(VII) ions move to?
  How can you tell?
- Do you think the manganate(VII) ions carry
  a positive or negative charge? (Remember that
  opposite charges attract.)
- Which way do you think the potassium ions move?

Ions carry their charge to the electrode

b) Copper chromate(VI)

Your teacher will set up this apparatus:
Leave the power on until you can see colours
near the electrodes.

Copper ions are **blue**.
Chromate(VI) ions are **yellow**.

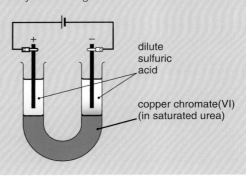

dilute
sulfuric
acid

copper chromate(VI)
(in saturated urea)

- Which colour can you see near the positive electrode?
- Which ion is attracted to the positive electrode?
- Which colour can you see near the negative electrode?
- Which ion is attracted to the negative electrode?

In general we can say that:

metal ions are positive

non-metal ions are negative

For example, Na^+, K^+, Ca^{2+}, Al^{3+}
(exceptions: H^+, hydrogen ions
and NH_4^+, ammonium ions)

For example, Cl^-, Br^-, I^-, O^{2-}
(exceptions: complex transition metal ions, like the ones in
the experiment above, MnO_4^-, CrO_4^{2-})

Cations and anions

We know that metal ions are positively charged.
We also know that opposite charges attract.

Therefore, as we have seen on the previous page:

> **metal ions are always attracted to the negative electrode (cathode).**

Positive ions are sometimes called **cations**.
This comes from the word cathode – a negative electrode.

Negative ions are attracted to
the positive electrode (anode).
These ions are sometimes called **anions**.
Why are they called anions?

Cat-ions are 'pussytive'!
Cations go to the cathode.
When they get there they
receive electrons.
(See page 87.)

Why are ions charged?

Lithium fluoride is made up of ions.
Let's look at the difference between the atoms
of lithium and fluorine and their ions:

CATions → CAThode
ANions → ANode

Can you remember your work on atomic structure
in Chapter 3?
Atoms have the same number of protons (+) as electrons (−).
Therefore atoms have no charge. They are neutral.

The charge in an atom is balanced

Lithium **atom**, Li		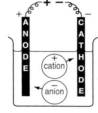
number of protons = 3+		
number of electrons = 3−		
overall charge = 0		

Fluorine **atom**, F		
number of protons = 9+		
number of electrons = 9−		
overall charge = 0		

However, in ions there are different numbers of
protons and electrons. Metals lose electrons
and non-metals gain electrons. You can read about
why this happens on page 248.

Now count the electrons (−) around each ion below:
Look at the positive charge in each nucleus.
Can you see why lithium is positively charged?

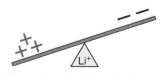

The charge in an ion is un-balanced.
Can you draw see-saws for F and F⁻?

Lithium **ion**, Li⁺		
number of protons = 3+		
number of electrons = 2−		
overall charge = 1+		

Fluoride **ion**, F⁻		
number of protons = 9+		
number of electrons = 10−		
overall charge = 1−		

Magnesium ions have a 2+ charge.
- How many more protons than electrons does an Mg^{2+} ion have?

► Electrolysis of copper chloride

We know that ions are attracted to
oppositely charged electrodes.
But what happens to the ions when they
get to the electrode? The next experiment
will help you to find out.

Experiment 7.4   Breaking down copper chloride

Set up the circuit as shown in a
fume-cupboard or well-ventilated room.
Watch what happens at each electrode.
Test the gas given off above the positive electrode
with damp litmus paper.
Turn off your power pack as soon as
you have detected the gas.

- What happens to your litmus paper?
- Which gas is this the test for?

Lift the negative electrode from the beaker.

- What does it look like?
- What do you think has formed on this electrode?

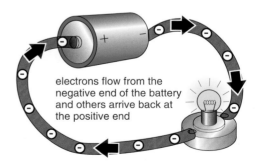

chlorine gas
copper chloride solution

4V
bulb
carbon
electrodes
copper
metal forms
copper
chloride
solution

damp litmus
paper is
bleached

Copper metal forms on the negative electrode (cathode).
Chlorine gas is given off at the positive electrode (anode).
The copper chloride is **electrolysed**.
It is broken down into its elements, copper and chlorine.

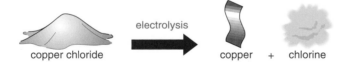

copper chloride   electrolysis   copper   +   chlorine

Explaining electrolysis

To understand how copper chloride is broken down,
we need to know how electricity flows around a circuit.
An electric current is the flow of electrons.
Look at the diagram opposite:

Remember that electrons are tiny negative particles.
As they are 'pushed out' of the negative end of the battery,
others are attracted back into the positive end.

During electrolysis, the electrons from the battery
go to the negative electrode. Here they meet the
positive ions. The **ions** then change into neutral **atoms**.

In the experiment above, copper ions changed into
copper atoms.

This carbon electrode has been coated with copper metal

electrons flow from the
negative end of the battery
and others arrive back at
the positive end

Electrons flow through the wires and bulb. This is an electric current.

Negative ions gather at the positive electrode.
The ions are negatively charged because they carry
extra electrons. (Look back to the example
of the fluoride ion on page 85.)
Their extra electrons are removed at the positive electrode.
These get 'sucked back' into the battery. In our experiment
we saw bubbles of chlorine gas on the positive electrode.

Look at the cartoon opposite:
Electrons leave the negative end of the battery.
Electrons arrive back at the positive end of the battery.
Therefore we have a complete circuit.
A bulb in the circuit will light up.

However, no free electrons pass through the electrolyte.

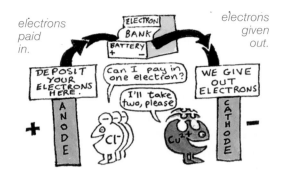

Electrolysis works like a bank. The ions are the customers. The electrons are the money. Electrons flow to and from the battery but not between the electrodes.

Ions carry charge through electrolytes. No free electrons pass through.

Half-equations

We can describe what happens at the electrodes
using **half-equations**.
These show what happens to the ions during electrolysis.

Let's look at the electrolysis of copper chloride
in more detail:

Copper ions have a 2+ charge. The original copper atoms have
lost 2 electrons. (See the example of the lithium ion on page 85.)
At the negative electrode the copper ions gain 2 electrons.
The ion is no longer charged. It becomes a copper atom.
This can be shown by a half-equation:

$$Cu^{2+} + 2e^- \longrightarrow Cu \quad \text{(at the cathode)}$$

At the positive electrode, the chloride ion, Cl^-,
loses its extra electron:

$$Cl^- - e^- \longrightarrow Cl$$

A chlorine atom is formed.
However, chlorine does not exist as Cl atoms,
but as Cl_2 molecules. So two Cl atoms join together
and we see chlorine gas:

$$Cl + Cl \longrightarrow Cl_2$$

These changes at the positive electrode (anode)
can be shown like this:

$$2\,Cl^- - 2e^- \longrightarrow Cl_2 \quad \text{(at the anode)}$$

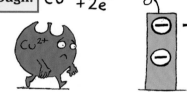

AT THE CATHODE (−)
The copper ion is feeling blue. It's lost two electrons. But help is available at the cathode.

Copper's in the pink. It gets two electrons and changes from an ion to an atom.

AT THE ANODE (+)
electrons go back
to the
battery

Two Cl⁻ ions each lose their extra electron and make a Cl_2 molecule

► Electrolysis of molten compounds

In Experiment 7.1 we saw that **solid** zinc chloride does not
conduct electricity. However, it does conduct when we **melt** it.
Can you remember why?
The **ions must be free to move** to the electrodes
before we get electrolysis.
In the next experiment you can look more closely at
the electrolysis of a molten compound.

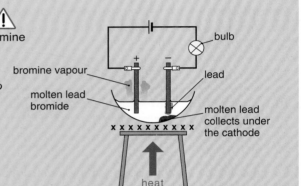

Demonstration 7.5   Electrolysis of lead bromide
Your teacher will set up the apparatus as shown
in a fume-cupboard.

- When does the bulb light up?
- What do you see at the positive electrode (anode)?

After a few minutes, carefully pour off the molten
lead bromide using tongs.

- What is left in the bottom of the dish?
- What has lead bromide turned into?

The electrolysis of lead bromide starts as soon as it melts.
The ions are then free to move between the electrodes.
The lead bromide breaks down into its elements:

lead bromide   electrolysis   lead   +   bromine

At the cathode ($-$) $Pb^{2+} + 2e^- \longrightarrow Pb$	**At the anode ($+$)** $2\,Br^- - 2e^- \longrightarrow Br_2$
Lead ions, Pb^{2+}, gain 2 electrons. They form lead atoms.	Two bromide ions, Br^-, each lose their extra electron. They form Br_2 molecules.

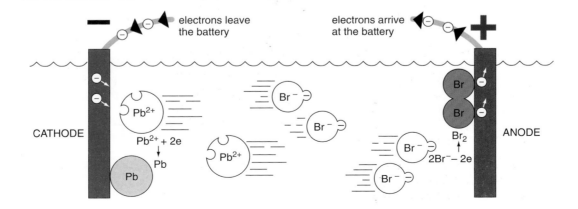

▶ Extraction of aluminium

As you know, highly reactive metals are extracted by electrolysis.
The most important of these metals is **aluminium**.
Aluminium has many useful properties.
It conducts heat and electricity well. It has a low density
for a metal. It also resists corrosion. That's because of the tough
layer of aluminium oxide on its surface which protects it.
Look at the pictures opposite :

- Explain why aluminium is chosen for each
  of these uses.

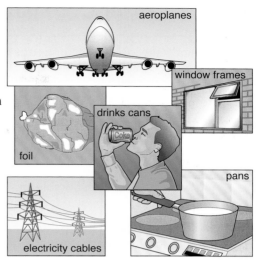

Getting the ore

Aluminium is extracted from its ore **bauxite**.
This is mainly **aluminium oxide**.
The impurities are removed before aluminium
is extracted.
Large amounts of bauxite are dug up in Jamaica.
Look at the photo opposite :

- What effect does the aluminium industry have
  on our environment ? Will recycling help ? (See page 90.)

Open-cast mining of bauxite in Jamaica

Electrolysis of aluminium oxide

The aluminium oxide is melted, then electrolysed.

- Why do we need to melt the aluminium oxide ?

Energy is saved by dissolving the oxide in molten cryolite.
This lowers its melting point.
The electrolysis is done in cells like the one shown :

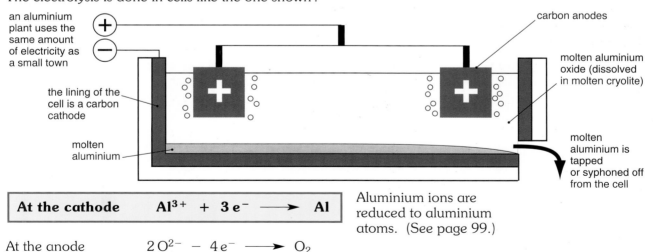

an aluminium plant uses the same amount of electricity as a small town

the lining of the cell is a carbon cathode

molten aluminium

carbon anodes

molten aluminium oxide (dissolved in molten cryolite)

molten aluminium is tapped or syphoned off from the cell

At the cathode	$Al^{3+} + 3e^- \longrightarrow Al$

Aluminium ions are reduced to aluminium atoms. (See page 99.)

At the anode $\qquad 2\,O^{2-} - 4e^- \longrightarrow O_2$

The oxygen gas reacts with the hot carbon anodes.
It makes carbon dioxide gas. This burns away
the anodes, which must be replaced quite often.

► Chemistry at work : Recycling metals

On page 89 we looked at some uses of aluminium
and on page 78 we saw how steel is used.
Have you ever recycled aluminium drinks cans?
How can you tell if a can is made from steel or aluminium?

Both aluminium and iron require large amounts of energy
to extract the metals from their ores.

- Can you think of all the ways in which energy is needed
  in a Blast Furnace and in making steel? (See pages 75–6.)
  And what about extracting aluminium from bauxite?
  (See page 89.)

Steel cans are separated from aluminium
cans and other household waste by magnets

There is a 95 % saving in energy if you compare recycled
aluminium to aluminium extracted from bauxite.
So this is good news for global warming as we burn
less fossil fuels, so less carbon dioxide is released into the air.
We also reduce other pollutant gases in the extraction process.

There is also the benefit of preserving the ore supplies
and less open cast mining means fewer scars on the landscape.
The aluminium that would have been thrown away
would also take up precious space in our landfill sites.

The process of recycling aluminium is shown below:

Recycling reduces pollution from the
extraction of the metal

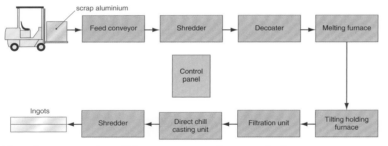

Besides saving landfill space, recycling steel also saves energy
(about 50% of the energy needed to make steel from iron ore) and
our natural resources.
For example, each year in the USA, steel recycling saves
enough energy to provide electricity for about 18 million homes!
Then there are the savings in coal (to make coke), the iron ore,
and the limestone needed in a Blast Furnace. Recycling also
means less greenhouse gases released into the air.

The recycling rate for steel cans is about 60 %, having risen from
15 % in 1988. However, there is still a lot we can all do to help
our environment in the years ahead.

Many people recycle aluminium and
steel

- Why do you think it is important to recycle metals?

► Chemistry at work : Cars for scrap

People have made a living from scrap cars for many years.

Each year about 2 million cars come to the end of their useful life. At present around 95% of the **metal** in a car is reused or recycled. But that still leaves almost 500 000 tonnes of other materials that get dumped in landfill sites.

This takes up valuable space, so new European targets have been set. By 2015 we will be required to recover 95% of all materials used to make a car.
Look at the metals we find in a normal car :

Material	Average mass (kg)	% mass
Ferrous metal (steels)	780	68.3
Light non-ferrous metal (mainly aluminium)	72	6.3
Heavy non-ferrous metal (for example lead)	17	1.5

Other materials used include **plastics, rubber and glass**. Car makers are trying to ensure these are all recyclable.

- What is the average mass of metal in a car?
- What percentage of a car's mass is made up of **non-metallic** materials? Discuss the problems in recycling these materials.

A car at the 'end of its life' can go to dismantlers and/or shredders. The dismantlers take off all the things that they can sell on. Shredders will recycle any of the rest that they get money for. However, they are worried that the new targets cannot be met while still making money out of scrap.

People can make money from scrap cars, but new recycling targets will prove difficult to meet

- What is the main metal found in most cars? Which of its properties allows it to be separated easily from other materials in the scrap?
- Discuss who should pay for meeting the European targets on recycling cars.

Aluminium is a less dense metal than steel. As it is lighter, a car made from aluminium uses less fuel than a traditional steel-bodied car.

However, aluminium is much more expensive than steel.

So, at the moment, only 'top of the range' manufacturers are making aluminium cars.

Aluminium parts, though, are getting more common as fuel economy becomes more and more important. This will help to reduce pollution from cars.

Aluminium cars should also last longer than steel cars because they will not corrode.

This aluminium car is expensive to buy!

► Electrolysis of solutions

To get a highly reactive metal from its compounds
we must melt the compound before we can electrolyse it.
Don't you think it would be cheaper and easier
to just dissolve the compound in water, then electrolyse it?
After all, solutions have free ions, just like molten compounds.
However, you can't get any **reactive** metals
from their solutions by electrolysis.

When electrolysing solutions, we have to think about
the **water** molecules in the electrolyte.
Water (H_2O) can affect what is formed at each electrode.
Let's test some of the products we get from
the electrolysis of solutions:

*You can't extract sodium from sea-water
(sodium chloride solution)*

Experiment 7.6   Electrolysing solutions

Set up the apparatus as shown:
See if you can identify any of the products shown below.
Do not breathe in any gas given off.

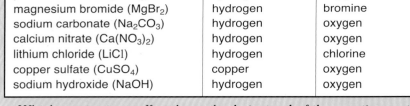

Solution	Cathode (−)	Anode (+)
potassium iodide (KI)	hydrogen	iodine
potassium nitrate (KNO_3)	hydrogen	oxygen
magnesium bromide ($MgBr_2$)	hydrogen	bromine
sodium carbonate (Na_2CO_3)	hydrogen	oxygen
calcium nitrate ($Ca(NO_3)_2$)	hydrogen	oxygen
lithium chloride (LiCl)	hydrogen	chlorine
copper sulfate ($CuSO_4$)	copper	oxygen
sodium hydroxide (NaOH)	hydrogen	oxygen

- Which gas is given off at the cathode instead of the reactive metal?
- Where do you think this gas has come from?
- What happens with a less reactive metal, like copper?

Look at the table above:
You can see that the **reactive metals stay in solution**.
Hydrogen gas is given off instead. The hydrogen comes from H_2O.

> **When you electrolyse a solution of a highly reactive metal,
> hydrogen – not the metal – is given off at the cathode.**

The half-equation is:
$$H^+ + e^- \longrightarrow H$$
A hydrogen ion, H^+, gains an electron to form a hydrogen atom, H.
Then two H atoms join together to make an H_2 molecule:
$$H + H \longrightarrow H_2$$
These 2 steps are usually shown as:

> $$2\,H^+ + 2\,e^- \longrightarrow H_2$$

*Hydrogen (not sodium) is given off when
we electrolyse sodium chloride solution*

Why is hydrogen given off?

Let's take a solution from the last experiment as an example. Let's look at calcium nitrate solution:

In any solution there are a few water molecules which have split up:

$$H_2O(l) \rightleftharpoons H^+(aq) + OH^-(aq)$$

They form hydrogen ions, H^+, and hydroxide ions, OH^-.

Let's think what happens at the cathode (−).
Both calcium ions and hydrogen ions are attracted towards the cathode.
However, calcium ions are more stable than hydrogen ions.
Remember that calcium reacts with acid (page 70).
Calcium can displace H^+ ions from an acidic solution.

So given a choice between H^+ and Ca^{2+} at the cathode, it's H^+ ions that have to leave the solution!

At the cathode	$2\,H^+(aq) + 2\,e^- \longrightarrow H_2(g)$

What happens at the anode?

Hydroxide ions, OH^-, and nitrate ions, NO_3^-, are both attracted to the anode (+).
But which of the two ions is 'kicked out' (or **discharged**) from the solution?

Look at the 'order of discharge' opposite:
If the negative ion is above the hydroxide in the list, it stays in solution. So nitrate (NO_3^-) ions stay in solution. Hydroxide ions from water are discharged.
When this happens, we get oxygen gas (O_2) given off.
Here is the half-equation:

Order of discharge

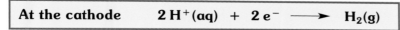

carbonate, CO_3^{2-}	these ions stay
nitrate, NO_3^-	in solution;
sulfate, SO_4^{2-}	oxygen is given off
hydroxide, OH^- ions from water	
chloride, Cl^-	OH^- stays in solution;
bromide, Br^-	the halogen is
iodide, I^-	given off

At the anode	$4\,OH^-(aq) - 4\,e^- \longrightarrow 2\,H_2O(l) + O_2(g)$

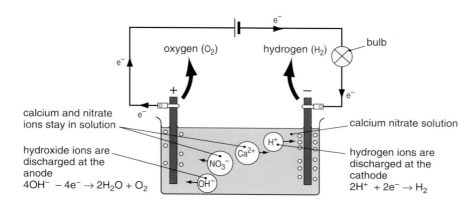

calcium and nitrate ions stay in solution

hydroxide ions are discharged at the anode
$4OH^- - 4e^- \rightarrow 2H_2O + O_2$

oxygen (O_2)   hydrogen (H_2)   bulb

calcium nitrate solution

hydrogen ions are discharged at the cathode
$2H^+ + 2e^- \rightarrow H_2$

► Active electrodes

We have now seen how electrolysis can be used to extract highly reactive metals, like aluminium. The other uses of electrolysis depend on electrodes which react – **active electrodes**.

Platinum is used to make inert electrodes. Why do you think we've used carbon but not platinum electrodes in our experiments?

In all our experiments so far, we have used carbon electrodes. These are called 'inert' electrodes. They take no part in the electrolysis. The carbon electrodes just carry electrons to and from the electrolyte.

However, some metal electrodes do take part in electrolysis.
These are called **active** electrodes.

Let's look at copper as an example of an active electrode. We will electrolyse copper sulfate solution.

These metals form 'active' electrodes!

Experiment 7.7   Electrolysis of copper sulfate solution using copper electrodes

Make sure your copper electrodes are clean and shiny.
Use a pencil to mark one electrode + and the other −.
Weigh them with an accurate balance.

Set up the apparatus as shown :

After 10 minutes, rinse the electrodes with distilled water.
Dry them by dipping in propanone and letting it evaporate.
When dry, re-weigh the electrodes.
- What do you find?
- Why would it be a good idea to repeat your experiment?

Look back to the electrolysis of copper sulfate solution on page 92.
- What difference do you notice at the anode (+)?

4V

copper anode loses mass

copper sulfate solution

copper cathode gains mass

When we electrolyse copper sulfate solution using copper electrodes, the mass of both electrodes changes. Remember that metals are always formed at the cathode (−). So not surprisingly, the cathode gets heavier.

The strange thing is that the anode loses mass.
The copper anode is an *active electrode*.
We find that :

loss in mass at the anode = gain in mass at the cathode

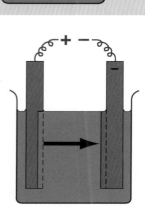

The anode's loss is the cathode's gain

Comparing inert and active electrodes

Let's compare the electrolysis of copper sulfate solution using carbon and copper electrodes:

	Carbon electrodes	Copper electrodes
At the cathode (−)	$Cu^{2+}(aq) + 2e^- \longrightarrow Cu(s)$	$Cu^{2+}(aq) + 2e^- \longrightarrow Cu(s)$
	the reactions are the same	
At the anode (+)	$4OH^-(aq) - 4e^- \longrightarrow 2H_2O(l) + O_2(g)$ oxygen gas is given off	$Cu(s) - 2e^- \longrightarrow Cu^{2+}(aq)$ no oxygen is seen; copper atoms from the anode lose 2 electrons and enter the solution as Cu^{2+} ions

The copper anode is an active electrode.
As copper atoms form on the cathode,
copper atoms are lost from the anode.
As copper ions leave the solution at the cathode,
they are replaced in the solution at the anode.

Copper ions, Cu^{2+}, make copper sulfate solution blue.
• What do you think would happen to the colour
  of the copper sulfate solution using carbon electrodes?
• Why won't the colour fade using copper electrodes?

Getting pure copper

Have you ever wired a plug or seen inside an electric cable?
The metal wire is copper.
Copper is an excellent conductor of electricity. However,
it must be very pure to do its job well.

Copper extracted from its ores is not pure enough
to be used for electrical wiring.
It can be made pure using electrolysis.

Look at the diagram opposite:
In industry a thin cathode of pure copper is used.
The anode is made of impure copper.
• What do you think happens at each electrode?

The cathode slowly gets bigger as the anode
gets smaller and smaller. (See table above.)

The impurities from the anode drop
to the bottom of the cell. This sludge contains
valuable metals, like gold and silver.

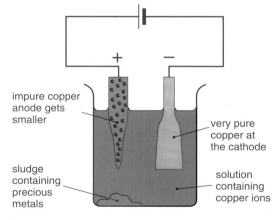

*As copper ions enter the solution at the anode,
copper ions leave the solution at the cathode*

impure copper anode gets smaller

sludge containing precious metals

very pure copper at the cathode

solution containing copper ions

*Many of these cells operate at the same time in industry.
The cathodes are removed about every two weeks.*

► Extracting copper

On the previous page you saw how we purify copper.
This is especially important for electrical copper wires.
But where does the copper anode come from? Where do we get
the impure copper from?

1. Recycling copper

High quality copper from wires can be recycled by melting and/or
re-using it. With just a few impurities, the copper can be made the
anode and purified by electrolysis (as on page 95).

Some copper scrap can be recycled to pure copper

However, a lot of old copper is mixed with other metals in alloys.
These are not so easy to recycle. You have to analyse the alloy to
see what percentage of each metal is in it. You might then just melt
the mixture of metals and re-use as the alloy.

If the percentage of copper is low, you can't use the alloy as the
anode to purify it. The process does not work.
You might be able to adjust the percentages of metals to get
another useful alloy. Or you could add more copper to the impure
copper by melting them together.
This raises the percentage of copper in the alloy so then you can
purify it electrically.
But this makes the copper more expensive to recycle.

2. Smelting copper ores

Some copper ores can be heated to leave copper by itself. (See page 72.)
In this experiment you can extract copper from one of its ores.
Malachite ore contains a form of copper carbonate.

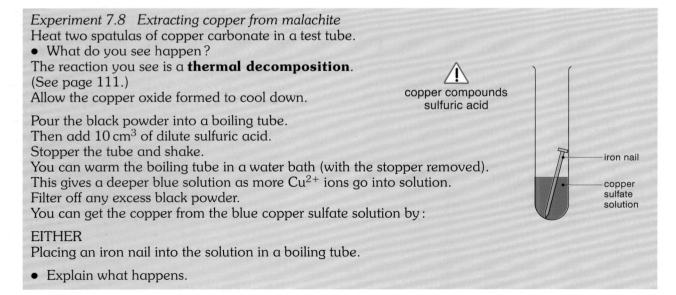

Experiment 7.8 Extracting copper from malachite
Heat two spatulas of copper carbonate in a test tube.
• What do you see happen?
The reaction you see is a **thermal decomposition**.
(See page 111.)
Allow the copper oxide formed to cool down.

⚠️
copper compounds
sulfuric acid

iron nail

copper
sulfate
solution

Pour the black powder into a boiling tube.
Then add 10 cm^3 of dilute sulfuric acid.
Stopper the tube and shake.
You can warm the boiling tube in a water bath (with the stopper removed).
This gives a deeper blue solution as more Cu^{2+} ions go into solution.
Filter off any excess black powder.
You can get the copper from the blue copper sulfate solution by:

EITHER
Placing an iron nail into the solution in a boiling tube.

• Explain what happens.

OR
Electrolysing the solution in a small beaker. Combine your copper sulfate solution with a couple of other groups and pass electricity through it.
Use carbon electrodes so you can see where the copper forms clearly.

- Explain what happens.

Smelting copper ores still provides the main source of copper.

3. Rock-eating bacteria (bio-leaching)

The extraction of copper involves open-cast mining that scars the landscape. (See next page.)
The sites have large waste tips of rock that contain too little copper to make it worthwhile extracting by smelting. Fortunately, there are bacteria that feed on the iron and copper compounds found in the 'low grade' ore.

By a combination of biological and chemical processes, the copper ions in the ore dissolve in sulfuric acid. Then the solution of copper ions runs off the slag heaps.

The solution can be collected and the copper extracted from it (by displacement or by electrolysis as in the previous experiment).
So now we can extract copper from waste and from new deposits of low grade ores that were thought to be uneconomic. At present about 20 % of copper is produced by leaching out copper with bacteria.

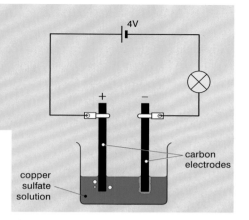

Advantages of bio-leaching

- The cost of setting up this bio-leaching process is low. It does not need huge open-cast mines that scar the landscape.
- As mentioned, bacteria can get the copper from low grade ores or waste. It uses a lot less energy and does not produce sulfur dioxide gas as in smelting sulfide ores.
- It can be used on ores that would pollute the air when smelted. For example, you can use copper ores that contain toxic arsenic.

Disadvantages of bio-leaching

- Bio-leaching does not recover precious metals. (See page 95.) So this reduces profits for the copper refiners.
- Sometimes acid has to be added to the ore – which costs money.
- Sometimes the bacteria themselves produce too much acid – which has to be neutralised. Again this costs more money.
- It can take years to extract only 50 % of the copper from the ores. So research continues to speed-up the action of the bacteria. However, as sources of high grade copper ore run out, bio-leaching will get more and more important.

Bio-leaching in action. By using waste rock from old copper mines we reduce the area of land scarred by the open-cast mining of high-grade ores. (See next page.)

▶ Chemistry at work : The costs of exploiting metal ores

Metals play a vital role in our technological society.
Without metals the electrical devices we depend on would be useless.

However, whenever we extract the metal found in the Earth's crust there is a price to pay. That price is not just the money we pay out. There is usually a cost to the environment.

Mining the ore

As you saw on page 89, open-cast mining is often used to get the metal ore from the ground. As well as aluminium, 90% of copper comes from open-cast mines.

This is an open-cast mine in Zambia.

These huge pits scar the landscape. They also destroy the habitats of plants and animals. The mining produces large unsightly slag heaps of waste rock.

The mining companies can help to solve these problems once a mine has been stripped of its useful ore. They can landscape the site, planting trees and perhaps creating a lake.
The hole can also be filled with our rubbish before covering it with earth. This is useful as it is difficult to find new land-fill sites.

The water in an area can also be affected by mining, especially if the ores contain sulfides. Rainwater seeps through the exposed ore and waste. The water and oxygen (plus the action of bacteria) turn the metal sulfides to sulfuric acid. This runs down into the groundwater and acidifies it.

Processing the ores

Sulfide ores are heated strongly in the process of smelting. This releases sulfur dioxide gas into the air, causing acid rain. (See pages 72 and 148.)

Strict controls are now enforced by law. Many smelters can stop more than 99% of the sulfur dioxide they make from escaping. Almost half the cost of producing copper goes in controlling pollution. However, the high cost of updating smelters has meant many have had to close down, resulting in thousands of job losses.

In developing countries, pollution control is not as tight and labour costs are cheaper. So the copper industry in richer countries struggles to compete.

This mine is in Zambia. Many African nations have rich mineral resources. Multi-national companies often set up mines and processing plants there.
● *What issues does this raise?*

- Discuss the issues facing the people who live in a town that relies on extracting copper. The town is in Arizona, USA where the Environmental Protection Agency insists on strict pollution control measures.

► Redox reactions

You have already seen how we extract metals from ores.
Often a metal oxide is *reduced*.
Look back to page 74.

Whenever a substance is reduced, something else is *oxidised*.
We call the reaction a **redox reaction**.
Can you see why?

Look at an example from the blast furnace on page 75:

iron oxide + carbon monoxide $\longrightarrow$ iron + carbon dioxide
$$Fe_2O_3 \ + \ 3\,CO \longrightarrow 2\,Fe \ + \ 3\,CO_2$$

iron oxide is reduced   carbon monoxide is oxidised

This blast furnace is about 50 m high.
Inside it, iron oxide is reduced to iron.

Iron oxide is reduced by carbon monoxide. It *loses oxygen*.
The carbon monoxide itself gets oxidised. It *gains oxygen*.
The carbon monoxide is called a reducing agent.

Can you remember how we extract highly reactive metals?
Carbon or carbon monoxide cannot reduce their oxides.
We have to use electrolysis.
Look at the extraction of aluminium on page 89.
The aluminium oxide is reduced.

At the cathode (−)
$$Al^{3+} \ + \ 3e^- \longrightarrow Al$$

The Al^{3+} ions gain electrons. They are *reduced*.
So we can say that:

> **Reduction is the *gain of electrons*.**

The Al^{3+} ions have been reduced to Al atoms at the cathode.

Oxidation and reduction are chemical opposites.
So what do you think happens to the number of
electrons in an ion that gets oxidised?

> **Oxidation is the *loss of electrons*.**

For example, $2\,O^{2-} - 4e^- \longrightarrow O_2$
We can also write this as:
$$2\,O^{2-} \longrightarrow O_2 \ + \ 4e^-$$
So oxide ions are oxidised at the anode.

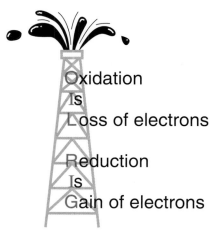

Oxidation
Is
Loss of electrons

Reduction
Is
Gain of electrons

OIL-RIG will help you to remember!

Investigation 7.9 Investigate electrolysis!

You can now use your ideas about electrolysis to find out:

Which factors affect the rate of electrolysis?

You can investigate copper sulfate solution.
Think about these things:
- How can you speed up the electrolysis of copper sulfate?
- Use your ideas about ions to explain why it is speeded up.
- How can you measure or judge how quickly it happens?
  (Hint: an ammeter might be useful.)
- How can you make it a fair test?
- Is your plan safe?
- How can you collect **reliable** data that you can trust?

Show your plan to your teacher before you start.

Summary

- **Electrolysis** is the breakdown of a substance using electricity.
  For example, copper chloride $\longrightarrow$ copper + chlorine
- **Ions** are charged particles. Metal ions are positive. Non-metal ions are negative.
- The negative electrode is called the **cathode**. The positive electrode is the **anode**.
- Solid compounds made from ions cannot be electrolysed.
  They must be melted or dissolved in water first.
  The ions then become free to move to the electrodes.
- Metals form at the cathode ($-$).
  Non-metals form at the anode ($+$).
  For example, in the electrolysis of copper chloride:
  at the cathode $Cu^{2+} + 2e^- \longrightarrow Cu$
  at the anode $2Cl^- - 2e^- \longrightarrow Cl_2$
- In solutions, reactive metals are not formed at the cathode.
  They stay in the solution and **hydrogen** (from the water) is given off:
  at the cathode $2H^+ + 2e^- \longrightarrow H_2$
- Reactive metals are extracted from **molten** compounds.
  For example, aluminium is extracted by electrolysing molten aluminium oxide.
- We also use electrolysis to purify copper.

▶ Questions

1. Copy and complete:
   When we break down a substance using
   electricity it is called . . . . .
   The . . . . electrode is called the cathode.
   The . . . . electrode is called the anode.
   The ions in a . . . . are not free to move.
   However, when the substance they are in is
   molten or . . . . in water, electrolysis can take
   place.

   Metals form at the . . . . , and non-metals at the
   . . . . . For example,
   at the cathode $Pb^{2+} +$ . . . . $\longrightarrow$ Pb
   at the anode $2Br^- - 2e^- \longrightarrow$ . . . .
   When solutions of reactive metal compounds
   are electrolysed . . . . is formed at the cathode,
   not the metal.
   Electrolysis is used in industry to extract
   reactive metals, and to purify . . . . .

2. Liz and Jabeen set up the experiment below:

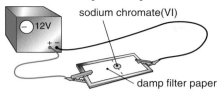

sodium chromate(VI)

12V

damp filter paper

They left it for 15 minutes.
Sodium ions are positively charged. They are colourless.
Chromate(VI) ions are negatively charged. They are yellow.
a) What did they **see** happen?
b) Explain the results of their experiment.

3. What do these words mean:
 a) cathode, b) anode, c) cation, d) anion?

4. Complete this table:

Substance electrolysed	What is formed at cathode?	What is formed at anode?
molten lead bromide	a) ?	b) ?
c) ?	potassium	chlorine
calcium nitrate solution	d) ?	e) ?
copper chloride solution	f) ?	g) ?
h)?	aluminium	oxygen
sodium iodide solution	i) ?	j) ?

5. Copy and complete these half-equations at the **cathode**:
 a) $Na^+ + \ldots \longrightarrow Na$
 b) $Li^+ + e^- \longrightarrow \ldots$
 c) $Mg^{2+} + \ldots \longrightarrow Mg$
 d) $\ldots + 2\,e^- \longrightarrow Ca$

6. Copy and complete these half-equations at the **anode**:
 a) $2\,Cl^- - \ldots \longrightarrow Cl_2$
 b) $\ldots - 2\,e^- \longrightarrow Br_2$
 c) $2\,I^- - 2\,e^- \longrightarrow \ldots$
 d) $2\,O^{2-} - 4\,e^- \longrightarrow \ldots$

7. Jo and Tim set up this experiment:

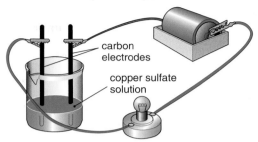

carbon electrodes

copper sulfate solution

a) Draw a circuit diagram of their apparatus.
b) What did they **see** at the cathode?
c) Complete this half-equation at the cathode:
  $Cu^{2+} + \ldots e^- \longrightarrow Cu$
d) What did they **see** at the anode?
e) Jo and Tim did the experiment again, but this time they used **copper** electrodes instead of carbon. What differences did they notice between their two experiments?

8. Look at the experiment below:

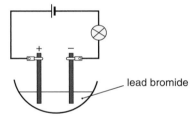

lead bromide

a) The bulb does not light up when you dip the electrodes into the solid lead bromide. Explain why not. (Use the word **ions** in your answer.)
b) You want to electrolyse the lead bromide. Which apparatus is missing from the diagram?
c) Copy and complete:
 .... moving through the wires make the bulb light up, whereas .... carry the charge through the lead bromide.

9. Explain why reactive metals can't be extracted from their solutions by electrolysis.

10. a) What difficulties do we have recycling scrap that contains copper?
 b) Why will bacteria become increasingly important in the extraction of copper?

Further questions on page 119 to 121.

SALT

Do you like a little salt on your food?
Even if you don't add it yourself, just look at the labels
on the packets and tins at home.
Salt is added to many foods to bring out the flavour.

It can also help to preserve food. Early explorers,
like Captain Cook, took meat rubbed in salt
on their long voyages.

> **The chemical name for salt is sodium chloride.
> Its formula is NaCl.**

Have you ever been swimming in the sea? If you have,
you will know how salty sea-water is. Imagine how
much salt must be dissolved in the world's oceans!
It's no wonder that chemists call salt '**common salt**'!

In hot countries salt is extracted from sea-water.
The sea-water is left in huge shallow lakes.
Then the Sun's energy evaporates the water. (See page 131.)
Why don't we get much salt this way in Britain?

Fortunately, we have found thick layers of salt
under the ground in this country. Most salt lies under
the Cheshire countryside. How do you think it got there?
Salt is a very useful raw material. It has attracted
the chemical industry into Cheshire.

So how do we get the salt up to the surface?
There are two methods:

1. *digging from mines*, and
2. *pumping up salt solution* (brine).

First of all, let's look at mining salt.

Salt mining

The salt is found about 300 m below ground.
In its natural, impure form it is called **rock salt**.
It is drilled and blasted out, then brought to the surface.
The underground caverns formed do not cave in
if pillars of rock salt are left in place.

The rock salt is sold to councils around the country.
It is used to grit roads in winter.

- Why is rock salt put on roads? What is the
  disadvantage for car owners? (See page 77.)

People think that millions of years ago there was an
inland sea where Cheshire is today. It
dried up leaving salt behind. The seam is up to
2000 metres thick in places!

Rock salt can be dug from underground

▶ Brine (salt solution)

The second way that we can get salt from
the ground is **solution mining**.
This method relies on salt dissolving in water.
Look at the diagram below:

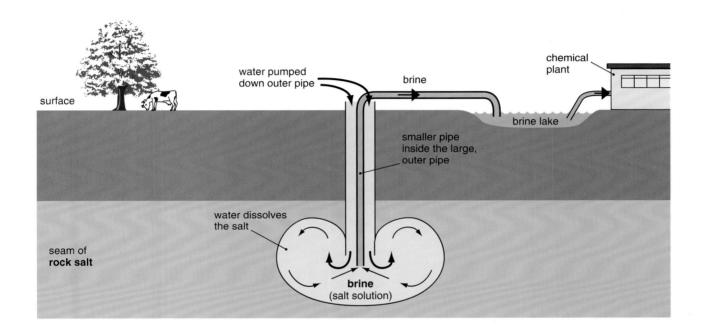

Hot water is pumped down the outer pipe.
It dissolves the salt. The salt solution is then
forced up the smaller, inner pipe by
the pressure of water.

The salt solution is called **brine**.
The brine gets stored in lakes until it is needed.
It is then pumped straight to the chemical plant.

When mining in an area has finished,
the holes left under the ground must be filled.
If not the land above can collapse and slide
into the old mines. This is called subsidence.
Look at the old photograph opposite:

Subsidence used to be more of a problem in
Cheshire. Nowadays the holes are more carefully
spaced out.

Subsidence was a problem in Cheshire

► Electrolysis of brine

We have now seen how brine (salt solution) is brought to the surface.
At the chemical plant it is electrolysed (broken down by electricity).
You can try this yourself in the next experiment:

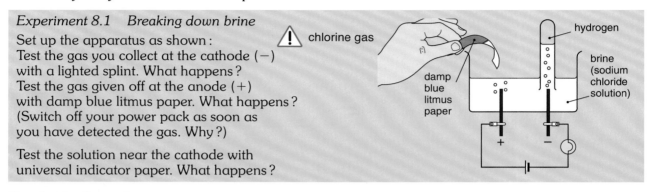

Experiment 8.1   Breaking down brine
Set up the apparatus as shown:
Test the gas you collect at the cathode (−) with a lighted splint. What happens?
Test the gas given off at the anode (+) with damp blue litmus paper. What happens?
(Switch off your power pack as soon as you have detected the gas. Why?)

Test the solution near the cathode with universal indicator paper. What happens?

Products from brine

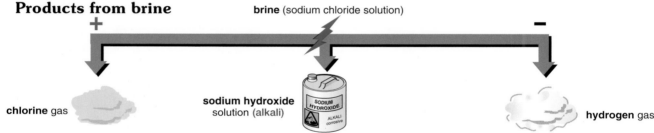

chlorine gas

sodium hydroxide solution (alkali)

hydrogen gas

You can read about the uses of chlorine, hydrogen and sodium hydroxide on pages 54, 105–7, 177 and 200.

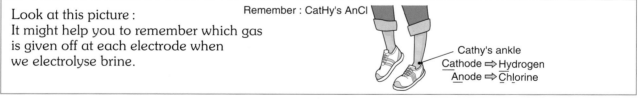

Look at this picture:
It might help you to remember which gas is given off at each electrode when we electrolyse brine.

Remember: CatHy's AnCl

Cathy's ankle
<u>Cat</u>hode ⇒ <u>Hy</u>drogen
<u>An</u>ode ⇒ <u>Cl</u>orine

In industry

There are several types of cell used in industry.
The most modern is called a membrane cell.
Look at the diagram opposite:
A plastic membrane divides the cell in two.
This keeps the chlorine and hydrogen gas apart and stops chlorine reacting with the sodium hydroxide.
The gases leave from the top of the cell.

Sodium ions can pass through the membrane.
Hydroxide ions can't get through. This gives us sodium hydroxide on the negative side of the cell.

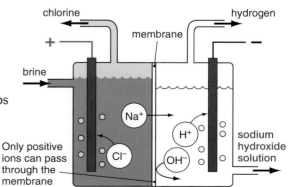

Explaining the electrolysis of brine

Let's see if we can work out how we get hydrogen, chlorine and sodium hydroxide from brine.

You might have read about the electrolysis of solutions on page 92. It explains why reactive metals, such as sodium, are not freed from solutions during electrolysis. Remember that brine is sodium chloride solution. Instead of sodium we get hydrogen forming at the cathode ($-$).

Can you recall where the hydrogen comes from? H^+ ions come from the water (H_2O) in the solution. A few water molecules split up into hydrogen ions (H^+) and hydroxide ions (OH^-):

$$H_2O(l) \rightleftharpoons H^+(aq) + OH^-(aq)$$

Therefore, when we electrolyse a solution of sodium chloride we can think of 4 different ions:

sodium Na^+, hydrogen H^+, chloride Cl^-, and hydroxide OH^- ions.

The H^+ and Cl^- ions are discharged at the electrodes. **Hydrogen** comes off at the cathode ($-$):

$$2\,H^+(aq) + 2\,e^- \longrightarrow H_2(g)$$

Chlorine comes off at the anode ($+$):

$$2\,Cl^-(aq) - 2\,e^- \longrightarrow Cl_2(g)$$

The Na^+ and OH^- ions stay in solution. They form **sodium hydroxide** solution in the cell. Sodium hydroxide is a very important alkali.

Look at the diagram:

rayon fibres
(using sodium hydroxide)

paper
(using chlorine)

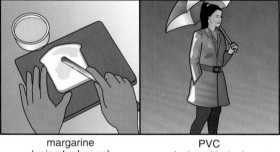

soap
(using sodium hydroxide)

bleach (using chlorine
and sodium hydroxide)

margarine
(using hydrogen)

PVC
(using chlorine)

These are some everyday products made using salt as a raw material

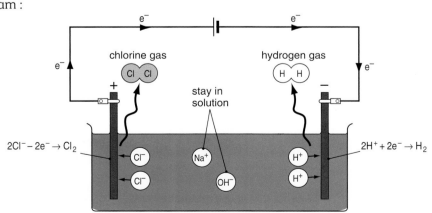

▶ Chemistry at work : Products from salt

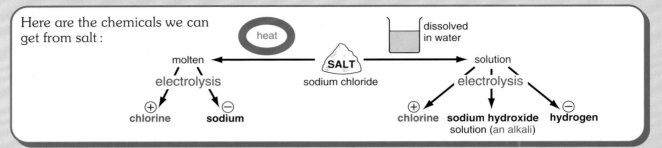

Here are the chemicals we can get from salt :

heat

dissolved in water

molten ← **SALT** → solution
sodium chloride

electrolysis

⊕ chlorine   ⊖ **sodium**

electrolysis

⊕ chlorine   **sodium hydroxide**   ⊖ hydrogen
solution (an alkali)

Uses of sodium

Street lamps
Have you ever noticed the yellow glow at night
in a lit-up area?
We can thank sodium for this effect.
A small amount of sodium vapour is inside
the yellow street lamps.

Sodium vapour is used in street lamps

Heat transfer
Sodium is also used in some nuclear reactors.
You might think it is strange to use
such a reactive metal in a nuclear power plant.
However, sodium has some useful properties.
It is a good conductor of heat and has a low melting point.
This is why it is used to transfer heat
from the reactor to the steam generators.
It is pumped around the power plant in sealed pipes.

- Why is it important to keep the pipes sealed?
- Why can't a safer, less reactive metal, such as copper,
  be used to transfer the heat?

Sodium is used in nuclear reactors

Uses of sodium hydroxide

The salt industry is known as the **'chlor-alkali'** industry.
The alkali, sodium hydroxide, is the most important
of the products from salt.
Look at the range of
things it is used to
make :

ceramics

rayon and acetate fibres

detergents
and
soaps to
remove
grease

BLEACH

paper

SODIUM HYDROXIDE

ALKALI
corrosive

making new
chemicals

purifying bauxite to extract aluminium

Using hydrogen in fuel cells

Hydrogen is used in space-craft to power **fuel cells**. Fuel cells are a very efficient way to use the chemical energy stored in fuels. The conventional way to transfer the energy stored in fuels is to burn the fuel. It reacts with oxygen and releases energy as heat. Basically, in a fuel cell the same reaction happens, but the chemical energy in the fuel is transferred directly into electrical energy.

Fuel cells are used in space-craft to transfer chemical energy into electrical energy

Advantages of fuel cells over conventional ways of generating electricity :

- They are more efficient because there are fewer stages in releasing the stored chemical energy.
- There is less pollution as the only waste product is water. However, the hydrogen used in a fuel cell must be produced from somewhere, e.g. from electrolysis of water. This requires energy which is often supplied by burning fossil fuels, so there can still be indirect pollution of the atmosphere.

The hydrogen fuel cell

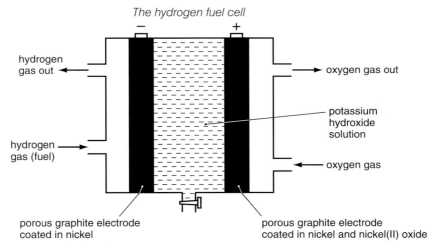

porous graphite electrode coated in nickel

porous graphite electrode coated in nickel and nickel(II) oxide

This cell uses hydrogen as the fuel :

at the negative electrode :

$$2H_2(g) + 4OH^-(aq) \longrightarrow 4H_2O(l) + 4e^- \text{ (hydrogen is oxidised)}$$

at the positive electrode :

$$O_2(g) + 2H_2O(l) + 4e^- \longrightarrow 4OH^-(aq) \text{ (oxygen is reduced)}$$

So overall we get :

$$2H_2(g) + O_2(g) \longrightarrow 2H_2O(l)$$

- Discuss how hydrogen fuel cells could be made to reduce pollution from cars.

Summary

- Common salt is **sodium chloride, NaCl**. It is found in the sea or in underground seams.
- Salt is dug up from mines as **rock salt**. It is also pumped to the surface as a solution. Salt solution is called **brine**.
- When *molten* sodium chloride is electrolysed we get:
  sodium at the cathode (−), and **chlorine** at the anode (+).
- When a *solution* of sodium chloride (brine) is electrolysed we get:
  hydrogen at the cathode (−),
  chlorine at the anode (+), and
  sodium hydroxide solution (*an alkali*) formed.

▶ Questions

1. Copy and complete:
The chemical name for common salt is
.... Its chemical formula is ..... When
molten sodium chloride is electrolysed, ...
metal forms at the .... (−) and .... gas is
given off at the anode (+).
Salt solution is called .... When it is
electrolysed .... gas forms at the cathode (−)
and .... gas at the anode (+). The solution
becomes an .... as sodium .... is left.

2. Copy and complete this flow diagram. It
shows the chemicals we can get from salt:

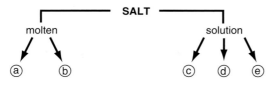

3. This diagram shows the solution mining of salt.

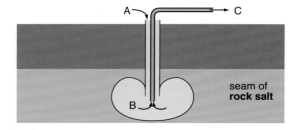

Copy the diagram and explain what happens
at A, B and C.

4. Sodium metal is extracted in a Down's Cell.

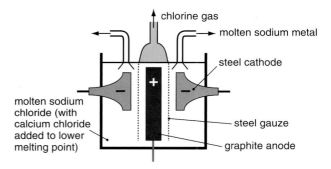

a) Complete this word equation to describe
what happens in the Down's Cell:

sodium chloride $\xrightarrow{electrolysis}$ .... + ...

b) It takes a lot of energy to melt the sodium
chloride.
 i) Why can't we use sodium chloride
solution to get the sodium metal?
 ii) Which substance is added to the sodium
chloride to lower its melting point?
Explain why sodium would cost more to
buy if this substance was not added.

5. Compare the use of hydrogen fuel cells with
electricity produced by burning fossil fuels.

6. a) Describe how brine is electrolysed in
industry. Include a diagram in your answer.
b) Imagine you are a sodium ion, Na⁺, in a
seam of rock salt. Describe your
adventures on the way to finding yourself
as sodium hydroxide in an oven cleaner.

Further questions on page 122.

chapter 9

Calcium carbonate

Have you ever seen the 'white cliffs of Dover'?
If you have, you've seen lots of **calcium carbonate**.
Chalk is one form of **limestone**, made from around
98 % calcium carbonate.

You might also find other rocks containing calcium carbonate
around your school.
Limestone itself is a common building material. It might
be spread on the school roads or roofs as chippings.

Another source of calcium carbonate is **marble**.

A school built from limestone

> **The chemical formula of calcium carbonate is $CaCO_3$.**

Many rocks with calcium carbonate in them were formed from
the shells of sea creatures that died millions of years ago. The
sediments built up on the seabed and slowly formed limestone.

Marble is formed from limestone crushed by powerful movements
in the Earth's crust and/or by very high temperatures near molten
rock (magma) underground.

*Limestone is a raw material for
making concrete and glass*

Limestone quarries

Have you ever visited the Peak District or the
Yorkshire Dales? If you have, you will have
enjoyed the beautiful views from the limestone hills.

You might have been pot-holing or visited the caves there.
These caves have been weathered out of the limestone
by rainwater. Rainwater is slightly acidic because
carbon dioxide dissolves in it as it falls.
You can see how limestone reacts with acid on page 204.

Limestone is a common rock in Britain. It is blasted
from the hillsides in huge quarries.
Look at the photo opposite:

- Why do you think that many people are against
  any plans for new quarries?

These arguments have to be balanced against the benefits.
Limestone is a vital raw material for industry.
Look at its uses on pages 112 and 113.
New jobs and money are created for the area.

- What are your views on quarrying? (See page 117.)

A limestone quarry 'scars' the landscape

▶ Looking at limestone

Once limestone is quarried, it is changed into many useful materials.

Is your home made from bricks? If it is, they are fixed in place by mortar. Mortar is mixed using cement, which we get from limestone.

Or perhaps you live in a building made mainly from concrete? The raw material for concrete is also limestone.

In this experiment you can make some new substances from limestone:

Limestone and its products are used a lot in the building industry

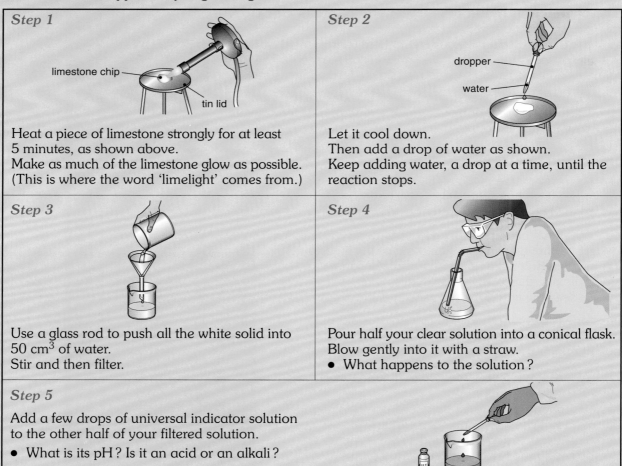

Experiment 9.1 Changing limestone

Follow the steps shown below.

● Record what happens as you go along.

Step 1	Step 2
limestone chip — tin lid	dropper — water
Heat a piece of limestone strongly for at least 5 minutes, as shown above. Make as much of the limestone glow as possible. (This is where the word 'limelight' comes from.)	Let it cool down. Then add a drop of water as shown. Keep adding water, a drop at a time, until the reaction stops.

Step 3	Step 4
Use a glass rod to push all the white solid into 50 cm³ of water. Stir and then filter.	Pour half your clear solution into a conical flask. Blow gently into it with a straw. ● What happens to the solution?

Step 5

Add a few drops of universal indicator solution to the other half of your filtered solution.

● What is its pH? Is it an acid or an alkali?

▶ Reactions of limestone

The reactions in the last experiment can be shown like this:

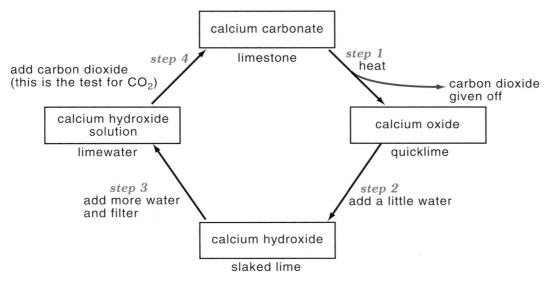

Equations

step 1

$$CaCO_3(s) \longrightarrow CaO(s) + CO_2(g)$$

step 2

$$CaO(s) + H_2O(l) \longrightarrow Ca(OH)_2(s)$$

step 4

$$Ca(OH)_2(aq) + CO_2(g) \longrightarrow CaCO_3(s) + H_2O(l)$$

The lime kiln

In industry the limestone is heated in lime kilns.
The limestone is broken down by heat in the kiln.
This reaction is called **thermal decomposition**.

calcium carbonate $\xrightarrow{\text{heat}}$ calcium oxide + carbon dioxide

$$CaCO_3(s) \longrightarrow CaO(s) + CO_2(g)$$

It's like Step 1 in Experiment 9.1 on a large scale.
The kiln is lined in fire-bricks. They have to
withstand very high temperatures (about 1500 °C).

Rotating kilns are used to turn limestone into **cement**.
The powdered limestone is heated with clay.
Most cement goes on to make **concrete**.
You can read more about these and other uses
of limestone on the next few pages.

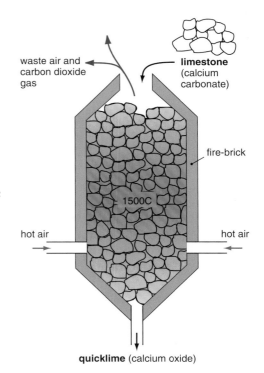

▶ Uses of limestone

Look at the uses of limestone below:

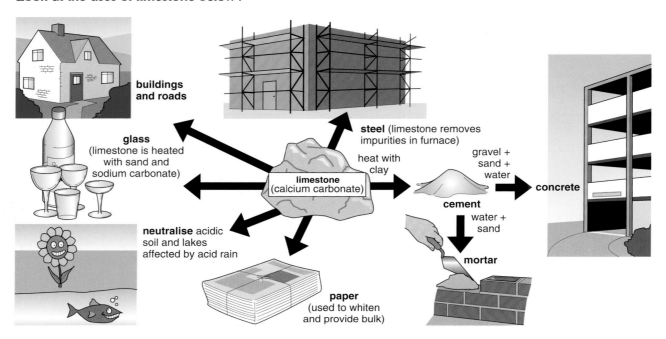

Cement

Have you ever watched brick-layers at work? They use a mixture of cement and sand, with water added to make a thick paste. This is called **mortar**. It takes a lot of skill to set the bricks in the right position. What do you think would happen if their mortar was too runny? What if it was too dry?

Cement is made by heating limestone and clay (or shale). They are ground up and heated in kilns which rotate. A little gypsum (calcium sulfate) is added. This stops the cement from setting as soon as you add water.

Cement sets as it reacts with water. This is called hydration. The reactions are complex. We can think of it as 'fingers' of crystals growing out from each grain of cement. These interlock and bind the mixture together. Mortar will set overnight. However, it carries on reacting and getting stronger for several months.

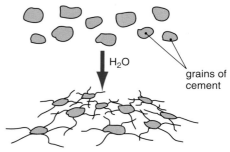

Crystal fingers (of calcium hydroxide) hold the structure together as it sets. Calcium hydroxide can react with CO_2 in the air to form calcium carbonate:

$$Ca(OH)_2 + CO_2 \longrightarrow CaCO_3 + H_2O$$

Investigation 9.2   Mortar

Why don't you try some brick-laying yourself! Experiment with different mixtures of cement and sand. See which mixture makes the best mortar.

⚠ cement

Concrete

What are the paths around your school made from?
Some will be made from concrete.
Concrete is made by mixing cement, sand and small stones.
If you look closely, you will see the small stones set in position.

Concrete is the most widely used building material in the world. That explains why we produce about 1000 million tonnes of cement in the world every year!

Concrete is delivered to customers ready-mixed in trucks. Why does the large drum on the back rotate continuously?

Investigation 9.3   Concrete

⚠ cement

You can test different mixtures of cement, sand, and gravel to see which makes the strongest concrete.
You can make your mixtures in yoghurt pots.
• Evaluate the reliability and validity of your investigation. (See page 6.)

Reinforced concrete

Concrete is a relatively cheap building material.
As a material itself it is hard, but a beam will snap if put under tension.
We can make it stronger by making **reinforced** concrete.

To make reinforced concrete, the mixture is allowed to set around a steel support. Its increased strength and flexibility means that we can use it to make buildings and bridges. You've probably seen lots of concrete on motorway bridges. The bridges would not be able to span across gaps without the embedded steel.

However, corrosion of the steel inside the concrete can be a serious problem, weakening a structure.

Reinforced concrete being laid

Investigation 9.4   Comparing concrete with reinforced concrete

Use the same concrete mix to make two small beams. In one beam embed some straightened paper clips.
• Test their strength.

▶ Glass

Limestone is one of the materials that we use to make **glass**.
Glass is important in all our lives. Think of all the things
we use every day that are made from glass.
Look at the photos:

- List some other things made from glass.

What is glass?

Glass is a strange material. Can you remember the 3 states of
matter? (See page 19.)
Some people call glass the 4th state of matter. It's like a 'solid liquid'!
The raw materials (see below) are heated to 1500 °C. At this high
temperature, they melt and react to form molten glass. As it cools
down, the glass turns into a solid. However, the particles don't form a
regular pattern.
It's as if the particles in the molten glass are frozen in place.
Look at the diagram:
Can you see how the structure is jumbled up?

*The disorderly structure in
borosilicate glass*

What's in glass?

Did you know that glass is made mainly from sand?
Glass has been around for a long time. It was probably discovered
by accident in the sand underneath an ancient fire!
The first glass object has been dated at about 4500 BC.
The Egyptians used glass containers around 3000 BC.

As well as **sand** (SiO_2), the other raw materials are
limestone ($CaCO_3$) and **sodium carbonate – soda –** (Na_2CO_3).
You have seen some of the uses of common salt (sodium chloride)
on page 106. Sodium carbonate is another useful product made
from salt.
Re-cycled glass (cullet) is becoming more important.
It can make up to 30 % of some glass-making mixtures.

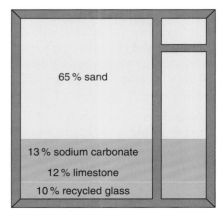

65 % sand

13 % sodium carbonate

12 % limestone

10 % recycled glass

The composition of glass

Different types of glass

Have you ever seen a car window that's been smashed?
The bits of glass look very different from the pieces
you get when a bottle smashes.
Bottles and car windows are made from different types of glass.

There are many different types of glass. Scientists have
tried changing the glass-making mixture. They've also
found ways of treating the glass to change its properties.
For example, windscreens are usually made like a
glass sandwich – with a thin sheet of plastic as the filling!
This is called laminated glass.
How does this help if a stone hits your windscreen?

Look at the table below:
Do all these types of glass have any properties in common?

This glass had not been laminated

Type of glass	Use
soda-lime	windows
boro-silicate	test-tubes, beakers, etc. (heat-proof, chemical resistant)
lead-crystal	decorative glasses, bowls and vases
glass fibres	fibre optics, fibreglass
optical glass	lenses in spectacles, cameras, projectors, etc.
glass ceramic	opaque oven-ware

*These glass fibres carry information
in the form of light.
A single fibre can carry 10 000
phone calls at once.*

Coloured glass

Have you taken any glass to be recycled at a bottle-bank?
The bottles are sorted out into different colours.
What are the most common colours of glass bottles?
Green and brown bottles get their colour from iron impurities in the
sand that it is made from.

On page 47 we found out that the **transition metals**
form coloured compounds. Adding small amounts
of their oxides makes glass coloured.

copper

nickel

manganese

Experiment 9.5   Make your own glass

1. Sugar, like sand, is made of crystals.
   You can make sugar-glass (toffee) by melting some sugar.
   Heat it until its colour is pale brown. (Do not touch!)
   Then cool it quickly to make a clear, hard toffee.

2. Heat some borax on a wire loop as shown:
   This melts to form a structure like glass.
   When molten, dip it in a *tiny* amount
   of a transition metal oxide.
   This will make a bead of coloured glass when cool.

⚠ molten borax
metal oxides

borax

glass or wooden
handle

wire loop

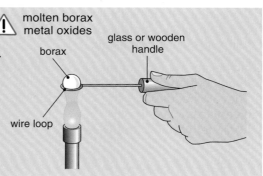

► Chemistry at work : New developments in concrete and glass

New types of concrete

Some form of mortar has been in use for thousands of years.
Do you remember how you investigated the effect of varying the mixture to see how it affected the strength (page 112)?
Then on page 113 you investigated reinforced concrete.

People have always tried out new ways to improve the properties of mortar and concrete. For example, around 400 BC the Romans were trying out animal fat, milk and even blood in their mortar. The experiments continue today with all the support of modern science to help. Concrete was first reinforced in France in 1867.

They used a wire mesh to strengthen it. Nowadays we can use:

- glass fibres
- carbon fibres
- steel fibres
- poly(propene), nylon, polyesters and Kevlar.

Some of the latest research uses pulp from wood, plants and recycled paper. The materials are shredded and sliced into small pieces before adding them to the concrete mixture.
We call these **reinforced concrete composites**. A 'composite' is a combination of materials, mixed to get improved properties for a certain use.

A little recycled paper can improve concretes resistance to cracking, impact (making it tougher) and scratching!

Using industrial waste as admixtures ('additives') to concrete makes economic sense. New developments use granulated slag from a blast furnace or ash from coal-fired power stations to make High Performance Concrete.

Glass in concrete

We can even substitute ground-up waste glass instead of sand in concrete. However, the cost of **cullet** (recycled glass) does not make this worthwhile at present.

However, research at the University of Sheffield, has shown that there could be a market for 'ConGlassCrete'.
They embedded larger pieces of glass and polished the surface to get a very attractive finish.
Look at the photo opposite:

The latest High Performance Concretes make new, creative designs possible

A scientist testing a new concrete composite

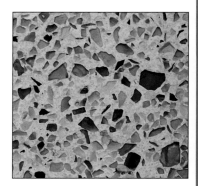

A decorative concrete composite made with scrap glass

New uses of glass

The glass we collect from recycling can now be used for:

- abrasives
- laying foundations in place of gravel and small stones
- filtering and for the drainage of water
- transparent bricks
- insulation foams made with glass fibres.

By bonding glass to other materials, such as wood or metal, we also get interesting composites. The new material made is stronger than the original materials we started with.

- Design a marketing leaflet for either the latest developments in glass or concrete. Carry out your research using the Internet.

Glass is finding many new uses

▶ Chemistry at work : Quarrying limestone

The limestone for all these uses is quarried out of the ground. Have you ever seen a quarry? They are not a pretty sight! They produce a huge scar on the landscape.

Imagine a limestone company want to build a quarry in a beautiful piece of countryside.

Look at some of the arguments below:

This is a limestone quarry. Workers use explosives to dislodge the limestone from the rock face.

The dust will settle on my crops and they won't grow as well

The lorries carrying limestone will go right through the village past our primary school

I think I'll get a lot more business from the workers at the quarry – and there will be more jobs for youngsters like you

The quarry will destroy the habitats of birds and animals

We might get that by-pass we've been asking for these last 10 years

Yes, but I don't know if my heart will stand those explosions when they blast the rock out !

We'll be able to supply limestone for the glass, steel and cement industries in this region now

- Discuss the issues involved in gaining planning permission for a new quarry.

Investigation 9.6   Thermal decomposition

You have seen on page 110 what happens when we heat calcium carbonate. The reaction is called thermal decomposition. Now investigate this problem.

Which carbonates break down most easily?
You can investigate some of the carbonates of:
calcium, magnesium, sodium, potassium, copper, and zinc.
Design a fair test to find out which of these decomposes most quickly.
● How can you make it safe?
Ask your teacher to check your plan before you try it.

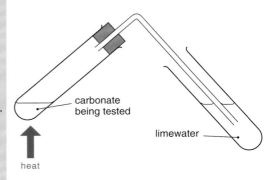

Summary

● Calcium carbonate ($CaCO_3$) is mainly found in nature in limestone, chalk and marble.
● It is broken down by heat (thermal decomposition), to give calcium oxide and carbon dioxide.
● Calcium oxide (quicklime) reacts with water to form calcium hydroxide (slaked lime).
  This can be used to raise the pH of acidic soil in a neutralisation reaction.
● Cement is made from crushed limestone and clay in a rotary kiln.
● Mortar is made from cement and sand, mixed into a thick paste with water.
● Concrete is also made from cement and sand, but also has small stones (or gravel) mixed in.
● Glass is made by heating limestone, sand and sodium carbonate (soda).

▶ Questions

1. Copy and complete:
   The chemical name for limestone is . . . .
   Its formula is . . . . .
   When heated, limestone breaks down in a reaction called . . . . decomposition.
   The products formed are . . . . and carbon dioxide.

   Quicklime is another name for calcium . . . .
   It makes . . . . lime when we add water, and this is used to . . . . the pH of acidic soil. Glass is made by . . . . a mixture of limestone, ... and . . . . .

2. Draw a table with 2 columns. Use it to show the advantages and disadvantages of starting a limestone quarry in a National Park.

3. Limestone is roasted in a lime kiln.
   a) Write a word and symbol equation for the reaction in a lime kiln.
   b) What do we call this type of reaction?
   c) Draw a diagram of the apparatus you could use to show that carbon dioxide gas is given off when calcium carbonate is heated.

4.

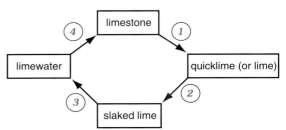

Match the numbers 1 to 4 with the letters A to D below:
A $Ca(OH)_2(s) \xrightarrow{+H_2O} Ca(OH)_2(aq)$
B $CaCO_3(s) \xrightarrow{heat} CaO(s) + CO_2(g)$
C $Ca(OH)_2(aq) + CO_2(g) \longrightarrow$
    $CaCO_3(s) + H_2O(l)$
D $CaO(s) + H_2O(l) \longrightarrow Ca(OH)_2(s)$

5. a) Describe how we turn limestone into cement.
   b) What is the difference between mortar and concrete?

6. Imagine that you work for an advertising company. You are given a job to design a poster to show the public how important limestone is. Use information from this chapter and other sources to create your poster.

▶ **Extraction of metals and Electrolysis**

1. Many iron ores contain iron(III) oxide.
Iron can be extracted from iron(III) oxide by heating it with carbon.
   a) Copy and complete this equation for the reaction.
   $2Fe_2O_3 + \ldots C \longrightarrow \ldots Fe + \ldots \ldots$ [3]
   b) This reaction is described as a reduction of the iron(III) oxide.
   Explain why. [1]
   c) Name two other metals that are obtained by a similar reaction. [2]

2. Look at this list of metals:
   aluminium   copper   iron   zinc
   a) Which of these metals cannot be extracted from its ore by reduction with carbon? [1]
   b) Explain why reduction with carbon cannot be used to extract this metal. [1]
   c) What method is used to extract this metal? [1]

3. The table gives information about some metals.

Name of the metal	Cost of one tonne of the metal in December 2003 (£)	Percentage of the metal in the crust of the Earth (%)
Aluminium	883	8.2
Platinum	16 720 000	0.000 000 1
Iron	216	4.1
Gold	8 236 800	0.000 000 1

   a) Use information in the table to suggest why gold and platinum are very expensive metals. [1]
   b) Aluminium and iron are made by **reduction** of their ores.
   i) Name the element that is removed from the ores when they are **reduced**. [1]
   ii) Use the reactivity table on page 70 to suggest a metal that would reduce aluminium ore. [1]
   c) Aluminium is made by the reduction of molten aluminium ore, using a very large amount of electricity.
   i) How is iron ore reduced in a blast furnace to make iron? [2]
   ii) Suggest why aluminium is more expensive than iron. [1]

4. Tungsten is used for the manufacture of filaments for light bulbs. Tungsten may be obtained by heating its oxide, WO_3, in hydrogen. The equation for this reaction is:
   $WO_3(s) + \ldots H_2(g) \longrightarrow W(s) + \ldots H_2O(g)$
   a) What is the chemical symbol for tungsten used in the above equation?
   b) Balance the above equation.
   c) What does $H_2O(g)$ in the equation suggest about the temperature of the reaction?
   d) What is **reduced** in this reaction?
   e) What is **oxidised** in this reaction? [6]
   f) Do you think that the above method would be suitable for the extraction of potassium from potassium oxide? Explain why.
   [2]   (EDEXCEL)

5. a) A student was trying to extract the metals from lead oxide and aluminium oxide. She heated each oxide with carbon in a fume cupboard:

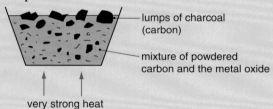

lumps of charcoal (carbon)

mixture of powdered carbon and the metal oxide

very strong heat

   She was able to extract lead from lead oxide but not aluminium from aluminium oxide.
   i) Explain the results of these experiments. [4]
   ii) Complete this word equation for the reaction between lead oxide and carbon.
   lead oxide + carbon $\longrightarrow$ .... + ...   [1]
   b) Copper can be extracted as shown below.

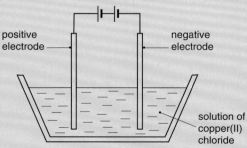

positive electrode

negative electrode

solution of copper(II) chloride

   Copper chloride is an ionic compound. State where the copper would collect and explain your answer fully.   [2]   (AQA)

6. Copper metal can be extracted from the ore malachite.
  a) What is an ore? [1]
  b) Malachite contains copper(II) carbonate. This is roasted in air to obtain copper(II) oxide.
  How is copper obtained from this copper(II) oxide? [2]
  c) The copper obtained by this process is not pure.
  What method can be used to purify the copper? [1]

7. It is thought that ancient Britons first made a useful substance by accident, when they mixed some malachite – an ore of copper with formula $Cu_2CO_3(OH)_2$ – with some charcoal, and heated them together.
  a) What useful substance was made? [1]
  b) What might this substance have been used for? [1]
  c) What type of reaction took place during heating? [1]

8. In which one of the following sets are all the metals so reactive that they all have to be extracted by electrolysis?
  A Ca   Cu   Pb
  B Ca   Mg   Na
  C Cu   Mg   Zn
  D Cu   K   Mg
  E Mg   Pb   Zn   [1]   (OCR)

9. Magnesium metal is manufactured by electrolysis of the salt magnesium chloride. This salt is obtained from sea water.
  Before electrolysis, magnesium chloride is mixed with sodium chloride and the mixture is heated to melt it.
  Adding sodium chloride lowers the melting point of the magnesium chloride without affecting the products of the electrolysis.
  a) i)   What is the state of magnesium chloride in sea water? [1]
      ii) What is the state of magnesium chloride during the electrolysis process? [1]
  b) i)   What **type** of particle is present in magnesium chloride which allows it to be electrolysed? [1]
      ii) At which electrode will magnesium metal be formed during electrolysis? [1]
      iii) Name the substance formed at the other electrode. [1]
  c) Sodium chloride lowers the melting point of magnesium chloride so less energy is needed to melt the magnesium chloride. Suggest **two** advantages of using less energy. [2]   (EDEXCEL)

10. Copper sulfate solution can be electrolysed. What forms at the cathode?
  A Copper
  B Hydrogen
  C Oxygen
  D Sulfur   [1]   (AQA)

11. The following apparatus is used to purify a sample of copper.

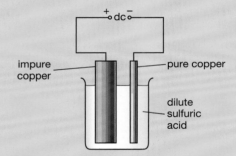

During the electrolysis copper leaves the positive electrode and joins onto the negative electrode.
  a) Write a half equation for the reaction at the positive electrode (anode). [2]
  b) Write a half equation for the reaction at the negative electrode (cathode). [2]
  c) During the electrolysis the negative electrode gains less mass than the positive electrode loses when the electrodes were weighed before and after the experiment. Explain why. [2]

12. Window frames made from iron corrode faster than those made from aluminium. This observation does **not** follow the order of the reactivity series. Explain the reasons for this. [3]   (AQA)

13. The diagram shows a method used to purify copper.

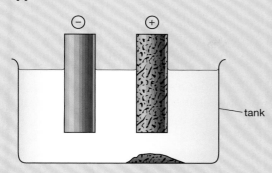

tank

Choose words or phrases from the box to complete the sentences below.

> **bottom of the tank   cracking
> displacement   electrolysis
> negative electrode   positive electrode
> surface of the solution**

This method of purifying copper is called . . . . .
The impure copper is made the . . . . and the pure copper collects at the . . . . .
The impurities collect at the . . . . .   *[4]*

14. The table shows information about different types of steel.

Type of steel	Percentage of carbon	Description
Mild steel	0.25	can easily be bent into shape
Medium steel	0.45	harder than mild steel
High carbon steel	0.80	hard and brittle
Cast iron	3.00	very brittle and easily cast

a) What is the effect of increasing the percentage of carbon in steel?   *[1]*
b) Impure iron from the blast furnace is called pig iron.
   Explain how this is changed into mild steel.   *[3]*
c) Mild steel is a mixture of substances but is not an alloy steel.
   Explain this statement.   *[3]*   (EDEXCEL)

15. The table gives information about five metals.

Metal	Density (g/cm³)	Tensile strength (MPa)	Resistance to corrosion
Aluminium	2.7	195	good
Copper	8.9	224	good
Mild steel	7.8	450	poor
Titanium	4.5	460	very good
Cast iron	7.8	150	poor

a) Aluminium is used for overhead power cables. These cables have a steel core.
   i) Aluminium and copper are both good electrical conductors.
       Explain why aluminium rather than copper is used for overhead cables.   *[1]*
   ii) Why do the cables have a steel core?   *[1]*
b) The properties of aluminium and mild steel can be modified by alloying.
   What is meant by alloying?   *[1]*
c) Cast iron can be converted into mild steel.
   Why is mild steel more useful than cast iron?   *[1]*
d) The following is an extract from an article in a science magazine.

> Titanium is fantastic. It is lighter than steel yet strong and tough enough to survive the extremes of space or the corrosive salts and pressures of the deep ocean with hardly a blemish. In fact, just about the only drawback with the material is its price. But this looks set to change, with the discovery of a new way to extract titanium metal that requires little more than black sand and electricity. Best of all the process promises to lower the price of the metal by up to three-quarters.

Titanium is the ninth most abundant element in the Earth's crust with an estimated 10 000 years supply of ore. Until now, the cost of extraction has made the metal very expensive.
   i) Why is it likely that the discovery of the new method of extracting titanium will increase demand for the metal?   *[1]*
   ii) What are the benefits to society of this new method of extraction of titanium?   *[1]*   (EDEXCEL)

▶ Salt

16. Look at the diagram.
It shows how sodium hydroxide solution can be manufactured. It is manufactured by the electrolysis of sodium chloride solution.

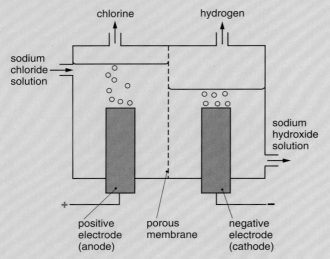

a) What is the meaning of the word 'electrolysis'? [2]

b) Sodium hydroxide is produced when sodium chloride solution is electrolysed. Give the names of the other **two** products. [2]

c) Sodium chloride solution contains particles called ions.
Explain how sodium chloride solution conducts electricity. [1]

d) The cost of making sodium hydroxide includes the cost of the raw materials used. Write down **three** other costs involved in making sodium hydroxide. [3]

e) The electrolysis of sodium chloride solution can also be carried out in the laboratory.
Describe what you would observe at the positive electrode. [1]   (OCR)

17. Rock salt is an important raw material for the chemical industry.
Rock salt is extracted by dissolving the salt in water to form brine. The brine is pumped out and converted into chlorine, hydrogen and sodium hydroxide.

a) Name the type of process used to obtain chlorine, hydrogen and sodium hydroxide from brine. [1]

b) Describe a test for chlorine. [2]

c) Give a use of sodium hydroxide. [1]
(EDEXCEL)

18. The diagram below shows the electrolysis of sodium chloride solution, in the laboratory.

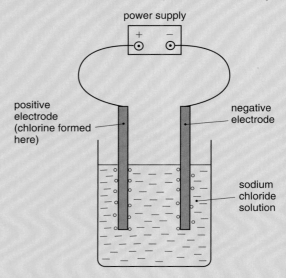

a) Which gas forms at the negative electrode? [1]

b) Explain why chlorine gas forms at the positive electrode. [2]

c) State **one** use of chlorine gas. [1]   (AQA)

19. What are the advantages and the disadvantages of using hydrogen fuel cells to power motor cars? [4]

▶ Limestone

20. The diagram at the top of page 123 shows a lime kiln. This is used to make calcium oxide from limestone (calcium carbonate).

a) Use information from the diagram to help you to write a word equation for the reaction which takes place in the lime kiln. [1]

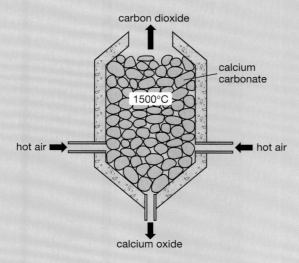

carbon dioxide

calcium carbonate

1500°C

hot air → ← hot air

calcium oxide

b) The diagram below shows another process that uses limestone.

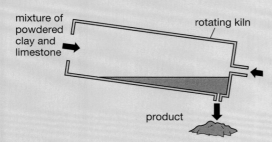

mixture of powdered clay and limestone

rotating kiln

product

Name the product made in this process.
[1]   (AQA)

21. This table shows the composition of some types of glass.

	SiO$_2$	Sodium carbonate	Lead oxide	Other oxides	Other
Soda lime glass	75%	15%			10%
Lead glass	60%	15%	25%		
Borosilicate glass	75%	5%		20%	

a) What is the common name for the chemical SiO$_2$?
b) Which type of glass contains the least amount of sodium carbonate?
c) What might be included in the 'other' category for soda lime glass?

d) Why is lead oxide added to some types of glass?
e) Borosilicate glass is often found in the kitchen. What is it used for?   [5]

22. a) Limestone is a very useful material which is readily available in the United Kingdom. When it is heated, it is broken down into simpler substances. What is the name for this type of chemical reaction?   [1]
b) Limestone can be used for many purposes. Some uses are shown below.

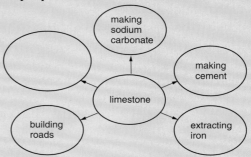

making sodium carbonate

making cement

limestone

building roads

extracting iron

Name **one** more use of limestone.
[1]   (AQA)

23. Limestone (made mainly of calcium carbonate – CaCO$_3$) is heated with sand (silicon dioxide) and soda ash (sodium carbonate) to make glass.
a) When limestone is heated it decomposes to make quicklime, which is calcium oxide, CaO, and carbon dioxide. Write a balanced equation for this reaction.   [2]
b) Quicklime is reacted with water to make slaked lime, which is calcium hydroxide, Ca(OH)$_2$.
Slaked lime is often added to soils that are too acidic.
How does the slaked lime cure this problem?   [1]
c) Limestone (calcium carbonate) was formed in the tourist area now called the Peak District, about 300 million years ago. Today limestone is mined near Buxton, in one of the largest limestone quarries in the United Kingdom.
Suggest **three** social or environmental issues which might be involved in the mining of limestone in the Peak District.
[3]   (OCR)

chapter 10

ACIDS AND ALKALIS

Acids all around

What do you think of when you hear the word **acid**?
Most people think of a fuming, corrosive liquid
which is very dangerous.

However, not all acids are like this. Most of us
like a little acid on our fish and chips!
Look at the cartoon on page 102.

● Which acid does vinegar contain?

Vinegar has the sharp, sour taste of acids.
(The Latin word for 'sour' is *acidus*.)
Citric acid gives oranges and lemons their sharp taste.
These fruits also contain ascorbic acid.
We know it better as vitamin C!

Did you know that even rainwater
is slightly acidic. (See page 291.)
Carbon dioxide gas dissolves in the rain as it falls.
This is also the gas in fizzy drinks, such as
cola, beer and sparkling mineral water.
Why do you think these drinks taste tangy?

These are all examples of **weak acids** that we meet every day.

These teenagers are enjoying a little ethanoic acid on their chips!

Acids in the lab

You will have used acids in your science lessons.
The 3 common acids found in school are:

hydrochloric acid	**HCl**
sulfuric acid	**H_2SO_4**
nitric acid	**HNO_3**

These are **strong acids**.

Did you know that we all have one of
these strong acids in our stomachs?
Hydrochloric acid helps to break down our food
into smaller molecules and kills bacteria.

You can read more about sulfuric acid on page 134.
Nitric acid is on page 235.

Fizzy drinks are acidic

What's the formula?

3 strong acids we should know,
Learn this rhyme and have a go!
Nitric *acid's first we see*
*Its formula's **HNO_3***
The next one you should learn as well
*Is **hydrochloric** – **HCl***
Sulfuric *leaves us just one more,*
*That famous **H_2SO_4***

▶ Neutralisation

Have you ever had indigestion? The 'burning' feeling
comes from too much hydrochloric acid in your stomach.
You can cure the pain quickly by taking a tablet.
The tablet contains an *alkali* (or base) which gets rid of the acid.

Acids and alkalis are chemical opposites.
They react together and 'cancel each other out'.

If we mix just the right amount of a strong acid and
a strong alkali together, we get a neutral solution.

*We all have hydrochloric acid
in our stomachs*

> **The reaction between an acid and an alkali is called neutralisation.**

The pH scale

You have used universal indicator before to measure pH.
Do you know the pH of a neutral solution?
Here is a reminder of the pH scale:

*These antacid tablets
can neutralise excess
acid in your stomach*

0	1	2	3	4	5	6	7	8	9	10	11	12	13	14

◀ more acidic ▬▬ neutral ▬▬ more alkaline ▶

In the next experiment you can neutralise some hydrochloric acid.
You will react the acid with an alkali, sodium hydroxide:

Experiment 10.1   Get the balance right!

Collect $10 \, cm^3$ of sodium hydroxide solution in a small flask.
Add 5 drops of universal indicator solution.

- What colour is the solution? What is its pH?

Measure out $9 \, cm^3$ of dilute hydrochloric acid in
a measuring cylinder. Add it to the alkali in the flask.
You can also use a burette to add the acid.

- What colour is the solution now?

As you can see, the solution is still strongly alkaline.
You will need to add the next hydrochloric acid
a drop at a time. Use a dropper or a burette.
Swirl the flask as you add each drop of acid.

Try to get a *neutral solution*.
Look at the pH scale above.
What colour are you aiming for?

If you add too much acid, you don't have to start again.

- What can you add, a drop at a time, to neutralise
  the acid? It's a bit like balancing a see-saw!

You can keep your neutral solution for the next experiment.

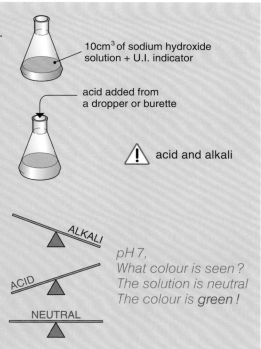

$10cm^3$ of sodium hydroxide
solution + U.I. indicator

acid added from
a dropper or burette

⚠ acid and alkali

ALKALI

ACID

NEUTRAL

*pH 7,
What colour is seen?
The solution is neutral
The colour is green!*

▷ Salts

In the last experiment you neutralised an acid with an alkali.

If you ever get stung by a bee, you might be grateful for that reaction!

A bee's sting is said to be acidic. So, in theory, it can be neutralised by an alkali. The pain is certainly eased by treating it with bicarbonate of soda (a weak alkali).

● Why shouldn't you use sodium hydroxide on the sting?

But what is made when an acid and alkali react together?
In general we can say:

A bee's sting is said to be acidic. A wasp's sting is alkaline. How would you treat a wasp sting? (See page 136.)

Let's look at the equation for the reaction in Experiment 10.1:

hydrochloric acid + sodium hydroxide ⟶ sodium chloride + water
$HCl(aq)$ + $NaOH(aq)$ ⟶ $NaCl(aq)$ + $H_2O(l)$

The salt, sodium chloride, is dissolved in water.
Can you think how we can get it from its solution?

You can try this out in the next experiment:

Experiment 10.2   Preparing sodium chloride

Have you got your neutral solution from the last experiment?
If not, follow the method for Experiment 10.1.

Add a spatula of charcoal powder to the green solution.
Stir it with a glass rod.
The charcoal takes the colour out of the solution.

Filter the mixture.
You should get a clear, colourless solution.
Put the solution into an evaporating dish.
Now heat it on a water bath as shown:

Stop heating when you see some white crystals around the edge of the solution.
Leave your evaporating dish for a few days.
The rest of the water will evaporate off slowly.
Slow evaporation gives bigger crystals.

● What is the chemical name for the salt you have made?
● Write the equation for the reaction to make the salt.
● What do we call this type of reaction?
● What shape are your salt crystals?
● Can you think of any reactions that we use in everyday life to neutralise acids? List any you know.

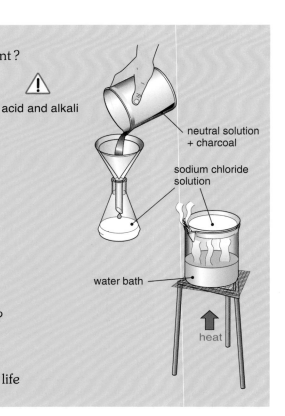

acid and alkali

neutral solution + charcoal

sodium chloride solution

water bath

heat

What is a salt?

You already know a lot about one salt. We have looked at common salt, sodium chloride, in Chapter 8. We have also made some crystals of it in the last experiment.

But chemists use the word **salt** to describe any *metal compounds that can be made from acids.*

Can you remember the formulas of the strong acids from the start of the chapter?
Look back to page 124 if you can't think of them. Which element do they all contain?

We find that, *all acids contain hydrogen.*

> **When we replace some or all of the hydrogen in an acid by a metal, we get a salt.**

Sodium chloride is a salt from hydrochloric acid, HCl.
- *Why would we never make sodium chloride by adding sodium to hydrochloric acid? (See page 70.)*

▷ Naming salts

Each acid has its own salts. Look at this table:

Acid	Its salts	Example
hydro**chlor**ic acid, HCl ⟶ **chlorides**		sodium chloride, NaCl
sulfuric acid, H_2SO_4 ⟶ **sulfates**		copper sulfate, $CuSO_4$
nitric acid, HNO_3 ⟶ **nitrates**		potassium nitrate, KNO_3

Exception to the rule
Ammonium salts do not contain a metal. They are made when an acid reacts with ammonia (an alkali). Examples include ammonium nitrate (NH_4NO_3) and ammonium chloride (NH_4Cl).

Naming a salt is like naming a person.
Look at the examples in the table above.
A salt gets its first name from the metal, and its surname from the acid.

- How does the name of the salt tell us that:
  a) there is just one other element combined with the metal?
  b) there is oxygen present in the salt?

Some salts can be made by **direct combination**. The chlorides of sodium, potassium, iron and aluminium are examples.

Look at the photo opposite:

Here iron is heated and placed in chlorine gas:

iron + chlorine ⟶ iron(III) chloride
$$2\,Fe(s) + 3\,Cl_2(g) \longrightarrow 2\,FeCl_3(s)$$

An acid turns into a salt! Hydrogen in the acid is replaced by a metal in the salt.

Iron and chlorine can make the salt iron(III) chloride by direct combination

▶ Preparing salts

You have now seen how to make a salt by
reacting an acid with an alkali.
But alkalis are part of a larger group of compounds
called **bases**.

Look at the diagram opposite:
An alkali is just a base that can dissolve in water.

Do you think bases will react with acids, as alkalis do?
The general equation is:

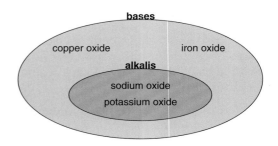

acid + base ⟶ a salt + water

For example,

sulfuric acid + zinc oxide ⟶ zinc sulfate + water
$H_2SO_4(aq)$ + $ZnO(s)$ ⟶ $ZnSO_4(aq)$ + $H_2O(l)$

As in all neutralisation reactions, a salt is made.
- What is the name of the salt in the reaction above?

All metal oxides are bases.
The soluble ones are alkalis.

Salts have many uses. This farmer is treating his vines
with copper sulfate to kill pests.
Why must grapes be washed well before we eat them?

Experiment 10.3   Making a salt from an acid and a base

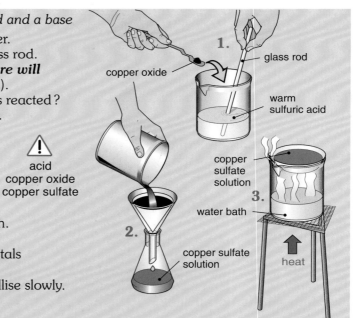

1. Put 25 cm³ of sulfuric acid in a small beaker.
   Add black copper oxide and stir with a glass rod.
   Keep adding the copper oxide until **no more will
   dissolve (react).** Warm gently (do not boil).
   - How do you know when all the acid has reacted?
   - Work out what is formed in the reaction.

2. Filter the mixture in your beaker.
   You should get a clear, blue solution.
   - What should you do if you still have
     some black powder in your solution?

   ⚠ acid
   copper oxide
   copper sulfate

3. Pour your solution into an evaporating dish.
   Then heat it on a water bath as shown:
   Stop heating as soon as you see some crystals
   around the edge of the solution.
   Leave the solution for a few days to crystallise slowly.
   - What is the name of your salt?
   - What do the crystals look like?

The sulfuric acid is neutralised by the base, copper oxide:

sulfuric acid + copper oxide ⟶ copper sulfate + water
$H_2SO_4(aq)$ + $CuO(s)$ ⟶ $CuSO_4(aq)$ + $H_2O(l)$

More soluble salts

Acids can also be neutralised by reacting them with carbonates. You have already seen how limestone (calcium carbonate) reacts with acid. In general we can say:

acid + a carbonate ⟶ a salt + water + carbon dioxide

- How is this different from the equation for an acid reacting with a base?
- How could you test for the gas given off?

We can use the same method as Experiment 10.3 to prepare a salt from a carbonate. This is because most carbonates are insoluble in water.

This lake has become acidic. It is being neutralised by calcium carbonate released from a helicopter.

Experiment 10.4 Preparing a salt from an acid and a carbonate

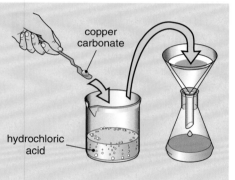

Collect 25 cm³ of hydrochloric acid in a small beaker.
Add a spatula of copper carbonate.
- What happens? What is the gas given off?

Add more copper carbonate until it stops fizzing.
Filter to remove the un-reacted copper carbonate.
Repeat step 3 from the end of Experiment 10.3 to get some crystals of your salt.
- What is your salt called?
- What do its crystals look like?

acid
copper carbonate

The hydrochloric acid is neutralised by copper carbonate:

hydrochloric acid + copper carbonate ⟶ copper chloride + water + carbon dioxide
$$2\,HCl(aq) \quad + \quad CuCO_3(s) \quad \longrightarrow \quad CuCl_2(aq) \quad + \quad H_2O(l) \quad + \quad CO_2(g)$$

Metals and acid

We have already seen how some metals react with acid.
Can you remember which gas is given off?
Do all metals react with acid? (See the table on page 70.)
If a metal does react, the general equation is:

acid + a metal ⟶ a salt + hydrogen

You can use the method from Experiment 10.3 to prepare the salt.
- Why would it be dangerous to use this method to prepare potassium chloride?
- Which acid and metal would you start with to make magnesium sulfate?

Here is magnesium reacting with dilute sulfuric acid.

- *How can you remove any excess magnesium after the acid has been neutralised?*

▶ Acids need water!

All the acids we have used so far have been solutions.
The acid molecules are dissolved in water.
Acids need water before they can show their acidic properties.
Let's look at citric acid (the acid in oranges) as an example:

Experiment 10.5   Testing citric acid

You can compare solid citric acid (which has
no water added to it) to a solution of the acid.   acid
Write down your results in a table like this:

Test	Results for pure citric acid	Results for citric acid solution
blue litmus paper sodium carbonate magnesium ribbon does it conduct electricity ?		

- How does water affect the properties of citric acid?
- Think of another test to show if a substance is acidic.

Without water, substances are not acidic.
Look at these properties of a typical acidic solution:
- turns litmus red, and has a pH less than 7,
- reacts with a base,
- fizzes with a carbonate, giving off carbon dioxide,
- fizzes with a metal, such as magnesium, giving off hydrogen.

Let's see what difference the water makes:
In water, the acid molecules split up.
We can look at hydrogen chloride, HCl, as an example:
Hydrogen chloride is a gas. It only shows
its acidic properties when we dissolve it in water.
We call the solution formed hydrochloric acid:

$$HCl(g) \xrightarrow{\text{dissolve in water}} H^+(aq) + Cl^-(aq)$$

Remember that all acids contain hydrogen.
When dissolved in water, the hydrogen is released
into the solution as **H$^+$(aq) ions**.

> **It is H$^+$(aq) ions that make a solution acidic.**

Explaining neutralisation

When we neutralise an acid, the H$^+$(aq) ions
are removed from the solution.
Alkaline solutions contain hydroxide ions, OH$^-$(aq):

> **H$^+$(aq) + OH$^-$(aq)** $\xrightarrow{\text{neutralisation}}$ **H$_2$O(l)**

The H$^+$ and OH$^-$ ions 'cancel each other out', forming water.

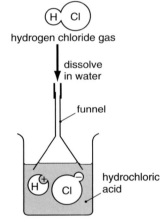

*The hydrogen chloride gas splits up (dissociates) in
water into ions. You can read more about this on
page 226. The solution is called hydrochloric acid.
(The gas is very soluble in water.)*
- *Why is a funnel used when dissolving the gas?*

*Acid plus alkali gives salt and water,
H$^+$ ions are led to the slaughter!*

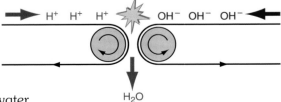

▶ Solubility of salts

Do you remember how we get a soluble salt
from its solution?
If we heat the solution, water evaporates off.
Small crystals of the salt start to form.
This is called the **point of crystallisation**.

We now have a **saturated** solution.
No more salt can dissolve in the solution.
As more water evaporates, the crystals
of the salt grow bigger.

*You can get the salt from
sea-water in hot countries.
As the water evaporates, the
solution becomes saturated.
Then salt crystallises out and
is collected.*

Solubility curves

We measure how soluble a salt is by seeing
how much dissolves in 100 g of water.
(100 cm^3 of water has a mass of 100 g.)
For example, the solubility of copper sulfate
is 19 g per 100 g of water, at 15 °C.

Notice that the value for the solubility above
is given at 15 °C.
Do you think that temperature affects
how soluble a salt is?

Look at the graph:
● Which of the 3 salts is most soluble at 20 °C?
● Which salt is most soluble at 60 °C?

We call the lines on the graph **solubility curves**.
They show us how much salt dissolves
at different temperatures.

You can find out more about solubility curves
on page 295.

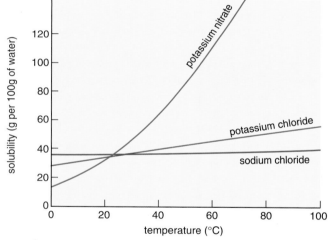

Solubility curves (also see page 295)

Investigation 10.6   Which salt is most soluble?
You can test the solubility of these salts:
copper sulfate, sodium chloride, calcium chloride and zinc sulfate.

Design a fair test to see which of these salts is most soluble.
Use 10 cm^3 of water for each test.
● What other variables must be kept constant for a fair test?
● How can you tell when your solutions are saturated?
● How will you measure how much of each salt dissolves?
Your teacher will check your plan before you start.

Investigation 10.7   Which factors affect the solubility of copper sulfate?
Your teacher will check your plan before you start.
● Why is it difficult to collect precise and accurate data?

⚠ copper sulfate

▶ Making insoluble salts

The salts we have made so far are all soluble in water.
To get the salt, we crystallise it from its solution.
However, some salts are **insoluble**.

An example of a salt which does not dissolve well in water is calcium sulfate.

We can prepare an insoluble salt by a precipitation reaction.
Do you know what a precipitate is?
Sometimes, when we mix two solutions we get an insoluble solid formed.
The solid is a **precipitate**.

Look at the cartoon below:
It explains what happens in a precipitation reaction.

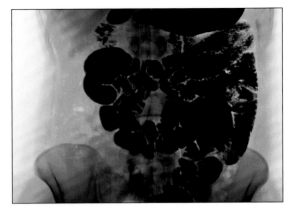

Barium sulfate is a salt which does not dissolve well in water. It is used in a 'barium meal' which enables doctors to see your digestive system using x-rays.

A and D have swapped partners and formed a precipitate

Experiment 10.8   Making an insoluble salt by precipitation

1. Take 5 cm³ of lead nitrate solution in a test tube.
   Add 10 cm³ of sodium iodide solution.   ⚠ lead salts
   - What happens?
   - What is the precipitate called?
   - How can we separate the precipitate from the solution?

2. Filter your mixture.
   The solid lead iodide is left on the filter paper.
   Rinse it with distilled water.
   - Why must we rinse the lead iodide?

3. Scrape the salt made on to some fresh filter paper.
   Leave it to dry.

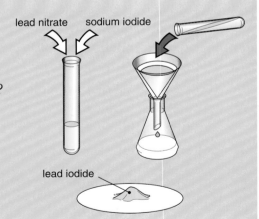

The lead nitrate and sodium iodide react to make a precipitate, lead iodide.

lead nitrate (aq) + sodium iodide (aq) $\longrightarrow$ lead iodide (s) + sodium nitrate (aq)

The lead iodide must be rinsed with water to wash away any soluble salts.

▶ Precipitation

In the last experiment we made an *insoluble* salt
from two *soluble* salts.
To help us choose which salts to mix,
we need to know which salts dissolve in water.

$$AB(aq) + CD(aq) \longrightarrow AD(s) + CB(aq)$$
A precipitation reaction

Look at the table below:

Salt	Solubility
chlorides	soluble (except for lead chloride, silver chloride)
sulfates	soluble (except for barium sulfate, calcium sulfate, lead sulfate)
nitrates	all soluble

Any salts of lithium, sodium or potassium are also soluble.

- Name two salt solutions you could mix to make barium sulfate.
- Which acid would you add to barium nitrate solution to make barium sulfate?

Look at the word equation for the last experiment.
The balanced equation is:

$$Pb(NO_3)_2(aq) + 2\,NaI(aq) \longrightarrow PbI_2(s) + 2\,NaNO_3(aq)$$

We can also show what happens like this:

$$Pb^{2+}(aq) + 2\,I^-(aq) \longrightarrow PbI_2(s)$$

This is called an **ionic equation**.

It shows us which ions stick together to form the solid
in the precipitation reaction.
The sodium ions, $Na^+(aq)$, and the nitrate ions, $NO_3^-(aq)$,
stay in the solution. They are not changed, so they
don't appear in the ionic equation.
They are called **spectator ions**.

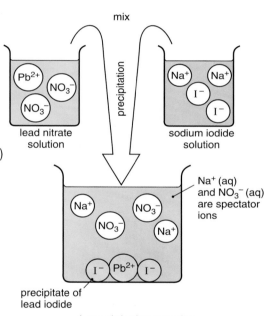

A precipitation reaction

The centrifuge

The solids in precipitation reactions
are formed as very small grains.
The mixture is called a **suspension**.
A centrifuge is a machine which spins test-tubes
around at high speed.
What do you think happens to the fine particles
of the precipitate? Think about what happens to you
on a ride that spins around quickly at a fairground!

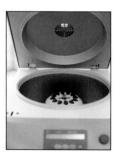

A centrifuge spins test tubes at high speed

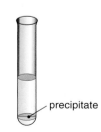

After centrifuging the solid is packed at the bottom. The solution above can be poured off (decanted). The precipitate is then washed and dried.

133

▶ Chemistry at work : Sulfuric acid, H_2SO_4

Uses of sulfuric acid

Sulfuric acid is one of the most important products of the chemical industry.
Look at its uses below :
Which products made from sulfuric acid have you used today ?

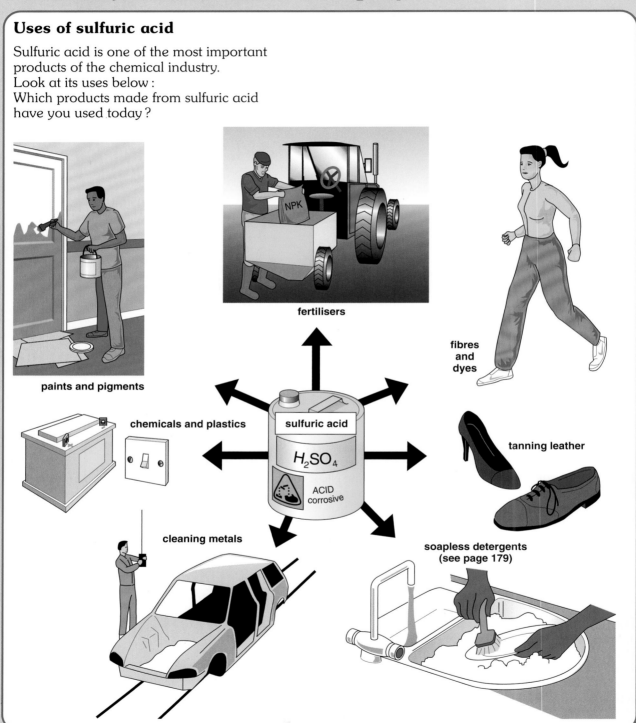

paints and pigments

fertilisers

fibres and dyes

chemicals and plastics

sulfuric acid

H_2SO_4

ACID corrosive

tanning leather

cleaning metals

soapless detergents
(see page 179)

▷ Chemistry at work : Sulfuric acid, H_2SO_4

The Contact process

Sulfuric acid is made in the Contact process.
The raw materials are sulfur, air and water.
Look at the diagram below :

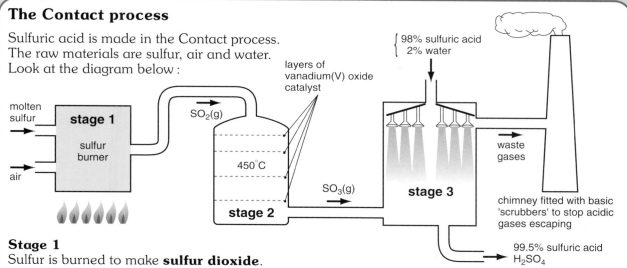

Stage 1
Sulfur is burned to make **sulfur dioxide**.
We can import the sulfur on ships from Poland or
the USA.
It can also be extracted from impurities in
crude oil or natural gas. However North Sea gas
contains very little sulfur, so it can't be used.
The sulfur burns in air :

$$\text{sulfur} \ + \ \text{oxygen} \longrightarrow \text{sulfur dioxide}$$
$$S(l) \ \ + \ \ O_2(g) \longrightarrow \ \ \ SO_2(g)$$

Stage 2
The sulfur dioxide is turned into **sulfur trioxide**.

$$\text{sulfur dioxide} \ + \ \text{oxygen} \rightleftharpoons \text{sulfur trioxide}$$
$$2\,SO_2(g) \ \ + \ \ O_2(g) \rightleftharpoons \ \ 2\,SO_3(g)$$

Notice that the reaction is reversible. You can read
more about this type of reaction in Chapter 17.
A catalyst, vanadium(V) oxide, is used to speed
up the reaction.
Why do you think there are 4 layers of catalyst ?

Stage 3
In the final stage, sulfur trioxide is changed into
sulfuric acid.
The sulfur trioxide gets absorbed into a mixture of
98 % sulfuric acid and 2 % water. In effect, the
sulfur trioxide reacts with the water :

$$\text{sulfur trioxide} \ + \ \text{water} \longrightarrow \text{sulfuric acid}$$
$$SO_3(g) \ \ \ \ + \ H_2O(l) \longrightarrow \ \ H_2SO_4(l)$$

Economic and environmental aspects

- The sulfur dioxide made in Stage 1 is
  supplied directly from smelting metal ores
  that contain sulfides. So smelting plants
  often make sulfuric acid at the same site.

- In Stage 2, 99.5 % of the sulfur dioxide gets
  converted into sulfur trioxide. Sulfur
  dioxide causes acid rain, so strict controls
  are needed on the waste gases.

- The reactions in each stage give out heat.
  As much energy as possible is conserved.

- If you used water to absorb the sulfur
  trioxide in Stage 3, a fine mist of sulfuric
  acid would be formed. This can't be
  condensed and would pollute the air. But
  chemists found that SO_3 is absorbed better
  into a mixture of 98 % sulfuric acid and 2 %
  water.

- Waste gases are passed over 'scrubbers'
  (such as basic calcium carbonate or calcium
  hydroxide) to stop acidic gases escaping into
  the air.

▶ Chemistry at work : Neutralisation

Acid soil

If soil is too acidic, most crops will not grow well.
Farmers can spread powdered limestone (calcium carbonate)
or slaked lime (calcium hydroxide) on the soil to raise its pH.

Experiment 10.9   Raising the pH of soil

Add 50 cm³ of distilled water to some
acidic soil. Stir it well, filter, then add a few drops
of universal indicator.
Add powdered limestone until you neutralise
the solution.
* How can you see when you have added enough ?

Sodium hydrogencarbonate

Baking powder helps cake-mix to rise.
It contains sodium hydrogencarbonate (known as
bicarbonate of soda) and a weak acid.
When water is added the two react. The reaction
is like an acid plus a carbonate.
Do you remember which gas is made ? (See page 129.)

*Baking powder
helps a cake rise
to the occasion !*

> **acid + a hydrogencarbonate ──▶ a salt + water + carbon dioxide**

For example with hydrochloric acid (but don't add to your cake mix!) :

$$HCl(aq) + NaHCO_3(aq) \longrightarrow NaCl(aq) + H_2O(l) + CO_2(g)$$

The carbon dioxide gets trapped in the cake-mix.
Self-raising flour has baking powder already added.
The sodium hydrogencarbonate also undergoes thermal decomposition
in the hot oven :

$$2NaHCO_3(s) \longrightarrow 2Na_2CO_3(s) + H_2O(g) + CO_2(g)$$

This also helps the cake to rise.
We also use sodium hydrogencarbonate in cleaning agents and in some
toothpastes. It neutralises acidic solutions.

Treating stings

Bee stings are said to be acidic, although this is debatable !
However, you can ease the pain with bicarbonate of soda
(sodium hydrogencarbonate) – a weak alkali.

Wasp stings are alkaline. You can neutralise them
with vinegar (ethanoic acid).

Summary

Acids form solutions which:

- turn blue litmus red
- have a pH less than 7
- react with an alkali (or base) to give a salt and water – this is called **neutralisation**
- react with a carbonate to give a salt, water and carbon dioxide
- react with metals to give a salt and hydrogen (the metal must be above copper in the Reactivity Series).

Salts are compounds made when we replace the hydrogen in an acid by a metal.
Hydrochloric acid (HCl) makes salts called chlorides.
Sulfuric acid (H_2SO_4) gives sulfates.
Nitric acid (HNO_3) gives nitrates.
Examples of salts include sodium chloride (NaCl), copper sulfate ($CuSO_4$) and silver nitrate ($AgNO_3$).

▶ Questions

1. Copy and complete:

   Acidic solutions turn blue litmus . . . . .
   Their pH is always . . . . than 7.
   Here are their common reactions:

   acid + base (or alkali) ⟶ a salt + ...

   acid + carbonate ⟶ a salt + water
   + ...

   acid + metal ⟶ a salt + ...

   When an acid and alkali react to 'cancel each other out', we call it a . . . . reaction.

   A salt is made when the . . . . in an acid is replaced by a . . . . .

Acid	Salt
hydrochloric acid ⟶	. . . .
. . . . acid ⟶	sulfates
nitric acid ⟶	. . . .

2. Copy and complete these word equations:

   a) sodium hydroxide + ... acid ⟶ sodium chloride + water
   b) copper oxide + sulfuric acid ⟶ . . . . + ...
   c) copper . . . . + hydrochloric acid ⟶ copper . . . . + water + carbon dioxide
   d) . . . . + sulfuric acid ⟶ magnesium sulfate + hydrogen
   e) sodium hydroxide + nitric acid ⟶ . . . . + ...

3. Joe tested some solutions with universal indicator paper.
   He wrote down their pHs:
   1, 5, 7, 14
   but forgot to write the names of the solutions.
   Can you help him by matching the pHs to the correct solutions?

Solution tested	pH
distilled water	
sulfuric acid	
sodium hydroxide	
vinegar	

4. Marie and Sakib want to make some crystals of copper sulfate.
   They have dilute sulfuric acid and black copper oxide powder.
   a) Describe how they can get copper sulfate crystals safely.
   b) How can they tell when the sulfuric acid has been neutralised?
   c) What type of substance is the black copper oxide powder – an alkali or a base?
   d) Write a word equation and a symbol equation for the reaction.

5. The table below shows the conditions some plants prefer:

Plant	pH
apple	5.0 – 6.5
potato	4.5 – 6.0
blackcurrant	6.0 – 8.0
mint	7.0 – 8.0
onion	6.0 – 7.0
strawberry	5.0 – 7.0
lettuce	6.0 – 7.0

a) Which plants grow well over the largest *range* of pH values?
b) Which plant can grow in the most acidic soil?
c) Describe how you can test the pH of a soil.
d) How can you raise the pH of an acidic soil?

6. Sulfuric acid is made in the Contact process.
a) What are the raw materials for the process?
b) Draw a flow diagram showing the steps in making sulfuric acid. Include word equations.
c) What is the catalyst used in the process?
d) In the last step, why don't we dissolve the sulfur trioxide in water?
e) Draw a spider diagram showing the uses of sulfuric acid.

7. Liz was investigating which of four indigestion tablets was most effective. She crushed each tablet and added it to 25 cm^3 of water and an indicator.
Then she added dilute hydrochloric acid 1 cm^3 at a time, until the indicator changed colour.
Here are her results:

Fizzo – 18 cm^3, Neutratabs – 17 cm^3, Soothers – 9 cm^3, Alkomix – 12 cm^3.

a) Put her results in a suitable table.
b) What type of graph would you use to show Liz's results?
c) Which tablet was the most effective?
d) How could she have made her results more reliable?

8. Look at the table below.
It shows the solubility of potassium nitrate at different temperatures.

Temperature (°C)	Solubility (g/100 g of water)
20	32
40	64
60	110
80	169
100	246

a) Draw a solubility curve for potassium nitrate. (See page 131.)
b) What pattern can you see from your graph?
c) What is the solubility of potassium nitrate at
   i) 50 °C   ii) 70 °C?
d) What does a 'saturated solution' mean?
e) How can you make a saturated solution of potassium nitrate? No solid should be present.

9. a) Vani has a solution which she thinks might be acidic. Describe 3 tests she could do to see if it is acidic or not.
b) Explain why water is needed before a substance can show its acidic properties.

10. Here is a list of soluble salts:

sodium nitrate – $NaNO_3$
potassium chloride – KCl
magnesium sulfate – $MgSO_4$
calcium chloride – $CaCl_2$
lead nitrate – $Pb(NO_3)_2$

a) Choose a pair of salts you could mix to make the insoluble salts:
   i) lead chloride ii) calcium sulfate.
b) i) How would you collect a *pure* sample of each salt made in part a)?
   ii) Write a word equation for each reaction.
   iii) Write a symbol equation for each reaction. Include state symbols.
   iv) Write an ionic equation for each reaction.

11. Look through this chapter (and others if you have time) and design a poster showing some uses of salts.

Further questions on pages 187 and 188.

PRODUCTS FROM

chapter 11

▷ Hydrocarbons

Imagine you could vote in a 'Molecule of the Century' competition. A vote for a **hydrocarbon** would stand a good chance of winning!

So what is a 'hydrocarbon'? And why are they so important?

As the name suggests:

> **A hydrocarbon is a compound containing only hydrogen and carbon.**

One of the world's most important raw materials is **crude oil**. Crude oil is a *mixture* of many hydrocarbons.

The hydrocarbons in crude oil are not only vital **fuels**. They are also the starting materials for many new products, such as **plastics**.

We get petrol from crude oil

The alkanes

There are lots of different hydrocarbons in crude oil. Most of them are called **alkanes**.

Natural gas, which is found with crude oil, is mainly methane. This is the smallest alkane, CH_4.

The alkane molecules have a 'backbone' of carbon atoms. Their carbon atoms are surrounded by hydrogen atoms.

Look at the pictures of some alkanes below:

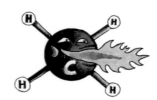

*My name is methane –
If you want a fire,
Just light the gas
And turn me higher!*

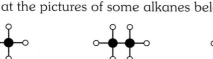

methane   ethane   propane   butane   pentane

Notice the bonds that join the atoms together: Carbons have 4 bonds. Hydrogens have 1 bond. (See page 257.)

You can see the names and formulas of the first few alkanes in this table.

Notice that all their names end in **-ane**.

- Can you see a pattern in their formulas?
- Can you work out the formula of hexane?
- Can you draw the structure of hexane?

● = carbon
○ = hydrogen

Name	Formula
methane	CH_4
ethane	C_2H_6
propane	C_3H_8
butane	C_4H_{10}
pentane	C_5H_{12}
hexane	

The general formula of an alkane is:

$$C_nH_{2n+2}$$

where *n* can be any number.

► Fossil fuels

Most of our common fuels are **fossil fuels**.
Can you name any?

Coal, **crude oil** (which gives us petrol), and **natural gas** are all fossil fuels.

Coal is the odd one out as it is not a hydrocarbon.
It does contain carbon and hydrogen.
However, there are also other types of atoms, such as oxygen, in some of its molecules.

Fossil fuels have taken millions of years to form.

Coal came from trees and ferns that died and were buried beneath swamps.

Crude oil was formed from tiny animals and plants which lived in the sea.

The plants and animals got their energy from the Sun.

So when you burn a fuel you are using energy that started off in the Sun!

Fern fossils are found in coal

the plankton in the sea which went on to make oil got their energy from the sun

the trees that made the coal got their energy from the Sun (in photosynthesis)

oil rig

coal mine

Scientists predict our supplies of crude oil could run out within your life-time (in about 50 years time). Coal could last another 300 years.

Fossil fuels are called **non-renewable** fuels. That's because once we use up our supplies on Earth they will be gone forever.

The story of oil begins

Crude oil was made from the bodies of tiny sea creatures and plants in plankton that died about 150 million years ago.

These were buried under layers of sand and silt on the sea bed. They did not decay normally as the bacteria feeding on them had little or no oxygen in these conditions.

As the pressure and temperature slowly increased, they were changed into oil.

Natural gas is usually found with the crude oil.

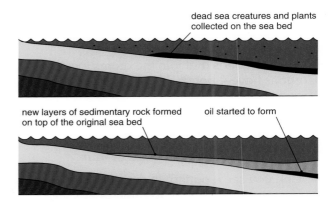

dead sea creatures and plants collected on the sea bed

new layers of sedimentary rock formed on top of the original sea bed

oil started to form

Finding oil

Oil supplies are scarce.
Oil companies, like BP and Shell, are looking for new oil all the time.
So how do they find their oil?

First of all scientists look for clues on the Earth's surface. Look at the diagram opposite:

Crude oil soaks into porous rock, just like water soaking into a sponge. The oil rises towards the surface but is usually stopped before it gets there by a layer of non-porous rock.

The oil is usually found under a dome shaped layer called a cap rock or anti-cline. (See page 316.)

Seismic survey

If an area looks good, scientists can find the structure of the rocks underground. They do this by a seismic survey.

Small explosions are set off on the surface.
These send out shock waves.
The waves that bounce back from each layer are picked up by sensors.

A computer helps to analyse these echoes. It then draws a map of the rocks beneath the surface.

If the layers are like the ones shown above, there might be oil. The scientists do not know for sure until they actually drill down. If rock containing oil is found, more wells are drilled. This helps the company see if it is worthwhile going into full production.
A mistake at this stage is very costly!

Transporting oil

There are two ways to move the oil from the oilfield to the refinery:
i) by pipeline or ii) by oil tanker.

Pipelines are often used when the oil is found reasonably close to the refinery. For example, North Sea oil is piped to Aberdeen in Scotland.

Giant ships called oil tankers take oil all over the world. However, accidents can happen at sea and the oil can be spilled.
The effects on wildlife are disastrous. (See page 338.)

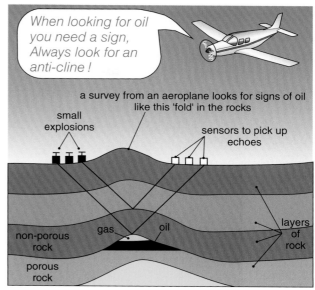

A seismic survey

An exploration rig

Oil tankers can be 400 m long. The crew sometimes use bicycles to get around the ship.

▶ Distilling crude oil

At the refinery

When crude oil reaches the refinery it is a thick black, smelly liquid. It is not much use to anyone.

You learned on page 139 that crude oil contains a mixture of different hydrocarbons. At the refinery these are sorted out into groups of useful substances called **fractions**.

These fractions are separated by **fractional distillation**.

An oil refinery at night

We can distil crude oil in the lab to help you understand what happens in the refinery.

- Where do you think would be the best place to build an oil refinery? Why?
- How could the refinery harm our environment?

Demonstration 11.1   Distillation of crude oil
The apparatus is set up as shown in a fume-cupboard.

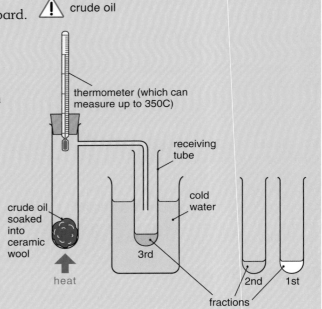

Heat the crude oil gently at first. Collect a few drops of the first liquid that collects.
Keep heating and change the receiving tube.
Now collect the liquid that distils up to 150 °C.
Again change the tube and collect the liquid when you heat strongly.

Pour each fraction collected onto a watch glass.

- Note its colour and how thick it is.

Try to light each fraction with a lighted splint.

- What happens?
- Why is the receiving tube kept in cold water?
- You need a fuel for a car. Why would you use the first fraction collected?
- Why would you use the last fraction to lubricate a car engine?

Q. What do you call a hydrocarbon who tells rude jokes?

A. Crude oil!

There was a young girl called Eve,
Who boiled oil with a friend called Steve.
Small molecules flew
Up out from the brew
As they were the first ones to leave.

Looking for patterns

Fraction	Size of molecules	Colour	Thickness (viscosity)	How it burns
low boiling point (up to 80 °C)	small	colourless	runny	lights easily (highly flammable), clean flame
medium boiling point (80–150 °C)	medium	yellow	thicker	harder to light, some smoke
high boiling point (above 150 °C)	large	dark orange	thick (viscous)	difficult to light (not very flammable), smoky flame

The fractions get **less volatile** (are harder to evaporate) as the molecules in their compounds get larger.

- Look down each column in the table above :
  What patterns can you see as the boiling point of the fraction gets higher ?

Explaining the distillation

If you have a plate of spaghetti, you find that the short strands are easier to pull out than the longer ones. This helps to explain the distillation of crude oil, Instead of spaghetti, think of molecules !

As the oil is heated, the small molecules boil off first. These gas molecules are then condensed (turned back to liquid) in the cold receiving tube.

Small hydrocarbons have lower boiling points than large ones. They are easier to separate from the mixture of molecules in crude oil.

As you carry on heating, the temperature of the crude oil rises. The larger molecules are then boiled off and are collected.

Remember :

The long pieces of spaghetti get tangled up. They are harder to separate out, just like the long molecules in crude oil.

> **Small hydrocarbons have lower boiling points than large ones.**

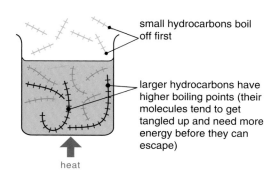

small hydrocarbons boil off first

larger hydrocarbons have higher boiling points (their molecules tend to get tangled up and need more energy before they can escape)

heat

> **The larger the molecules, the stronger the forces of attraction between molecules.**

▶ Fractional distillation in industry

In an oil refinery the crude oil is separated into its fractions. This happens in huge **fractionating columns**.

Remember that a **fraction** is a group of hydrocarbons with similar boiling points.

Just like distillation in the lab, they use the different boiling points of the hydrocarbons to separate them. But there is a difference.

In Demonstration 11.1 we boiled off each fraction in turn. In industry, they boil up all the fractions together. Then they condense them at difference temperatures at the same time.
You can see this in the column shown below.

A fractionating column

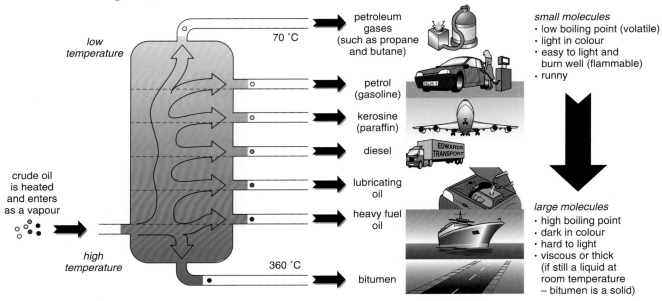

Explanation
The crude oil is heated up and evaporates. It enters the column as a vapour.

The fractionating column is hot at the bottom and cooler at the top. This means that the larger hydrocarbons, with the high boiling points, turn back to liquids nearer the bottom.

At the high temperatures there, the smaller hydrocarbons stay as gases. They rise up the column. The different fractions now condense at different levels.

At the top of the column there are hydrocarbons with low boiling points. At 70 °C these have still not condensed, but come out of the top as gases.

Fraction	length of carbon chain
petroleum gases	$C_1 - C_4$
petrol	$C_4 - C_{12}$
kerosine	$C_{11} - C_{15}$
diesel	$C_{15} - C_{19}$
lubricating oil	$C_{20} - C_{30}$
fuel oil	$C_{30} - C_{40}$
bitumen	C_{50} and above

▶ Cracking

After distilling the crude oil, the oil companies find that they have too many large hydrocarbons. We just don't need that much of the heavy fractions.

Yet the smaller hydrocarbons, like petrol, are in great demand. So scientists have found a way to change the larger, less useful molecules into smaller, more useful ones.

The reaction used is called **cracking**.

The big molecules are broken down by *heating* them as they pass over a **catalyst**. A catalyst helps to speed up a reaction. (See page 212.)

In the refinery this happens inside a **cracker**!

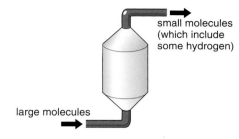

A cracker in an oil refinery

It's a cracker

Do-it-yourself cracker

Q. What do you call two mad scientists in an oil refinery?

A. Crackers!

Experiment 11.2   Cracking

In this experiment you can crack a large molecule (paraffin) into smaller ones.

Set up your apparatus as shown:

⚠ *You must take the end of the delivery tube out of the water before you stop heating.*
(You don't want cold water to be sucked back into your hot tube!)

Start by heating the aluminium oxide strongly. Then just move the flame onto the paraffin every now and again. This is just to make sure that some vapour is passing along the tube.
(Don't collect the first few bubbles that appear. This is hot air inside the tube expanding.)

Collect two test tubes of gas. The gas is **ethene**.

Test the ethene with a lighted splint.
● What happens?

Test the other tube by adding a little bromine water. Replace the bung quickly and shake.
● What happens?

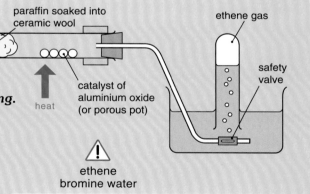

Remember:

Cracking is when we break down large hydrocarbons into smaller, more useful ones. This is done by passing the vapours over a hot catalyst or mixing them with steam at very high temperatures.
Cracking is an example of a thermal decomposition reaction.

▶ Ethene and the alkenes

You have seen how ethene can be formed in cracking. For example:

$$C_{10}H_{22} \longrightarrow C_8H_{18} + C_2H_4$$
decane   octane   ethene
(used in petrol)

Notice that ethene is a hydrocarbon. Like the alkane family on page 139, it contains only hydrogen and carbon atoms. However, they are joined together differently.

Ethene belongs to a family of hydrocarbons called the **alkenes**. Look at the table:

Their names start off like the alkanes, but end in **-ene** instead of -ane.

Remember the alkanes have only single bonds joining their atoms together. The alkenes have at least one **double bond** between carbon atoms.

(If you want to know more about bonds, read Chapter 20.)

double bond

an ethene molecule

A double bond makes a molecule reactive.

Ethene even reacts with itself to make poly(ethene). (See pages 153 and 154 for details.)

Saturated and unsaturated molecules

Ethene and the other alkenes are called **unsaturated** compounds. They have one or more double bonds between carbon atoms in their molecules.

On the other hand, the alkanes are said to be **saturated**. They contain only single bonds.

You can test to see if a compound is unsaturated:

> Unsaturated compounds (containing carbon–carbon double bonds) turn yellow **bromine water** colourless.

When we crack large hydrocarbons, we get a mixture of smaller saturated and unsaturated hydrocarbons.
- Look at the equation at the top of the page:
  Label each hydrocarbon as saturated or unsaturated.

Name	Formula	Picture of molecule
ethene	C_2H_4	H_C=C_H (H, H, H, H)
propene	C_3H_6	H_C=C–C–H (H, H, H, H)
butene	C_4H_8	

The table shows the first three members of the **alkene** family.
- *Can you see a pattern in their formulas?*
- *Can you guess the names and formulas of the next two alkenes?*
  (Look at the table of alkanes on page 139 for clues.)
- *Can you draw a molecule of butene?*
- *Why is there no alkene with just one carbon atom?*

The general formula of an alkene containing one carbon–carbon double bond is:

$$C_nH_{2n}$$

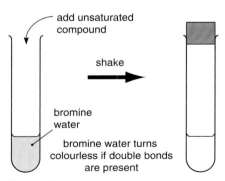

add unsaturated compound

shake

bromine water

bromine water turns colourless if double bonds are present

The test for an unsaturated compound

For example
$$C_2H_4 + Br_2 \longrightarrow C_2H_4Br_2$$
*This is called an **addition** reaction*

▷ Combustion

As you know, the fuels we get from crude oil are called hydrocarbons.
Petrol is a mixture of hydrocarbons.
Can you remember which 2 elements make up a hydrocarbon? Petrol contains octane. Can you guess how many carbon atoms there are in octane?

When we burn a fuel, the reaction is called **combustion**.
The fuel reacts with oxygen gas from the air and heat is given out.
But what is made in the reaction? You can find out in the next experiment:

Combustion is a useful reaction

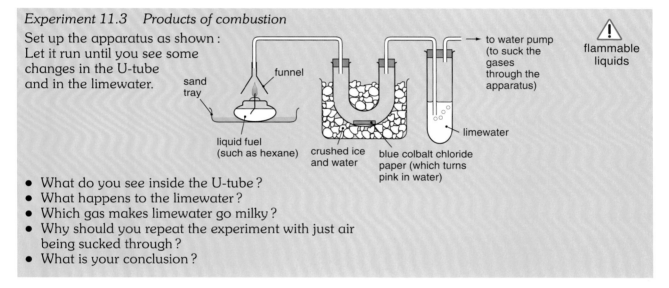

Experiment 11.3   Products of combustion

Set up the apparatus as shown:
Let it run until you see some changes in the U-tube and in the limewater.

to water pump (to suck the gases through the apparatus)

flammable liquids

sand tray

funnel

liquid fuel (such as hexane)

crushed ice and water

blue colbalt chloride paper (which turns pink in water)

limewater

- What do you see inside the U-tube?
- What happens to the limewater?
- Which gas makes limewater go milky?
- Why should you repeat the experiment with just air being sucked through?
- What is your conclusion?

When hydrocarbons burn, they are **oxidised**.
We get carbon dioxide and water formed (*never* hydrogen!):

| hydrocarbon + oxygen ⟶ carbon dioxide + water |

Incomplete combustion

There is only a small amount of oxygen inside a car engine.
There is not enough to turn all the carbon in the hydrocarbons into carbon dioxide. Some **carbon monoxide**, a toxic gas, is also made.
Catalytic converters on exhausts, once warmed up, can turn carbon monoxide into carbon dioxide.
Sometimes fuels burn with a smoky flame. Incomplete combustion means that **carbon** itself is given off as soot.

Particles from diesel are also polluting the air in our cities.
Unburnt hydrocarbons in **particulates** can cause cancer.

Carbon monoxide is given off from car exhausts. This gas reacts with haemoglobin. It stops your blood carrying oxygen around your body. This toxic gas can kill unsuspecting people when given off by faulty gas heaters.

▶ Acid rain

Burning fossil fuels causes **acid rain**.

Coal-fired power stations give off sulfur dioxide gas. This causes acid rain.

Most fossil fuels contain sulfur in compounds present as impurities.
When we burn the fuel, the sulfur is oxidised.
It turns into **sulfur dioxide** (SO_2) gas.

Power stations burning coal or oil give off most sulfur dioxide.
This is the main cause of acid rain. The gas dissolves in rainwater,
and reacts with oxygen in the air, to form sulfuric acid.

Cars also make our rain acidic. Most petrol is sulfur-free now.
However, car exhausts give off **nitrogen oxides**. In the high
temperatures inside an engine, even unreactive nitrogen gas will
react with oxygen.
The nitrogen oxides formed make nitric acid when it rains.

Experiment 11.4   Effect of sulfur dioxide on plants

Set up the apparatus as shown :
The sodium compound gives off sulfur dioxide
which is toxic.
Leave the seeds for a few days.
- Explain why we need to set up both tests.
- What effect does sulfur dioxide have on the growth of the seeds?

⚠ sodium metabisulfite sulfur dioxide

sodium meta-bisulfate added to make sulfur dioxide   cress seeds   cotton wool

---

Effects of acid rain

1. *Forests*   – Trees are damaged and even killed.
   Over half the forests in Germany are dead or dying.
2. *Fish*   – Hundreds of lakes in Norway and Sweden
   now have no fish left in them at all.
   Aluminium, which is normally 'locked' in the soil,
   dissolves in acid rain. It then gets washed into
   the lakes, where it poisons the fish.
3. *Buildings* – Acid rain attacks buildings and metal structures.
   Limestone buildings are most badly affected.

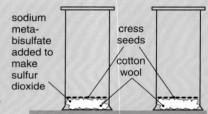
Acid rain damages buildings

What can be done?

We can burn less fossil fuels by using energy more efficiently.
Or we could use alternative forms of energy.

We are now starting to remove sulfur from fossil fuels
before we burn them. The sulfur can then be used to
make sulfuric acid. (See page 135.)

We can also get rid of the acidic gases before
they leave power stations. A mixture of limestone
and water neutralises any oxides of nitrogen and sulfur dioxide :
e.g. $CaCO_3(s) + SO_2(aq) \longrightarrow CaSO_3(s) + CO_2(g)$

*Nitrogen oxides are changed into harmless nitrogen gas by **catalytic converters** in car exhausts. The reaction is :*
$2CO(g) + 2NO(g) \longrightarrow N_2(g) + 2CO_2(g)$

▷ Greenhouse effect

The Earth's atmosphere acts like a greenhouse.
It lets rays from the Sun through to warm the Earth.
But gases, such as carbon dioxide and water vapour, absorb some of the heat waves given off as the Earth cools down.

We are lucky to have these natural 'greenhouse gases'.
Without them, the Earth would be about 30 °C colder!
How do you think that would affect life on Earth?

However, we are making more and more of these gases.
Whenever we burn a fossil fuel we make carbon dioxide.
And we are now burning up fossil fuels at an incredible rate.
This disturbs the natural balance of carbon dioxide. (See page 304.)

Although plants absorb carbon dioxide, we are cutting down huge areas of forest every day. The trees are often just burned to clear land for farming. This makes even more carbon dioxide.

More carbon dioxide, plus other 'greenhouse gases', such as methane from cattle, seem to be making the Earth hotter. We call this **global warming**.

Look at the graph opposite:
* What is the general pattern?

But some scientists are not sure if these changes are just part of the Earth's natural variations.
However, if temperatures go on rising, it will affect the Earth. Again, scientists disagree on the changes and how long they will take to happen.

Ice caps are melting. Warmer oceans could also expand, causing a rise in sea levels.
The shape of the world map would change as low lands flood. Climates could change around the world. However, scientists are not sure exactly how different places will be affected. An experiment on a global scale is not easy to predict, even using the most powerful computer models.

What can be done?

As with acid rain, burning less fossil fuels will help.
But cleaning the gases given off will not reduce the amount of carbon dioxide released into the air.
However, nitrogen oxides, which are also 'greenhouse gases', can be removed from car exhausts by catalytic converters. (See page 148.)

Planting trees to replace those cut down will help to restore the Earth's natural balance. (See Carbon Cycle, page 304.)

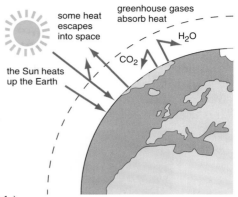

Carbon dioxide and water vapour are the main 'greenhouse gases'.
Methane from cattle and decomposing vegetation is another greenhouse gas.

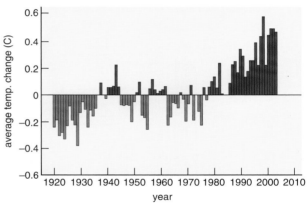

Changes from average temperatures, 1920–2005.

Like the glass in a greenhouse, the atmosphere lets light in but heat energy is trapped inside. However, scientists think that tiny solid particles in the atmosphere (from incomplete combustion of fuels) are also affecting climate. These stop energy from the Sun reaching the Earth's surface causing a cooling effect. This is known as **global dimming**.

▷ Chemistry at work : Naphtha

A source of new materials

You have already seen some uses of the fractions
we get from crude oil. (See page 144.)
You can also get a light fraction called **naphtha**.
It is used to make many new products.

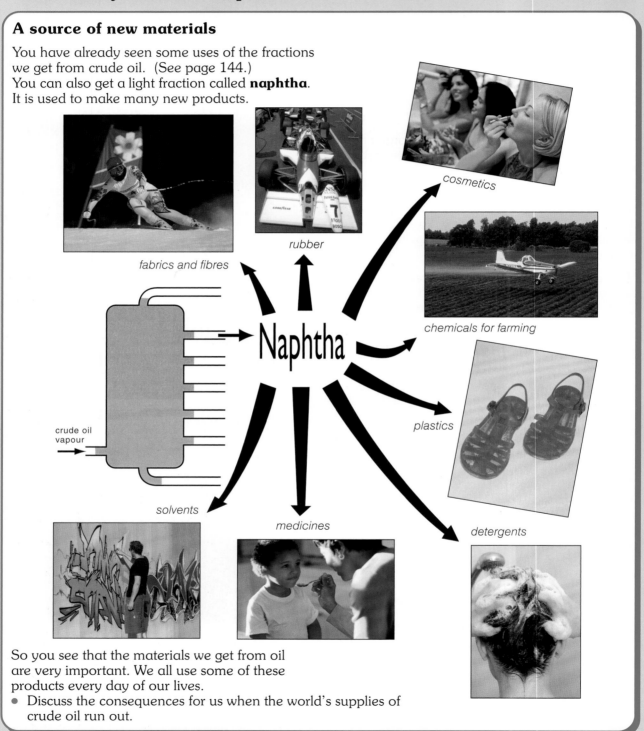

fabrics and fibres

rubber

cosmetics

crude oil
vapour

Naphtha

chemicals for farming

plastics

solvents

medicines

detergents

So you see that the materials we get from oil
are very important. We all use some of these
products every day of our lives.
● Discuss the consequences for us when the world's supplies of
crude oil run out.

Summary

- Crude oil contains a mixture of **hydrocarbons**. Hydrocarbons are compounds made of hydrogen and carbon only.
- Crude oil is separated into groups of useful substances (fractions) by **fractional distillation**. This works because different hydrocarbons have different boiling points.
- The small hydrocarbons
  - have lower boiling points,
  - are lighter in colour,
  - are easier to light, burning with a cleaner flame,
  - are thinner and more runny.
- The large molecules from crude oil can be broken down or **cracked** into smaller, more useful molecules. This is done at high temperature, using a catalyst.
- When a hydrocarbon burns in enough oxygen, we get carbon dioxide and water formed. The CO_2 is a **greenhouse gas**.
- If a hydrocarbon burns in a limited supply of oxygen, we also get **carbon monoxide** gas. Tiny particles of carbon and unburnt toxic hydrocarbons are also released. These particles can cause **global dimming**.

▶ Questions

1. Copy out and complete :
   a) A hydrocarbon is a compound containing .... and carbon atoms only.
   b) Crude oil is a .... of hydrocarbons. In an oil ...., the crude oil is separated into its fractions by .... The heated oil enters the fractionating .... as a gas. The hydrocarbons have different .... points and condense at different temperatures. The smaller hydrocarbons, with the lower boiling points, are collected near the .... of the column.
   c) Large hydrocarbons can be cracked by heating them with a ..... The new molecules made are .... and more useful.
   d) When a hydrocarbon burns in plenty of oxygen, it gives carbon dioxide and .... This reaction is called .... (or oxidation). In a limited supply of oxygen, we also get toxic carbon .... gas and .... particles.
   e) Burning fossil fuels causes .... rain because of .... dioxide gas given off from impurities. Scientists are also worried about the large volumes of .... dioxide gas produced. This has been linked to .... warming caused by the .... effect.

2. Carla distils some crude oil. She collects 3 fractions in different test tubes. But at the end of her experiment, the tubes get mixed up. Give 3 ways that Carla could find out which order they were collected in.

3. a) Copy and complete this table :

Alkane	Structure of molecule
methane (CH_4)	
butane (C_4H_{10})	
pentane (C_5H_{12})	

b) Petrol contains the alkane with 8 carbons. Can you guess its name ? Draw a diagram to show the structure of the molecule.

4. Imagine you are a small hydrocarbon molecule in crude oil. Describe what happens to you from the time you are discovered by an oil company, to the time you end up heating beans on a camping stove.

5. Design a leaflet for your parents which explains about the greenhouse effect and its possible consequences for our planet.

6. Look at the diagram of a **catalytic converter** below.

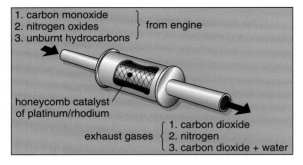

1. carbon monoxide
2. nitrogen oxides   } from engine
3. unburnt hydrocarbons

honeycomb catalyst of platinum/rhodium

exhaust gases {
1. carbon dioxide
2. nitrogen
3. carbon dioxide + water

a) Explain how a catalytic converter helps to reduce acid rain.
b) Does a catalytic converter help to reduce the greenhouse effect? Explain your answer.
c) Explain what happens to the carbon monoxide and nitrogen oxides in the catalytic converter. Include a word and symbol equation.

7. a) Finish off this equation:

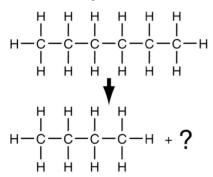

b) What is this type of reaction called?
c) What are the names of the 2 compounds made in the reaction?
d) What must you do to make this reaction happen?
e) Describe and explain a test and its results to distinguish between the products formed in the reaction.

8. Many power stations burn fossil fuels.
a) Name 3 fossil fuels. (See page 140.)
b) Which gas given off from power stations can cause acid rain?
c) Make a list of the things we can do to reduce the problem of acid rain.

9. Zara and Lee want to test the products formed when a hydrocarbon burns. They set up the experiment below:

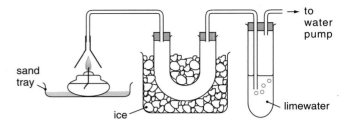

sand tray

ice

to water pump

limewater

a) Why do they put ice around the U-tube?
b) How can they test that the liquid formed in the U-tube is water?
c) The other product is carbon dioxide. What happens to the limewater?
d) There is a small amount of carbon dioxide in the air. How can they show that the carbon dioxide they test for is not just the carbon dioxide in the air?
e) Give the word and symbol equation for the complete cumbustion of methane, CH_4.

10. a) Complete this general formula for the alkanes:

$$C_nH_?$$

b) What is the formula of the alkane with 9 carbon atoms?
Look at the boiling points in this table:

Alkane (°C)	Number of carbon atoms	Boiling point
methane	1	−161
ethane	2	−88
propane	3	−42
butane	4	−0.5
pentane	5	
hexane	6	69

c) Draw a graph of their boiling points (vertical axis) against the number of carbon atoms (horizontal axis).
d) What is the general pattern you see from your graph?
e) Use your graph to predict the boiling point of pentane.

Further questions on pages 188 and 189.

Polymers

chapter 12

▶ Making plastics

Whenever oil companies crack large molecules into smaller ones, **ethene** is made.

Ethene is a very useful little molecule. It is the starting material for many **plastics**.

Plastics were first made on a large scale in the 1930s. They are now a very important part of our modern world.

Look at the picture of a toddler's bedroom from around 1900. Nothing in it is made from plastics.

● *What things would be made from plastics nowadays?*

Polymers

Plastics are huge molecules. They are usually long chains, made from thousands of atoms. These long chain molecules are called **polymers**.

(**Poly** meaning *many* e.g. a polygon is a many sided shape.)

Polymers are made by joining together thousands of small, reactive molecules called **monomers**.

(**Mono** meaning *one* e.g. a monorail is a railway with only one track.)

It's a bit like stringing beads together to make a necklace. But using real beads, your polymer necklace would be about half a kilometre long!

> **Lots of small, reactive molecules called monomers join together to make a polymer.**

You can also get the idea by using shapes :

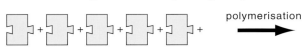

The reaction is called **polymerisation**.
The most common monomer is ethene.
When it joins together in a long chain it makes **poly(ethene)**. We know it better as polythene :

  lots of ethenes ⟶ poly(ethene)

> **monomers ⟶ polymer**

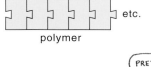

Poly-parrot?

▶ Polymerisation

You can now look at polymerisation in more detail. There are two types of reaction to make polymers. These are **addition** reactions and **condensation** reactions.

1. Addition reactions

The diagram of the shapes joining together on the previous page is an example of addition. Here is another with more detail added.

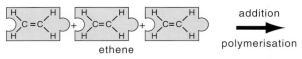
ethene

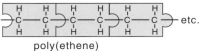
addition
polymerisation

poly(ethene)

In addition reactions the monomers have at least one double bond between carbon atoms.

The polymer is the only thing formed in the reaction.

The reaction of ethene to make poly(ethene) is the best known example, as shown above.

The reaction can take place at high pressure, using a catalyst. It can be shown like this:

$$n\,C_2H_4 \longrightarrow \left(C_2H_4\right)_n$$

where n = a large number.

Look at the diagram below:

The double bonds in ethene 'open up' and neighbouring molecules join end to end.

These supermarket trolleys make a polymer! What are the monomers in this polymer? What would you call the polymer? Poly(trolley)?

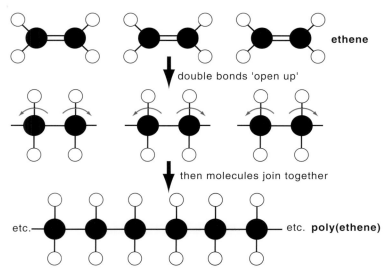

ethene

double bonds 'open up'

then molecules join together

etc. **poly(ethene)**

An addition reaction?

Other addition polymers

Some of the hydrogen atoms in ethene can be swapped for different atoms. You can then get a new polymer with new properties:

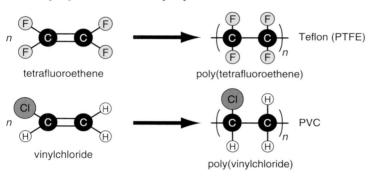

tetrafluoroethene → poly(tetrafluoroethene) — Teflon (PTFE)

vinylchloride → poly(vinylchloride) — PVC

Teflon is the non-stick lining on pans

2. Condensation reactions

This is the other type of polymerisation. Nylon is an example of a polymer made in a condensation reaction.

Nylon is very strong

> **Experiment 12.1 Making nylon**
>
> Put a thin layer of monomer A into the bottom of a very small beaker.
> Carefully pour a layer of monomer B on top of this.
> ● **What do you see happen?**
> Gently draw a thread out of the beaker using a pair of tweezers.
> Wind it around a test-tube as shown.
> Do not touch the nylon formed.   ⚠ monomers
>
> nylon
> monomer B
> monomer A

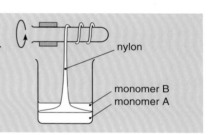

You see fumes given off as the two **different monomers** react together.
The fumes are hydrogen chloride (HCl) gas.
A small molecule is always given off in a condensation reaction.

The monomers have reactive parts at both ends of their molecules.
They join together, end to end, to make a long chain.

A condensation reaction?

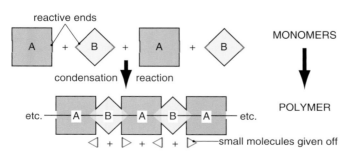

reactive ends

A + B + A + B   MONOMERS

condensation reaction

etc. — A — B — A — B — A — etc.   POLYMER

◁ + ▷ + ◁ + ▷ — small molecules given off

▷ Properties of plastics

Why are so many things made out of plastics?
They are certainly cheap to make.
However, lots of plastics do the job they are
designed for better than traditional materials.

- What advantages do:
  a) PVC gutters have over iron ones?
  b) melamine kitchen work surfaces have over
     wooden ones?
  c) poly(propene) milk crates have over
     metal ones?

We can make plastics that do their jobs well
in all sorts of ways.
For example, poly(styrene) is the plastic used
to make yoghurt pots. But if some gas is blown
into it during moulding you get 'expanded' poly(styrene).
This is the plastic that crumbles very easily.
It is used to hold hot drinks and 'fast food'.
The gas trapped inside makes it an excellent
heat insulator.

- What are the disadvantages of plastics in
  house fires or in rubbish dumps?

*The driver of this car escaped without serious
burns thanks to Kevlar. Racing drivers wear fire-
proof suits made from this new polymer.*

Structure of plastics

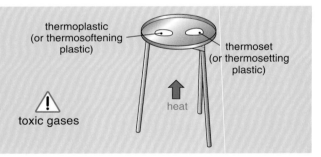

Experiment 12.2   Heating plastics

Heat a sample of a thermoplastic and a thermoset
on a tin lid in a ***fume-cupboard***.
- Which melts more easily?

When the thermoplastic softens, use a glass rod to
draw out a thread. You are remoulding your plastic.

- What eventually happens to your thermoset?

thermoplastic
(or thermosoftening
plastic)

thermoset
(or thermosetting
plastic)

⚠ toxic gases

heat

**Thermoplastics soften easily and can be
remoulded into new shapes. They are
sometimes called thermosoftening plastics.**

**Thermosets (or thermosetting plastics) do
not soften. If you heat them strongly
enough, they eventually break down and
char. They are hard and rigid.**

These different properties can be explained if you
look at the arrangement of the polymer chains:

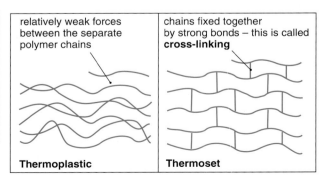

relatively weak forces
between the separate
polymer chains

chains fixed together
by strong bonds – this is called
cross-linking

Thermoplastic

Thermoset

► Changing the properties of plastics

On page 155 you have seen how using different monomers will produce a new polymer, with different properties. That's not surprising if you think about the different shapes of the monomers joining together.

Some plastics are made of polymers with very straight chains. Others have polymers with large groups of atoms sticking out at random. Straight chains can pack together neatly. This is also true of polymers with side-groups if the bits sticking out are the same and occur at regular intervals down the 'spine' of the polymer chain.

Regular packing of polymer chains makes a plastic denser than those containing polymers with random side-chains.
Look at the diagrams opposite:
We can make two forms of poly(ethene). Low density poly(ethene) is made up of polymer chains with random side-branches. But using ethene as the monomer with a special catalyst added in the reaction vessel, makes high density poly(ethene). The reaction takes place at lower pressure than used to make low density poly(ethene).

- Why do the two types of poly(ethene) differ?
- Which type of poly(ethene) will soften at a lower temperature?
- Which type will be stronger?

Mixing additives with plastics

As well as changing monomers and reaction conditions, we can also add things to modify the properties of plastics.
There is a wide range of additives to mix with a plastic.

Examples include:

- **Plasticisers** – these make a plastic easier to mould and more flexible. Their molecules get in between the polymer molecules and stop then lining up so neatly.
- **Preservatives** – these stop the plastic breaking down in harsh conditions or when it is being heated as it forms.
  Some stop the plastic reacting with oxygen (**antioxidants**).
  Others stop sunlight affecting the plastic (**UV stabilisers**).
- **Flame retardants** – these make it more difficult to ignite a plastic and more difficult for flames to spread.

Some substances are also added to make plastics impact resistant or harder wearing. For example, if you add sulfur to rubber, it becomes much more durable. That's because the sulfur forms **cross-links** between the rubber polymer chains.
Look at the diagram opposite:
By bonding neighbouring chains together, the rubber is now suitable for making tyres. We say that the rubber has been **vulcanised.**

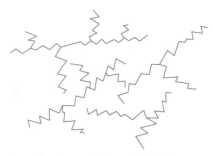

Low density poly(ethene) with its random branching on polymer chains

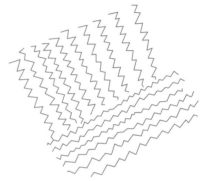

High density poly(ethene) with straight-chain polymers

We can get doors and windows made from uPVC (unplasticised PVC). Yet cling film is made from PVC with a plasticiser added.

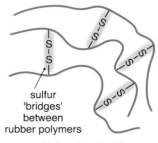

sulfur
'bridges'
between
rubber polymers

Vulcanised rubber

▶ Developing new polymers

Chemists can often predict how to make new polymers. They look at reactions that make similar polymers and change things slightly. But sometimes what they make surprises them!

a. In1938 Dr Roy Plunkett was researching substances to use in fridges to keep them cold. By accident, he found a waxy solid was blocking one of his gas cylinders. This turned out to be poly(tetrafluoroethene) or PTFE. (See page 155.) The American company he worked for marketed the new polymer as **Teflon.** This versatile polymer has special 'non-stick' properties. It is also very unreactive. These have led to a wide variety of uses, including Goretex. (See below.)

b. In the 1960s, the same American company were trying to discover a new polymer to make lighter weight tyres. This would help save fuel.
Stephanie Kwolek and her team worked on the problem. One day the chemicals she mixed formed a milky liquid, unlike the clear liquid she was expecting. But she didn't just throw the liquid away and start again.
She sent her discovery to the test lab. This stuff was incredible! It formed a plastic that was 9 times stronger than a similar mass of steel. Yet it was only half the density of fibreglass. Eventually it was marketed as **Kevlar**.
Its strength, low density and heat resistance led to many uses. These include bullet-proof vests, aeroplanes, motorcycle 'leathers' and tennis rackets amongst other things!

c. In 1970, Spencer Silver was trying to invent a super-strong glue. He did make a glue. However, the glue he came up with was even weaker than existing glues. But ten years later, the glue found world-wide fame as the sticky band at the top of **Post-its**. It is strong enough to stick to paper but weak enough not to damage the paper when it is removed.

Goretex

It is easy to produce a cheap waterproof coat from a plastic like nylon. However, if you exercise in the coat you soon get uncomfortable. Sweat condenses on the inside making you wet anyway!

Goretex is a material that solves this problem. It is made of expanded PTFE. (See page 155.) The PTFE polymer has gas blown into it as it forms. This produces fine fibres with tiny gaps between its strands. Look at the diagram opposite:

The 'pores' are 700 times the size of a water molecule. So drops of water can't get through, but water vapour can. That's why Goretex is described as '**breathable**'. You stay dry inside your waterproof protection.

Kevlar has saved millions of lives

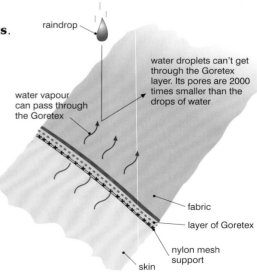

How Goretex works – it is waterproof but also 'breathable' (allowing water vapour to escape)

The breathable polymer is sandwiched between a nylon mesh support and an outer fabric.
Not surprisingly, Goretex is used widely wherever people need to work hard but stay protected against the weather.

Medical and dental applications

New polymers are now being designed to improve our medical care. Many of these are called 'smart' materials. These can respond to changes in their environment automatically.

More and more operations are now done through 'key-hole' surgery. But imagine trying to stitch inside a patient through the small incision.

Shape memory polymers are now being developed that can help. These can 'remember' their original shape when heated to a certain temperature. In this case the original shape of the fibre is the knot to be inserted in the stitch. Look at the diagram opposite:
The thread is pulled through the wound. Then, as it warms up to body temperature, it changes back to its knotted shape, applying just the right pressure. The polymer is biodegradable so that it just dissolves after a few days.

Hydrogels are polymer chains with a few cross-linking units. This matrix can trap water. (See Experiment 12.3, page 160.) These hydrogels have found use as wound dressings to allow the body to heal in moist, sterile conditions. They are useful for treating burns. The hydrogel can deliver tea tree oil as well as water to treat the burn. The polymer can also be used to grow new tissue on. The latest 'soft' contact lenses are also made from hydrogels.

To vary its properties, scientists can insert anywhere between 38% and 80% water in a hydrogel. Smart gels can shrink or swell up to 1000 times their volume due to changes in pH or temperature.

New polymers will eventually take over from mercury amalgams for filling teeth. Working with the toxic mercury every day is a potential hazard to dental workers.
Other developments include new softer linings for dentures, as well as implants that can slowly release drugs into a patient.

Active and intelligent packaging

Active packaging can change the conditions inside a pack.
Examples include :
- Ethene scavenging (to stop fruit ripening too quickly).
- Oxygen scavenging (to stop oxidation of a food).
- Humidity control (to remove water that encourages mould to grow).
- Microbial inhibitors (To stop bacteria, fungus and mould growing).
These can be included in sachets in the pack or built into the plastic packaging.
Intelligent packaging can sense its environment (possibly using microelectronics) and respond or pass on the information.
For example, diagnostic labels will change colour when a food goes off.
Thermochromic inks will also change colour at pre-set temperatures.

Goretex is used throughout the worlds of sport, work and the military

A
fibre knotted at 37°C

fibre for stitches made of shape memory polymers

B
surgeon inserts straightened fibre at room temperature

cut

needle

C
fibre warms up to body temperature (37°C) and takes up its knotted shape again

sides of cut pulled together

Smart polymers help surgeons to stitch in awkward operations

New soft contact lenses use a silicone polymer to trap water in the structure

Active and intelligent packaging is helping shoppers. Now we have the 'instant check-out'

▶ Everyday uses of plastics

All plastics are made by small molecules (monomers) reacting together to make very long molecules (polymers). The plastic is always named after its monomer. You just put poly- in front of the monomer's name. Look at these examples :

Monomer		Polymer
ethene	⟶	poly(ethene)
styrene	⟶	poly(styrene)
vinyl chloride	⟶	poly(vinyl chloride)

As you can see, the names can get quite long! So many plastics are known by their trade names. Well-known examples are Perspex, Teflon and Kevlar. (Nylon gets its name from two famous cities. Can you guess which ones?)

Here are some common plastics and their uses :

Special plastics have been injected into the timbers of the Mary Rose to preserve them. The ship sank in 1545 and was raised to the surface in 1982.

poly(ethene)

poly(styrene)

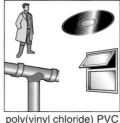
poly(vinyl chloride) PVC

poly(propene)

nylon

melamine

phenolic resins

- Which of these plastics are thermosetting plastics ?

You know that the world's crude oil will probably run out within your life-time. Look at all the uses of materials we get from oil :

- How would your life change without plastics ?

Experiment 12.3   Slimy polymer from PVA – poly(ethenol)

Gently heat $100 \, \text{cm}^3$ of water in a beaker. Do not boil. Slowly add 4 g of PVA glue to the water while stirring. When all the PVA has dissolved, stop heating. Then add a few drops of borax solution and stir. You may add a food colouring if you wish. Different groups can add different amounts of borax. Leave to dry.

- How does the amount of borax affect the polymer made ?
- Try to think of a use for your polymer.

- Look back at page 159 : What type of polymer is slime ?

▷ Plastic waste

Do you ever think about how much plastic
you throw away as waste? Whenever you go shopping
you notice how much packaging is used on everything
from foods to clothes.

Plastics have taken the place of many traditional materials
because they last a long time. They don't rot.
However, this becomes a disadvantage when it comes
to getting rid of the plastic. Most of our rubbish ends up
in land-fill sites. These are huge holes in the ground
that are covered with soil once the site is full.
Plastics take up much of the space available,
and many types take a very long time to rot away.

But we are running out of places to dump our rubbish.
So scientists are trying to help reduce the problems
we have disposing of plastics. We need to make sure
that new developments in plastics are **sustainable**.
Our environment must be protected.

Recycling

On page 156 we looked at the two types of plastics –
thermoplastics and thermosetting plastics.

We can melt thermoplastics and remould them into
new shapes. This means that we can recycle
many of the plastics we use. But we do need to sort
them into their different types.

Look at the symbols opposite:
Objects with these signs on them should be recycled.
Do you recycle plastics?
Do you have a recycling point near your home?
Some places have different bins for different types
of rubbish. What do you think of that idea?

Burning plastics

Burning plastics will reduce the volume of waste.
However, many plastics produce toxic gases as they burn.
We need very high temperatures in **incinerators** to break down
some of the polluting gases. Others need to be
removed in chimneys fitted with 'scrubbers'.
These contain a basic material, such as calcium carbonate,
which removes any acidic gases like sulfur dioxide. If the
plastic burnt contains chlorine, then poisonous hydrogen chloride
gas is given off. Hydrogen cyanide is produced when plastics
containing nitrogen burn in a limited supply of oxygen.

● Discuss the issues involved for a local council planning to build
an incinerator at a waste tip.

Space in land-fill sites is running out

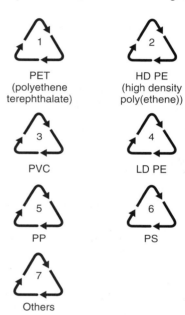

1 PET (polyethene terephthalate)

2 HD PE (high density poly(ethene))

3 PVC

4 LD PE

5 PP

6 PS

7 Others

Recycling symbols for plastics.
● *What do LD PE, PVC, PP and PS
stand for?*

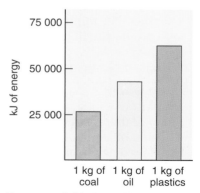

*Energy available from different fuels.
Switzerland burns most of its waste
to generate electricity.*

▷ Chemistry at work : Getting rid of plastic waste

Degradable plastics

More and more plastics are now designed so that they
do not last forever. They rot away, or **degrade**, naturally.
Normally a thermoplastic will break down over many years
and eventually decompose.
Now chemists have worked out various ways to help
plastics break down more quickly.

A degradable plastic

... by light

Chemists can design polymer chains that contain
certain groups of atoms that absorb light energy.
This energy splits the chain at those groups.
The plastic breaks down into small bits
which rot away more quickly.
This is used in the plastic mesh used to hold
some types of beer cans to each other.

... by bacteria

New plastics have been made by bacteria.
The bacteria are grown and they produce granules
of a plastic. The new plastic is totally **biodegradable**.
It breaks down in nature completely in about 9 months.
One problem is that the new plastic costs about
15 times as much as normal plastics.
- Would you pay more to be environmentally friendly?

*A biodegradable plastic at
different stages in its breakdown*

Scientists in the USA have used genetic engineering
to grow plants which make the new plastic themselves.
But the yield of plastic from crops, such as potato,
is still low.
- What do you think of research in this area?

Other plastics have been developed that have cornstarch
built into their structures.
The microbes that decompose rubbish in land-fill sites
feed on the starch and break down the plastic.
It is then easier and quicker for microbes to work on
the bits of plastic rather than large chunks.

... by water

Do plastics usually dissolve in water?
Many of their uses depend on their water-resistance.
- Think of some of these uses.
But now a new plastic has been designed
that is soluble in water.
By changing the reacting mixture, you can get plastics
that dissolve at different temperatures.
- Try to think of some uses for a soluble plastic.

*These plastic bags in hospital
dissolve in the hot water in a
washing machine.*
- *Why are they used?*

Summary

- Small, reactive molecules, called **monomers**, can join together to make very large molecules called **polymers**.
- The reaction is called **polymerisation**.
- The type of polymer we get depends on:
    - the monomer(s)
    - the conditions of the reaction
    - additives used.
- New polymers are still being developed, for example **smart** polymers. These can respond to changes in their environment.
- Degradable plastics can help us reduce the problems of plastic waste.

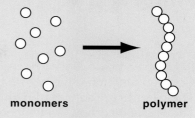

monomers   polymer

▷ Questions

1. Copy and complete this table :

Object	Traditional material	Plastic used
a) drain pipe	iron	
b) disposable cup	paper	
c) blouse	cotton	
d) disposable bag		poly(ethene)

What are the advantages of the plastic in each example ?

2. a) What happens in a **polymerisation** reaction ? Use the words *monomers* and *polymer* in your answer. Include equations if possible to help explain.
   b) What are the **monomers** called that make :
      i) poly(styrene) ?   ii) poly(propene) ?

3. a) How can you tell the difference between a **thermoplastic** and a **thermoset** ?
   b) How do their structures help to explain these differences ?
   c) What are the most important properties for the plastics used to make :
      i) a light switch ?
      ii) a fizzy drinks bottle ?
      iii) a pair of sandals for the beach ?
      iv) a wind shield for a motor-bike ?
   d) Carry out some research to find the uses and properties of Lycra and Thinsulate.

4. Look at this key to identify plastics :

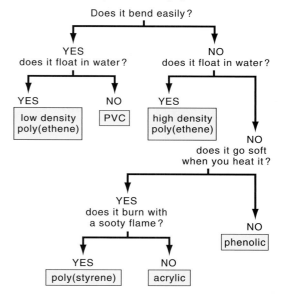

a) Which plastic bends easily and sinks in water ?
b) Which plastic does **not** bend easily, float in water or go soft when you heat it ?
c) Use the key to say as much as you can about acrylic.
d) Notice that in the key, it says that poly(styrene) does not float. However, it can be changed in its manufacture into a form which does float.
   i) What is this type of poly(styrene) called ?
   ii) How is it made ?
   iii) What are its useful properties ?
   iv) What is it used for ?

Further questions on page 190.

chapter 13

CARBON CHEMISTRY

In Chapter 11 we looked at crude oil and hydrocarbons.
Can you remember the names of two families of hydrocarbons?
The alkanes are saturated hydrocarbons.
They have a carbon 'skeleton' or 'back-bone', surrounded by
hydrogen atoms. These compounds, based on carbon and its
ability to form 4 bonds, are called **organic compounds**.
By replacing hydrogen atoms, we get lots of different organic
compounds. Organic compounds are very important because they
make up all living things.

The alkanes, and the alkenes, are examples of **homologous series**.
Members of the same series, or family, have similar
chemical properties. They react in a similar way.

Organic compounds form the basis of all living things

▶ Alcohols

If you replace a hydrogen atom in an alkane
by an –OH group, you get an **alcohol**.

```
    H   H
    |   |
H — C — C — O — H
    |   |
    H   H
```

ethanol is an alcohol
(its structure can be shown as
C_2H_5OH or CH_3CH_2OH)

The alcohols are another homologous series.
The alcohol shown above is the compound most of us
think of when we use the word 'alcohol' in everyday life.
Ethanol is the alcohol that we find in beer, wine and spirits.

Can you see how we name an alcohol?
The number of carbons gives us the stem from the matching alkane.
Then the –e at the end of the alkane is replaced by –ol.

You have probably heard of cholesterol. It is an alcohol and an essential steroid in our diet, but too much can 'fur up' your arteries and cause heart disease.

Look at the first five members of the alcohol family:

```
    H                  H   H                  H   H   H
    |                  |   |                  |   |   |
H — C — O — H      H — C — C — O — H      H — C — C — C — O — H
    |                  |   |                  |   |   |
    H                  H   H                  H   H   H
```
methanol, CH_3OH ethanol, C_2H_5OH propanol, C_3H_7OH

```
    H   H   H   H                  H   H   H   H   H
    |   |   |   |                  |   |   |   |   |
H — C — C — C — C — O — H      H — C — C — C — C — C — O — H
    |   |   |   |                  |   |   |   |   |
    H   H   H   H                  H   H   H   H   H
```
butanol, C_4H_9OH pentanol, $C_5H_{11}OH$

Physical properties of alcohols

The –OH group increases the boiling points of the alcohols compared to the alkanes with the same number of carbon atoms.
Look at the graph opposite :

The –OH groups also means that the smaller alcohols are soluble in water, unlike hydrocarbons. They form neutral solutions.

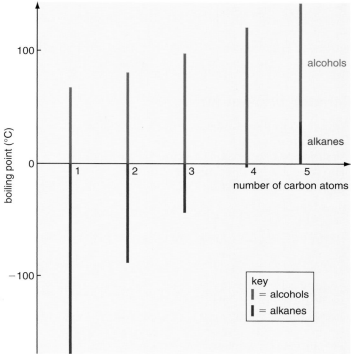

► Isomers

Look at the five alcohols at the bottom of the previous page :

The –OH group is at the end of each molecule. Would we have the same molecule if the –OH group was positioned somewhere else on the carbon chain ?
Look at propanol below :

propan-1-ol

propan-2-ol

Count the carbon, hydrogen and oxygen atoms in each molecule.
What do you find ?
The two molecules above are called **isomers**.

> **Isomers have the same formula, but their atoms are arranged differently.**

These isomers of propanol are both alcohols, but their physical properties are different. Look at the table opposite :
Notice the way the name of each isomer tells you where the –OH group is on the carbon chain.
The numbering always starts at the end nearest the –OH group.
Why are there only two alcohol isomers of propanol ?
Can you see that it doesn't matter if the –OH group is on the first or the last carbon ? It's the same molecule flipped over !
(Make a 'ball and stick' model to convince yourself !)

Isomer	Boiling point (°C)
propan-1-ol	97.5
propan-2-ol	82.5

● How many different positions can the –OH group occupy in butanol ?
● Draw an isomer of butanol with a branched carbon chain.

165

▶ Fermentation

People have used enzymes to make new products for thousands of years. Alcoholic drinks and bread are both made using the enzymes in **yeast**.

Making alcohol (ethanol)

Yeast is a type of fungus. Like all living things, it contains enzymes. Yeast feeds on sugars, in the absence of oxygen, turning them into alcohol. The reaction is called **fermentation**.
You can ferment a sugar (glucose) in the next experiment:

Experiment 13.1   Fermentation

Set up the apparatus as shown:
Put it somewhere warm.
(Near a radiator is a good place.)
Leave it for 10 minutes to start reacting.
• What happens inside the flask?
You can now leave your experiment until the next lesson.
• What has happened to the limewater?
• Which gas is given off during fermentation?

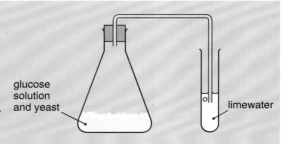

glucose solution and yeast

limewater

The enzymes in yeast break down the sugar, **glucose**, into **ethanol** (which we know as alcohol). Carbon dioxide is also given off:

glucose $\xrightarrow{\text{fermentation}}$ ethanol + carbon dioxide
(sugar)   (alcohol)

$$C_6H_{12}O_6 \xrightarrow{\substack{\text{enzymes in} \\ \text{yeast}}} 2\,C_2H_5OH + 2\,CO_2$$

The amount of ethanol in the fermenting mixture cannot rise above about 15 %. At this level, the ethanol poisons the yeast.
We have to **distil** the mixture to get purer ethanol.

Fractional distillation will separate the ethanol off. You can use the method on page 175, but you will need a fractionating column attached to the flask used in simple distillation.

Many people brew their own wine or beer. The air-lock at the top of the flask lets CO_2 gas escape, but won't let bacteria or oxygen from the air in. Why is this important?

*There was an unfortunate beast.
Who fermented some sugar with yeast.
When he swallowed the brew
His stomach, it grew
'Til BANG! CO_2 was released.*

Making bread

Have you ever wondered how the tiny holes get inside bread?
By 4000 BC, the Egyptians had found out that yeast makes bread rise.

The yeast feeds on sugar, as in brewing.
Which gas is given off in fermentation?
The carbon dioxide gas gets trapped in the dough.
This makes the bread rise.

Look at this recipe for making bread:

Can you see the holes where carbon dioxide gas was trapped in the dough?

Home-made bread

1. Mix dried yeast with some sugar in warm water.
2. When it has a froth, add it to flour with a little salt.
3. Mix and knead the dough.
4. Leave the mixture in a warm place for at least an hour. The dough should rise to about twice its volume.
5. Bake in a hot oven at 200 °C.

- Why must you add sugar to the yeast in step 1?
- What causes the froth in step 2?
- Why do you leave the mixture in a warm place for at least an hour before it goes into the hot oven? (Think about the effect of temperature on enzymes.)
- What do you think happens to the size of the gas bubbles in the bread as it is baked? (Think about the effect of temperature on gases.)

This baker is kneading dough. It is then left to rise, before being baked.

Experiment 13.2   Rising dough!

Put 20 g of flour in a beaker.
Add 1 g of sugar. Slowly pour in
25 cm^3 of yeast solution, stirring as you add it.

You should get a smooth paste.
Pour the paste into a large measuring cylinder, without letting it touch the sides.

Record the volume of your mixture every 2 minutes.
Stop after 30 minutes.

mixture of flour, sugar and yeast

Time (mins)	Volume of mixture (cm^3)

Plot your results as a line graph.
(Put volume of mixture up the side and time along the bottom.)

- Explain the pattern you see in your graph.
- Comment on the reliability of your data.

▶ Which method to make ethanol?

Do you know anyone who likes home brewing?
You can buy kits to help you make your own beer or wine.
The kits have brewers **yeast** in them. Yeast is a type of fungus.
You mix this with water, sugar and any flavourings you like.
The ethanol, which makes a drink alcoholic, is made
when the yeast feeds on sugar, in the absence of oxygen.

A chemical reaction takes place with enzymes in yeast.
Glucose (a type of sugar) is broken down to ethanol.
We call the reaction **fermentation**. (See page 166.)
Look at the equation below:

The alcoholic solution is syphoned off before bottling beer or wine

$$C_6H_{12}O_6(aq) \xrightarrow{\text{yeast}} 2\ C_2H_5OH(aq)\ +\ 2\ CO_2(g)$$
glucose   ethanol   carbon dioxide

> Fermentation
> of glucose

Ethanol can also be made in industry from ethene.
In Chapter 11 we saw how ethene is made when
we crack large hydrocarbons found in crude oil. (See page 145.)
Look at the reaction below:

$$C_2H_4(g)\ +\ H_2O(g) \xrightarrow[\text{high pressure}]{\text{phosphoric acid catalyst}} C_2H_5OH(g)$$
ethene   steam   ethanol

> Hydration
> of ethene

Making ethanol from ethene relies on crude oil supplies.
As you know, crude oil is a fossil fuel that is running out.
However, while supplies last, we make our 'lower grade'
ethanol from ethene. The process can be run continuously.
Although the reaction is reversible, unreacted ethene
is recycled back into the reaction vessel. The reverse reaction is
called **dehydration** (i.e. removal of H_2O):

Sugar cane grows well in Brazil but needs large areas of fertile land.
- *Why is this causing concern in the Amazon rainforest?*

$$C_2H_5OH(g) \xrightarrow{\text{dehydration}} C_2H_4(g)\ +\ H_2O(g)$$

Fermentation is always used to make the ethanol in alcoholic drinks.
In countries, like Brazil, with no natural oil supplies of their own,
ethanol is manufactured on a large scale by fermentation.
They have a climate that is ideal for growing sugar cane.
So although the fermentation process is a slow reaction,
it is the cheapest way for them to make ethanol used as a fuel.

The ethanol has to be made in batches and left to ferment.
The ethanol is distilled off, then used to run some Brazilian cars.
This helps preserve fossil fuel supplies, as petrol and ethene
come from crude oil.
It also reduces carbon dioxide released into the atmosphere.
- Why should we want to reduce CO_2 emissions? (See page 149.)

Ethanol does give off carbon dioxide when it burns,
but the sugar cane removes CO_2 gas as it grows!
(Remember photosynthesis!)

Cars in Brazil can be adapted to run on ethanol or a mixture of petrol and ethanol (called 'gasohol'). Ethanol produces about 70% of the energy released by the same volume of petrol.

► Chemistry at work : Uses of ethanol

Have you ever looked at the label on an *alcoholic drink*?
Spirits, such as gin or vodka, are made up of between 35% and
40% ethanol. You used to get an idea of the strength of a drink
from its 'proof' value.

The word 'proof' is used from olden days when tolls had to be paid
on alcoholic drinks. The 'proof' or test of whether or not the duty
was paid used gunpowder. If gunpowder soaked in the drink
still ignited when lit, this was 'proof' that the drink was alcoholic!

The copper vessels shown opposite are used to distil
alcohol when making whisky. This is needed to increase
the concentration of ethanol. The yeast in the initial brew
will not survive beyond about a 15 % solution of ethanol.
The whisky can be stored for years in oak barrels
which help give it its characteristic flavour.

Pure ethanol is poisonous. Do you know which organ
of the body removes harmful ethanol from your blood?
Too much alcohol over time will damage your liver
and 'binge' drinking can lead to alcoholic poisoning.

Alcohol is a legalised drug. It modifies the way your body
functions. A little can help people relax and relieve stress.
Excess dulls the senses and your reactions slow down.
This is why it is so dangerous to drink and drive.

Alcoholics are addicted to ethanol. This addiction leads
some people to drink meths. This is much cheaper
than an alcoholic drink because there is no tax to pay on it.
However, it has a purple dye added, as well as methanol.
Methanol is more toxic than ethanol. It is meant to stop
people drinking meths. Meths drinkers are often the poorest
members of society whose addiction can quickly lead to
blindness, liver failure and death.

Used properly, meths (or methylated spirits) is a useful *solvent*.
Have you seen any methylated spirits in your home?
● What is it used for?
Ethanol is used as a solvent in the perfume industry.

It can also be used as a *fuel* in spirit burners (or mixed with petrol
in some Brazilian cars – see previous page).
Ethanol burns with a 'clean', blue flame.
You can see the equation on the next page.

*Less than half of the volume of
this whisky is ethanol*

*Alcohol abuse can lead to addiction,
liver failure and anti-social behaviour*

▶ Reactions of alcohols

Combustion

Alcohols are flammable. They burn to produce carbon dioxide
and water. You can use the apparatus on page 147
to test for these products.

Experiment 13.3   Combustion of ethanol

Set up the apparatus as in the diagram on page 147
with ethanol in the spirit burner.

 ethanol

- How can you test for carbon dioxide?
- How can you test that water is formed?
- Is the flame clean or sooty?
- Write a word equation for the reaction.

Ethanol burning

The equation for the complete combustion of ethanol is shown below:

$$C_2H_5OH(l) + 3\,O_2(g) \longrightarrow 2\,CO_2(g) + 3\,H_2O(g)$$

It burns with a 'clean' blue flame.
Alcohols are sometimes added to petrol to make it easier to ignite.

Adding sodium

Demonstration 13.4 Sodium and ethanol
Watch your teacher add a small piece of sodium metal
to some ethanol on a dish.

 sodium
ethanol

- What do you see happen?
- Is this a more vigorous reaction than when sodium
  reacts with water?

The gas given off is hydrogen.

Sodium reacting with ethanol

Oxidation

Do you know what happens to the taste of beer or wine
if you leave it open overnight? Which gas in the air
do you think might react with the ethanol in the drink?

The sour, sharp taste results from a reaction with oxygen
caused by microbes in the air.
That's why you need an air-lock if you brew your own beer
or wine. The air-lock lets the bubbles of CO_2 escape,
but stops microbes and oxygen from the air entering the jar.

Can you remember which type of substances make things
taste sour or sharp? (See page 124.)
The ethanol has been *oxidised* to form **ethanoic acid**.
Where have you met this acid before?

oxidised
ethanol ⟶ **ethanoic acid**

*Can you see how these air-locks keep
air out from the fermenting mixture,
but let CO_2 escape?*

▶ Carboxylic acids

Ethanoic acid is the main acid in vinegar. It is also used to manufacture the fabric called rayon.
Look at its structure opposite :

ethanoic acid (CH_3COOH)

Can you see how it differs from ethanol?
And why we say that ethanol has been oxidised?
Ethanoic acid is one of the homologous series of carboxylic acids. Look at two other carboxylic acids opposite :

methanoic acid (HCOOH)

propanoic acid (C_2H_5COOH)

Other carboxylic acids include citric acid, found in oranges, lemons and limes.
These citrus fruits are also good sources of vitamin C.
This is another carboxylic acid called ascorbic acid.

The carboxylic acids are weak acids, but do show the typical reactions of acids.

Aspirin is a carboxylic acid used to relieve pain and reduce risk of heart attacks. (See page 330.)

Experiment 13.5   Reactions of carboxylic acids

1. Add a piece of magnesium ribbon to 5 cm³ of ethanoic acid solution.
   Test the gas given off with a lighted splint.
2. Add a spatula of sodium carbonate to 5 cm³ of ethanoic acid solution.
   Test the gas with limewater. (We use this as the test for carboxylic acids.)
3. Repeat test 2 with sodium hydrogencarbonate.

As with other acids, carbon dioxide is given off in reactions with carbonates and hydrogencarbonates.   A salt and water is also formed.
With magnesium, a salt is made and hydrogen gas is given off.

Esterification

Demonstration 13.6 Alcohol plus carboxylic acid

Add 2 cm³ of ethanol to 1 cm³ of concentrated ethanoic acid in a test tube. Add 3 drops of concentrated sulfuric acid to act as a catalyst (to speed up this reversible reaction).
Warm the reaction mixture gently in a hot water bath for 5 minutes.
Pour the reaction mixture into a small beaker containing sodium hydrogencarbonate solution, and stir well.
This gets rid of excess acid.

● How do the products smell compared to the reactants?

⚠ concentrated acid
ethanol

The new products include an **ester**.
In this case the ester is called ethyl ethanoate :

In general we can say :

sulfuric acid

ethanoic acid + ethanol ⇌ ethyl ethanoate + water
$$CH_3COOH + C_2H_5OH \rightleftharpoons CH_3COOC_2H_5 + H_2O$$

sulfuric acid
alcohol + carboxylic acid ⇌ ester + water

► Chemistry at work : Uses of esters

Food flavourings

Have you ever eaten sweets called pear drops?
If you have, you will know the fruity smell of esters.
The smaller esters are *volatile*. They evaporate easily.
Fruits, such as pineapple, pear, banana and strawberry,
owe their smell and taste to complex mixtures
that contain many esters.

Scientists can imitate the smells using simpler mixtures
of esters. It would cost too much to try to match
the exact mixtures we find in nature.
The esters are then used as artificial flavourings in foods.

Perfumes

You won't be surprised that we also find
the sweet-smelling esters in perfumes.
Cosmetic chemists create new scents by blending
esters along with other ingredients. They are made to
stimulate the sense cells in your nose.

The chemists have to design a mixture that not only smells
great, but also has the right volatility. If the perfume
evaporates off too quickly, the smell will not last long enough.
If it isn't very volatile, the smell will not be strong enough.
So they face a tricky balancing act to find the right mixture.

The cosmetic chemist must also make sure the perfume :
- is non-toxic (you don't want to be poisoned!)
- does not react with water (otherwise it would
  change when you sweat)
- does not irritate your skin (you must be able
  to apply it directly on to your skin without harm)
- is insoluble in water (so it is not washed off too
  easily).

Solvents

Do you remember the smell of ethyl ethanoate from page 171?
You might have recognised it from the smell of some glues.
The ester is used as a solvent to dissolve a plastic in the glue.
When the glue is applied, the ester evaporates off, and leaves
the solid plastic behind to stick objects together.

This ester is also found in nail varnish removers.
The ester molecules are strongly attracted to nail varnish
molecules, unlike water molecules. So they act as a solvent
and dissolve the nail varnish.

Many drinks have esters added for flavour. Plastic drinks bottles can be made of polyester. These can be recycled and made into new products.

Perfume contains esters

*There was a young compound named Ester
Who molecules would frequently pester.
Asked "What's the attraction?"
They replied in a fraction
"She smells sweet like a good perfume tester!"*

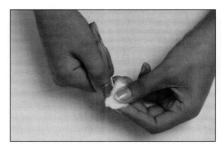

You might see 'ethyl acetate' written on the contents of nail varnish remover. This is the old name for ethyl ethanoate.

Summary

- **Ethanol** (C_2H_5OH) is a member of the **alcohol** family. It is made by fermentation of sugar using yeast in the brewing industry. Industrial ethanol is made by passing ethene gas and steam over a hot catalyst (phosphoric acid) at high pressures.

- Ethanol is oxidised to form ethanoic acid. Ethanoic acid is a **carboxylic acid**. These are weak acids, and show the typical reactions of acids. (See page 226.)

- An alcohol and a carboxylic acid react together to form an **ester**. Water is also formed in this reversible reaction. A catalyst of concentrated sulfuric acid is needed to speed up the reaction.

- Esters are used as perfumes, food flavourings and solvents.

Woof!

▶ Questions

1. Copy and complete:
   The alcohol we find in beer, wine and spirits is called .... It is made by f.................... sugar with .... (a type of fungus). During the process, ... gas is given off.
   We can also make this alcohol by reacting .... and steam in a continuous industrial process.
   If you leave an alcoholic drink in the air too long, it tastes .... This is because .... acid has formed in an .... reaction.

   Alcohols can react with .... acids to form sweet smelling compounds called .... A catalyst of concentrated .... acid is used in the reaction.

2. a) What are the factors that make a good perfume?
   b) Which compounds give perfumes their sweet smells?
   c) Describe how the smell of perfume spreads through a room. (See page 18.)
   d) Give two other uses of the compounds named in part b).

3. Draw spider diagrams to show all the reactions of both the alcohols and carboxylic acids in this chapter.

4. a) Name the carboxylic acid we find in vinegar.
   b) Give 3 sources of citric acid.
   c) What is ascorbic acid commonly known as?
   d) State two uses of aspirin.

5. a) Name this carboxylic acid :

   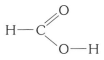

   b) Name and draw the structure of the carboxylic acid which has 3 carbon atoms.
   c) Which alcohol would you use to react with ethanoic acid to make methyl ethanoate?

6. Discuss the advantages and disadvantages of producing ethanol for use as a fuel by fermentation and by hydration of ethene using steam.

Further questions on page 191.

chapter 14

Plant oils

▶ Oils in plants

Do you enjoy olives on your pizza? Some people like olives, but others hate them. Most olives are used to make olive oil.
The olive trees grow well all round the Mediterranean. People who live in the warm countries of Southern Europe often say that it's the olive oil in their diet that keeps them healthy.
It provides nutrients as well as lots of energy.

Like animals, plants also store excess energy they take in as chemical energy in fats. The fats in plants are often liquids at 20 °C so we call them **oils**. Many of the compounds in plant oils are esters. (See pages 171–2.) This group of carbon compounds also contain hydrogen and oxygen atoms.

Lots of oils can be extracted from a wide variety of plants.
Fruits, seeds and nuts are good sources of oil, for example peanut oil. Palm oil is found in around 10% of products in supermarkets.
Some oils are used in aromatherapy treatments.
Some people think that these 'essential oils' have special healing properties.

People are not only interested in plant oils in their diet and in aromatherapy products. The oils are also being hailed as new sources of alternative fuels. (See page 183.)

Olives are the source of an important plant oil. Plant oils contain unsaturated oils that result in low-density cholesterol that is good for you. Saturated fats produce high-density cholesterol that clogs up your arteries.

Extracting the oil from plants

In order to get the oil from a plant, we have to break open its cells to release it. There are two main ways to break down the cell walls:
- **pressing** (applying pressure)
- **distillation** (heat using steam).
Then water and other impurities are removed.

Oils extracted from plants are used in aromatherapy

Pressing

In this method we squash the plant material. Then we collect the liquid. The oil and the water-based solutions are immiscible.
They do not dissolve in each other and will separate out into two layers.

This is the way we extract olive oil. After collecting olives from the trees, they are crushed. Then the 'mash' is spread out on thin mats in a press. When the pressure is applied, the liquid drains out. Then the oil and water are separated by **centrifuging** (spinning the mixture at high speeds) or just by **decanting** (pouring off).

Olives are crushed and 'pressed' to extract their oil. It is called virgin olive oil when no heat or chemicals are used, just 'cold pressing'

Experiment 14.1 Extracting oil by pressing

Take some finely chopped unprocessed nuts.
Place them between two pieces of filter paper.
Put the nuts and filter paper between two thick slabs of wood.
Then stand on the wood to 'press' the nuts.
Scrape the nuts off the filter paper.
- What do you see when you hold the filter paper up to the light?
- What does this show? (Remember your food tests in Biology.)

⚠ nut allergies

Distillation

We use simple distillation to get the oil from plant material:

To extract plant oils we can either:
- boil up the plant material with water, or
- pass steam through the plant material.

Then we condense the vapour and collect the mixture of water and oil.

We can separate the oil and water in another stage of the process. This can involve dissolving the oil in a solvent.

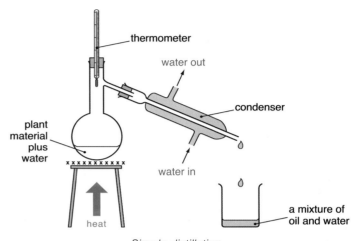

Simple distillation

Experiment 14.2 Extracting oil from an orange

You can try getting the oil out of the *zest*
(the outer layer of peel) from
an orange using micro-scale equipment.
Grate the zest from a quarter of an orange.
Set up the apparatus as shown opposite:
Heat the zest and water until a
couple of drops of liquid distil
over into the small receiver.

- Describe the distillate.
- What can you say about its smell?

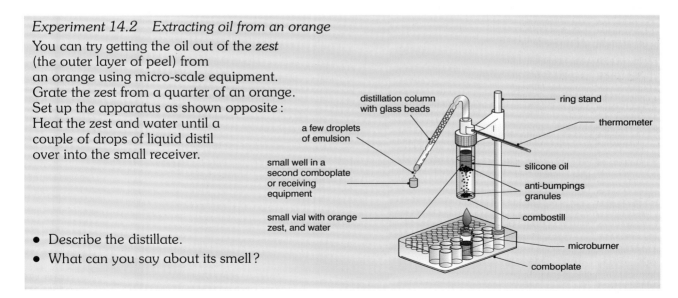

The liquid you collect is a mixture of water with tiny droplets
of the orange oil spread through it.
This is called an **emulsion**. (See next page.)

▶ Emulsions

As you saw in the last experiment, plant oils do not dissolve in water. We say that oil and water are **immiscible**.

Look at the salad dressing opposite:

The oil and water form two separate layers. The oil is less dense than the water so it floats on top.

When you shake the salad dressing, you get tiny droplets of oil spread (dispersed) throughout the water in vinegar. We call this type of mixture an **emulsion**.

Salad dressing: the oil and water are immiscible

When shaken, oil and water form an emulsion. Can you think of any other uses of emulsions?

An emulsion is a finely dispersed mixture of two or more liquids that do not dissolve in each other.

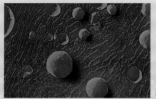

Experiment 14.3   Looking at an emulsion
Look at a drop of full fat milk under a microscope.
- What can you see?
- How does this differ from butter on a slide?

Milk under powerful magnification

Emulsifiers

Salad cream also contains oil and water. However, you don't need to shake it up to make an emulsion each time you use it. That's because salad cream has egg yolk added. This keeps the oil droplets mixed with the water. It stops the oil and water separating out into layers.
A substance that does this is called an **emulsifier**.

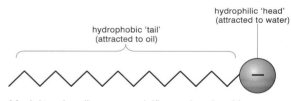

hydrophilic 'head' (attracted to water)

hydrophobic 'tail' (attracted to oil)

Model to visualise an emulsifier molecule with charged 'head'

The molecules of an emulsifier have a long chain made of carbon and hydrogen atoms. This part of the molecule dissolves well in oil. At one end of the molecule it has a charged part that dissolves well in water. Look at a model of an emulsifier molecule opposite:

So an oil droplet becomes 'studded' with emulsifier molecules. The droplets are called **micelles**. Look at the bottom diagram:

Because the surface of each micelle carries the same charge, they repel each other. This keeps the droplets of oil spread throughout the water, as an emulsion.
Lots of foods have artificial emulsifiers added. They make liquids thicker and smoother. They can then be coated on other foods more effectively. Emulsifiers also improve the texture and appearance of food, such as chocolate.
You can read more about emulsifiers on page 181.

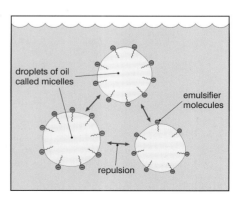

droplets of oil called micelles

emulsifier molecules

repulsion

The micelles repel each other and remain dispersed throughout the water, forming an emulsion

▶ Margarines

Margarines are made from plant oils. Sunflower oil is a popular choice for margarine manufacturers.

Margarines are often advertised as being 'high in poly-unsaturates'. This means that the hydrocarbon chain the oil molecules contain lots of carbon–carbon double bonds. Others are 'mono-unsaturated. As the name suggests, they have one double bond per molecule. You have met unsaturated compounds before on page 146 (the alkenes). Can you remember the test for them?

Experiment 14.4   Testing margarine and butter

Shake up some margarine with a little ethanol to dissolve it. Then add some bromine water.
- What happens?

Keep the test tube to compare with the same test carried out on butter.
- What can you say about the number of carbon–carbon double bonds in margarine and butter?

You can use solutions of bromine to test for unsaturated compounds.

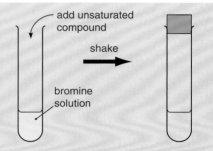

Bromine solution *is* **decolourised** *by unsaturated compounds. Food scientists also use a reaction involving iodine to find the* **iodine number** *of fats. The higher the iodine number, the more unsaturated the fat is.*

Unsaturated fats and oils are healthier for us than the saturated molecules we find in butter. Saturated fats can cause heart disease as fatty deposits block up arteries.

Hardening plant oils

The problem with making margarine is that the plant oils are liquids at room temperature. So they are too runny to spread on bread. They are less viscous than saturated oils or fats, such as butter. Saturated molecules have straighter chains that can pack together better. They have stronger forces between their molecules.

However, the oils can be made thicker. We can react them with hydrogen gas to saturate some of the double bonds. This is an **addition reaction**.
The reaction takes place at about 60 °C with a nickel catalyst:

plant oil + hydrogen $\xrightarrow[60\,°C]{\text{nickel}}$ margarine (less unsaturated than the oil)

As the molecules straighten, the oils get more viscous. The chemists have to add just the right amount of hydrogen. If they add too much, the margarine will be too hard to spread when it comes out of the fridge. If they add too little, the margarine will be too runny if left out of the fridge.
The soft, solid margarine is easier to make cakes, biscuits and pastries with, as well as spreading smoothly on bread.

Hydrogen reacts with plant oils (called 'hardening') to make margarine with the right consistency

► Chemistry at work : Detergents

Many detergents are made from the products of crude oil. Washing-up liquid is an example.

Detergents are substances which help the cleaning action of water.

Water is good at dissolving many things. However, it cannot dissolve oil or grease. This is where detergents help. They act as **emulsifiers**. They remove grease and keep it dispersed in water.

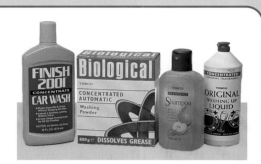

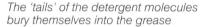

The 'head' of the molecule is strongly attracted to water. It is called 'hydrophilic' – water loving.

The 'tail' is a long hydrocarbon chain which dissolves in grease. It is called 'hydrophobic' – water hating.

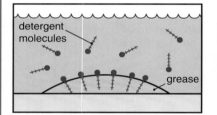

The 'tails' of the detergent molecules bury themselves into the grease

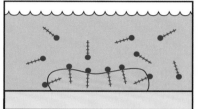

The 'heads' stick out and are pulled towards the water molecules

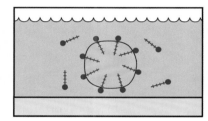

The grease then floats off into the water. The detergent molecules form 'micelles' with the droplets of grease (see page 176).

Experiment 14.5   Detergents are 'wetting agents'

Collect some water in a beaker.
Use a dropper to carefully put a drop of the water on to a flat piece of cotton.

Now add a few drops of detergent to the water in your beaker.
Again, place a drop on the cotton.

● What difference do you see?

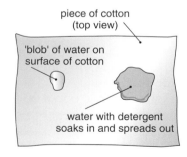

piece of cotton (top view)

'blob' of water on surface of cotton

water with detergent soaks in and spreads out

side view of cotton

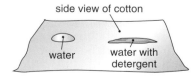

water   water with detergent

Detergents help water to soak into clothes when you wash them. They form a thin 'skin' on top of the water. This breaks down the strong forces of attraction between water molecules at the surface. Detergents reduce the water's surface tension.

The water can then spread out more easily. That's why detergents are called **wetting agents**.

▶ Chemistry at work: Detergents

Soaps

Experiment 14.6   Making soap

Add $10\,cm^3$ of castor (or olive) oil, $10\,cm^3$ of sodium hydroxide solution and $10\,cm^3$ of ethanol in a beaker ($250\,cm^3$).

Heat gently on an electric heater, stirring with a glass rod. When the mixture stops frothing, add 2 spatulas of sodium chloride. Stir with a glass rod.
Take care as the mixture can froth up again at this stage.
Allow the beaker to cool.
Pour off the liquid soap from the surface of the mixture into the wells on a spotting tile.
Let the soap solidify. Blot dry with filter paper or a paper towel.
Do not use the soap on your skin.
- Why not?
- What happens if you shake one of your soap tablets with some distilled water in a test tube?

⚠️
ethanol
sodium hydroxide
solution

We can summarise the reaction as:

oil/fat + sodium hydroxide ⟶ soap + glycerol
(fatty acid)

We say that oil is hydrolysed by the alkali. The soap formed is the sodium salt of the fatty acid.
Making soap is called **saponification**.

Which plant oils do you think this soap is made from?
Oils or fats are heated with sodium hydroxide to make soaps.
If potassium hydroxide is used to make the soap, we get a soap which is a potassium salt of a fatty acid. (Remember that all alkali metal salts are soluble in water)

Soapless detergents

Soaps are detergents. Traditionally, they are made from animal fats and plant oils.
Using soap in areas with hard water causes 'scum' to form. (See page 290.) The white bits of 'scum' can stick to clothes when they are being washed.

However, **soapless detergents** from crude oil do not have this problem. (See Experiment 25.5, page 338.)

Soapless detergents do not make 'scum' in hard water. No detergent is wasted reacting with the hardness, so you save money as well.
About 80% of all detergents made are soapless detergents.

Treating oil spills

As you know from page 141, crude oil is transported in giant oil tankers. If these ships have an accident, the crude oil can escape.
It floats on top of the sea, forming an oil slick.
Soapless detergents are used to clean up the mess.
The detergent breaks up the slick. Then the oil is spread out by the action of the waves.

This oil tanker was grounded on the rocks near Shetland

► Food additives

Do you ever read the list of ingredients on food
before you buy it? Have you noticed how many
E numbers we have in some foods?
Does it worry you?
Look at the label opposite from some sweets:

Permitted food additives are each given an **E number**.
Many are extracted from natural products, such as seaweed.
However, some people are concerned about their effects
on our health. So why do we add them at all?

Advantages of food additives

Additives can colour, preserve and add flavour to our foods.
There is a system used to give an additive its particular E number.
Look at the table below:

E number range	Use of additive
E100 – E181	colouring
E200 – E290	preservatives
E296 – E385	anti-oxidants and acids
E400 – E495	emulsifiers and stabilisers
E500 – E585	anti-caking agents and mineral salts
E620 - E640	flavour enhancers
E900 – E1520	others (such as wax glazes on fruit)

Let's look at each use in more detail:

Colourings

When we process foods much of the natural colour
is lost. The food does not look appetising.
So food chemists have searched for substances
to put the colour back into processed food.
Other compounds are used to add vivid colours.
For example, sweets and soft drinks are more attractive
to children if they are brightly coloured.

Preservatives and anti-oxidants

Fresh foods soon 'go off' as they are attacked
by bacteria or oxygen in the air. For example,
fats in food are oxidised to acids, making it rancid.
However, preservatives and anti-oxidants protect
the food. They have a longer 'shelf life' which is
better for shops and consumers.
E300 is ascorbic acid which is Vitamin C.
This is added by law to flour to replace
the vitamin lost in processing.

SHERBET FRUITS

An assortment of
fruit flavour sweets
with a sherbet
cracknel centre

INGREDIENTS
Glucose Syrup, Sugar,
Hydrogenated Vegetable Oil,
Flavourings, Colours:
E124, E141, E100, E110;
Contains 23% Sherbet

250g

*Permitted food additives are given
an E number*

Brightly coloured sweets

*This margarine contains preservatives
E200 (sorbic acid), E270 (lactic acid)
and Vitamin E*

Emulsifiers and stabilisers

We have met emulsifiers before in this chapter.
These help fats and oils to mix with water.
For example, a Caramel cake bar contains
emulsifiers E471 and E475.
Look at the label on the next bar of chocolate
you eat to find the emulsifiers used.

Flavourings

On page 172, we saw how chemists
use esters to flavour foods. We can't reproduce
the exact mixture of compounds that make up
natural flavours. For example, the smell of a strawberry
is made up from about 280 different compounds.
But the main compounds are identified
and made as artificial flavourings.

We also looked at flavour enhancers on page 45.
These act, like salt, to increase our taste sensations.
Can you remember some of the concerns
surrounding the use of monosodium glutamate (E621)?

Disadvantages of food additives

People are becoming more and more aware
of health issues. They are looking more closely
at the things they eat and want more information
about the additives in food.
We have seen the advantages of using E numbers,
but there are worries about some compounds used.

For example, tartrazine (E102) is a yellow colouring.
It is used in sweets, jams, drinks and snack foods.
However, it can provoke asthma attacks and hyperactivity
in children. There are also claims of other side effects.
E102 has now been banned in Norway and Austria.

Butylated hydroxyanisole (E320) is another controversial
compound. It was banned in Japan as early as 1958
and experts recommended its ban in the UK. However,
due to pressure from industry it was not banned.
Interestingly, McDonald's stopped using this additive
in America in 1986.

You might not recognise E954, but you've probably
heard of saccharin. It is an artificial sweetener.
It was banned in the USA in 1977, but was later reinstated
provided manufacturers print a health warning on labels.

- What do you think about these issues?

Chocolate and ice cream contain emulsifiers. But do the delicious textures made possible by emulsifiers encourage people to eat unhealthy foods?

Esters are used to add flavour

Some colourings have been linked with hyperactivity in children

Do you check the ingredients of the drinks you buy?

▶ Identifying additives in food

We can use **chromatography** to separate
small amounts of dissolved solids.
For example, you have probably separated the dyes in inks.

Now try to separate the dyes in food colourings:

Experiment 14.7   Chromatography

Set up a piece of chromatography paper
as shown:

Let the water soak up the paper until
you have separated the colours.
Leave it to dry and stick the paper,
called a chromatogram, into your book.

Make a table to show which colours
make up each food colouring.

- Which food colourings contain only one dye?

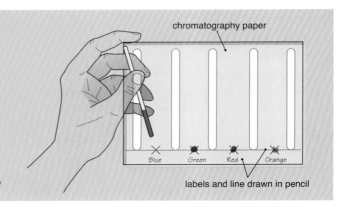

chromatography paper

Blue   Green   Red   Orange

labels and line drawn in pencil

The more soluble a substance, the
further up the paper it is carried by the water.
The soluble substance is called the **solute**.
Water is the **solvent** in Experiment 14.7.
Other solvents, such as ethanol, can also be used.
Do you think you would get the same results?

Chromatography can also be used to identify
unknown substances.
We can look at chromatograms for substances
we know and compare these with results for the
unknown substance.
Look at the example shown:

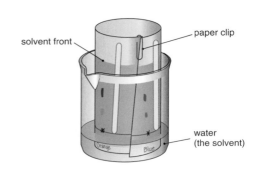

solvent front

paper clip

water
(the solvent)

The technique can also be used with
colourless substances. The spots are made
visible by spraying with a locating agent.

We can also use the latest analytical methods
to identify unknown additives in foods. These
machines are expensive but can detect and
identify the tiniest amounts of unknown
 substances. (See page 366.) They include
mass spectrometers and infra-red
spectrometers.

*This chromatogram shows that
the unknown food colouring C is a
mixture of A and B. It should not be
used in food because colouring B
has been banned.*

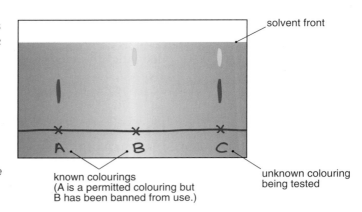

solvent front

A   B   C

known colourings
(A is a permitted colouring but
B has been banned from use.)

unknown colouring
being tested

▶ Bio-diesel

Have you seen the spectacular bright yellow fields as you drive past farms in spring and early summer? These are fields of the rape plant. Its seeds are one source of an exciting new development in fuels. We use oilseed rape to make **bio-diesel**. Sunflower oil is also used. Even recycled oil from chip shops can be used as the starting material for bio-diesel!

Plant oils are a real alternative to the diesel we get from crude oil. As you know, our supplies of fossil fuels are running out. So getting fuel from a source that can be re-grown every year will help to solve this problem. In other words, we can replace a non-renewable energy source with a renewable one.

The bio-diesel can be added to normal diesel to make it last longer. In France, about 5% of their diesel is made up from bio-diesel. Engines don't even have to be changed to run on bio-diesel.

Pollution is also reduced by burning bio-diesel. It still gives out carbon dioxide, a greenhouse gas, just like diesel. But this is largely balanced by the CO_2 absorbed as the plants grow. What's more, bio-diesel does not produce sulfur dioxide. The fuel does not contain impurities of sulfur compounds like those from crude oil. Therefore, it helps people with breathing problems and reduces acid rain. (See page 148.)

The particulates that diesel engines give out, containing volatile un-burnt hydrocarbons that cause cancer, are also reduced.

Many boats on inland waterways use diesel as their fuel. If there is an accident the diesel pollutes the water and endangers plant and animal life. But bio-diesel is biodegradable, so it breaks down if spilt in the water.

As diesel prices rise, bio-diesel is getting more attention as the cheaper option. People are even adding vegetable oil from the supermarket to their tanks of diesel. After paying tax, it has been worked out that this saves you about £10 per tank. Its fans say that their cars even run more smoothly on their 'bio-diesel mix'.

However, bio-diesel does have problems at high and low temperatures. It freezes at a higher temperature than normal diesel, turning to sludge. Then at high temperatures the oil molecules can oxidise and form polymers that can 'gum up' engines.

On a global scale, the use of large areas of farmland to produce fuel instead of food could pose problems if we start to rely on it. People are also worried about the destruction of habitats of endangered species. For example, orang-utans are under threat of extinction. Large areas of tropical forest where they live are being turned into palm plantations to extract palm oil.

The wonderful sight of oilseed rape growing could become more common as crude oil supplies run out

Bio-diesel is a 'cleaner' fuel than diesel from crude oil

The habitats of endangered species are under threat from plantations that supply the demand for plant oils

▶ Chemistry at work: Cooking

Kitchen chemistry!

Cooking could really be called 'kitchen chemistry'.
It's all about:
- making and separating mixtures (**physical** changes).
- making new substances (**chemical** changes).

In cooking, we often mix oil-based ingredients with water.
You have seen how emulsifiers help us achieve that.
Egg yolk is an excellent emulsifying agent.

But the chemistry really starts when you turn on the heat!
That's when we start changing one substance into another.
In cooking these changes are not possible to reverse – a sure sign
that a chemical reaction has taken place.

We cook foods to:
- kill bacteria in the high temperatures.
- improve the texture, flavour and taste.
- make them easier to digest.

We can cook food in many ways, including microwave, baking, boiling, steaming, grilling and frying

Cooking meat and eggs

Meat and eggs are rich sources of proteins. We need these to build
muscle and to help repair our bodies.
To understand what happens when we cook them, you need to
know a little bit about the structure of protein molecules.

Proteins are natural polymers made from amino acid monomers.
You have looked at monomers and polymers in Chapter 12.
The long protein molecules are held in shape by forces within and
between their chains. When we cook meat or eggs, their protein
molecules start moving around more vigorously.
Once the temperature gets high enough, they lose their original
shape. We say they have been **denatured**. (See page 215.)
Once denatured, the proteins cannot change back again.

A body builder's diet contains a higher proportion of protein than that of most people. What food might they eat?

In cooked meat, the proteins separate from each other and it
becomes tender. In a raw egg, the proteins in the egg white are
coiled up and floating around in a watery solution.
When fried or boiled, the proteins 'uncurl' and start forming bonds
between each other. The water in the egg white gets trapped
between the open structure that builds up.
It's a bit like the slimy polymer on page 160. The longer you heat
the egg, the more bonds form between chains. Then the less water
can be trapped. That's why overcooking makes the egg white more
rubbery!

I suppose boiling for 4 minutes might be better than 20 minutes, but this is more fun!

Boiling an egg is harder than you think!

▶ Chemistry at work : Cooking

Cooking potatoes

If you ask people what their favourite food is, many will reply 'chips' or 'French fries'. But doctors are warning people that too much of this type of 'fast food' is creating health problems. More people than ever are obese and at risk from heart disease, strokes and diabetes.

However, potatoes are a good source of carbohydrates. They provide us with the energy we need. Potatoes contain plenty of starch. Starch is another natural polymer. It is made up of lots of sugar (glucose) monomers.

We need to cook potatoes before we eat them to help us get at the starch. The starch is stored inside potato cells. Cooking helps to break down the tough cell walls and release the starch. Then we can digest the large starch molecules, breaking them down into glucose molecules.

Plant oils are used to fry potatoes, but doctors would prefer us to boil our potatoes.
Frying in oil means you can cook at a higher temperature (plant oils have much higher boiling points than water). Like all chemical reactions, the rate will increase as we raise the temperature. So cooking is quicker. Not only that, people like the taste better because of the flavour added by the oil. The crispy coating and the soft potato in the middle is another attraction.

However, the fried potatoes will absorb some of the oil. This increases the energy input of the potato significantly. That's what causes the link to be made with obesity. If you don't exercise enough to use the extra energy, your body will store it as fat.

Not only that, at the higher cooking temperatures, Swedish scientists have shown that acrylamide forms. This compound has been linked to cancer, although some argue that this has not been proved yet. The Swedish scientists did not find acrylamide in boiled potatoes.

Research like this can often indicate links (**correlations**) between variables. However, this is not necessarily a **causal link**. For example, people who eat chips regularly might be more likely to get a certain disease. But is it the chips or some other factor that causes the disease?

- Discuss the issue of eating potatoes as part of your diet. Point out the good points and the bad points. Then form your opinions.

French fries are very popular in 'fast food' outlets

Potatoes absorb some of the oil when we fry them. This can cause health problems. However, the high energy content of plant oils is an advantage in some uses, such as the bio-fuels on page 183.

Summary

- We can extract plant oils by pressing or distillation.
- Oil and water are immiscible. When shaken together they form an **emulsion**.
- **Emulsifiers** keep the oil and water from separating out into layers.
- Plant oils are **unsaturated** compounds. They contain carbon–carbon double bonds in their molecules. You can test for these with bromine or iodine solutions (which turn colourless).
- Plant oils can be hardened and turned into margarine by reacting them with hydrogen.
- Processed foods can contain permitted additives that are assigned E numbers.
- We can use chromatography to detect and identify colourings added to foods. We can also use modern analysis techniques using expensive machines, such as a mass spectrometer.
- Plant oils can also be used to make renewable **bio-fuels**.

▶ Questions

1. Copy and complete:
   We can get the . . . . from plants by . . . . or distillation.
   When we mix up plant oils and water we get a mixture called an . . . . .
   These oils can be converted to . . . . in a reaction with . . . . gas.
   Permitted food . . . . are given E numbers.
   We can separate the colours in a food dye by . . . . .

2. Here are the ingredients of French vanilla ice cream:

8 large eggs yolks
$\frac{3}{4}$ cup sugar
$2\frac{1}{2}$ cups heavy cream
$1\frac{1}{2}$ cups cold milk
Pinch of fine salt
1 vanilla bean

   Cream is an emulsion containing lots of fat/oil mixed with water. Milk is a much more watery emulsion with much less fat/oil mixed in. Cream and milk separate out into two layers in a bottle of full-cream milk.
   Using the information above where necessary:
   a) Give the name of an emulsion that contains mainly water.
   b) What is the missing word below?
      Two liquids that separate out into layers are said to be i . . . . .
   c) Ice cream does not separate out into layers like full-cream milk.
      Which ingredient in the list acts as an emulsifier?
   d) Which substances are used to improve the taste of the ice cream?

3. Plant oils can be used to make margarine.
   a) Which gas is added to the margarine?
   b) What conditions are needed for the reaction to take place?
   c) Explain why this can be called an 'addition reaction'.
   d) Explain why it is necessary to carry out this reaction and how it solves the problem.

4. A food scientist was testing a new type of sweet. She decided to make a chromatogram of the colouring used with some known colourings that were banned. Look at the chromatogram below:

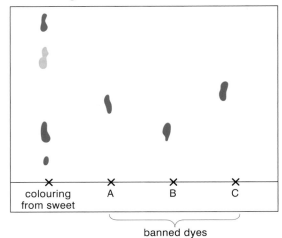

   What could a food scientist deduce from the chromatogram?

5. List the advantages and disadvantages of using plant oils to make bio-fuels.

Further questions on page 191.

▶ Acids and Alkalis

1. This question is about acids and alkalis.
   a) Copy and complete the following table. *[3]*

Colour for universal indicator	pH of solution
red	
	11
green	

   b) The reaction between an acid and an alkali can be summarised as:

   acid + alkali ⟶ salt + water

   i)   Name this type of reaction. *[1]*
   ii) Name the salt that would be formed by reacting dilute hydrochloric acid with potassium hydroxide solution. *[1]*
   iii) Name the acid and alkali needed to produce the salt sodium nitrate.

   [2]   (WJEC)

2. a) Clare made sodium chloride solution by adding sodium hydroxide solution, an alkali, to hydrochloric acid.
   i)   What name is given to substances, such as sodium chloride, formed from the reaction of an acid with an alkali? *[1]*
   ii) Name the process taking place in this reaction. *[1]*
   iii) What should Clare use to find out when she has added enough sodium hydroxide solution to react with all the hydrochloric acid? *[1]*
   iv) Describe how she should use this to make sure she added the correct amount of sodium hydroxide. *[2]*

   b) Sodium chloride can be separated from its solution by evaporation. Draw a labelled diagram to show how this is done. *[3]*
   c) Milk of Magnesia can cure indigestion. Explain how Milk of Magnesia works.
   [2]   (EDEXCEL)

3. Sodium carbonate reacts with acids.
   a) Complete the word equation: *[1]*

   sodium carbonate + hydrochloric acid ⟶ sodium chloride + ... + water

   b) Name the salt produced if sodium carbonate reacts with dilute nitric acid. *[1]*   (AQA)

4. An acid and an alkali react together to form a salt. These reactions are always exothermic.
   a) i) What name is given to the reaction between an acid and an alkali? *[1]*
      ii) What substance, other than a salt, is always formed when an acid reacts with an alkali? *[1]*
   b) Copy and complete the following table to show the salt formed on mixing the alkali and the acid.

Alkali	Acid	Salt
....	nitric acid	sodium nitrate
barium hydroxide	hydrochloric acid	....
....	....	potassium sulfate

   [4]

   c) Describe how you could prepare a solution of magnesium chloride starting with dilute hydrochloric acid. Include in your answer what you would do, what you would **see** and how you would make sure that all the hydrochloric acid had been used up.
   [5]   (EDEXCEL)

5. Complete the following word equations:
   a) zinc + sulfuric acid ⟶ .... + ... *[1]*
   b) sodium hydroxide + nitric acid ⟶ .... + ... *[1]*
   c) magnesium carbonate + hydrochloric acid ⟶ ..... + ... + water
   [1]   (AQA)

6. A student made a sample of zinc sulfate crystals.
   She added excess zinc metal to acid and warmed the mixture. When the reaction was finished, she filtered the mixture. She left the solution to allow crystals to form.
   a) Name the acid she used. *[1]*
   b) Name the gas formed by the reaction. *[1]*
   c) i)   Why did she use excess zinc metal? *[1]*
      ii) Why did she warm the mixture? *[1]*
      iii) Why did she filter the mixture? *[1]*
   (EDEXCEL)

7. The diagrams show what happens when an acid is added to an alkali.

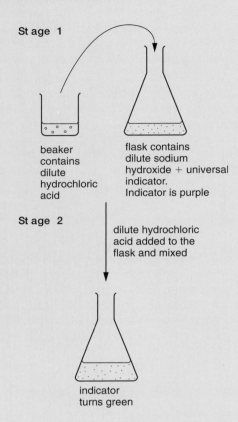

St age 1

beaker contains dilute hydrochloric acid

flask contains dilute sodium hydroxide + universal indicator. Indicator is purple

St age 2

dilute hydrochloric acid added to the flask and mixed

indicator turns green

a) What is present in the flask at stage 2, besides universal indicator and water? *[1]*

b) Write an ionic equation to show how water is formed in this reaction and state the sources of the ions. *[3]* (AQA)

8. a) Give the symbol for the ion which is produced when:

   i) any acid

   ii) any alkali

   dissolves in water. *[2]*

b) Copy and complete the following word equation.

calcium carbonate + hydrochloric acid ⟶ .... + ... + ...

[1] (WJEC)

▷ **Products from oil**

9. Crude oil is a mixture of hydrocarbons.
   a) Name the **two** elements found in every hydrocarbon. *[2]*
   b) To turn crude oil into useful products, it has to be separated into different fractions.
      i) Name the method used to separate different fractions in crude oil. *[1]*
      ii) Explain how this method of separation works. *[2]* (OCR)

10. The demand for the fractions containing smaller hydrocarbons, e.g. petrol, is higher than for those containing larger hydrocarbons. Smaller hydrocarbons are produced from larger ones by the process called cracking.
   a) State **two** conditions which are required for the cracking of large hydrocarbons. *[2]*
   b) Octane and ethene may be made from decane, $C_{10}H_{22}$, by cracking. Write a balanced symbol equation for this reaction. *[1]* (OCR)

11. The equation below shows the cracking of a hydrocarbon compound into two different compounds, A and B.

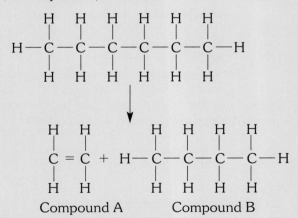

Compound A   Compound B

a) State **two** differences between the structures of compounds A and B. *[2]*
b) Why is compound A useful in industry? *[1]* (AQA)

12. Methane (CH_4) contains the elements carbon and hydrogen only. A student wanted to find out which new substances are produced when methane is burned. The student set up the apparatus shown below.

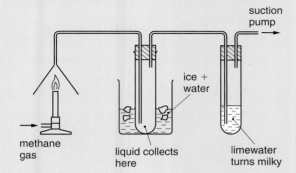

a) Which gas in the air reacts with methane when it burns? [1]
b) Name the liquid collected. [1]
c) Name the gas which turns limewater milky. [1]
d) When methane burns an exothermic reaction takes place. What is meant by an exothermic reaction? [2] (AQA)

13. Orimulsion is a **fossil fuel**.
It is a mixture of tar and water.
Orimulsion is much cheaper than coal.
When Orimulsion burns it gives out energy.
There is a lot of sulfur in Orimulsion.
a) i) Fossil fuels store energy. Where did this stored energy **first** come from? [1]
   ii) Will supplies of Orimulsion ever run out? Explain your answer in detail. [3]
b) Several gases are made when Orimulsion burns. Carbon dioxide and sulfur dioxide are two of the gases.
   i) Explain how the sulfur dioxide is made. [1]
   ii) Describe how sulfur dioxide can have a bad effect on the environment. [2]
   iii) Describe how carbon dioxide can have a bad effect on the environment. [2] (OCR)

14. The table at the top of the next column shows three of the substances which can be obtained from crude petroleum.

Name of substance	Formula	Structural formula	Use
methane	CH_4	H—C—H (with H above and H below)	fuel
a)	C_3H_8	b)	c)
ethene	C_2H_4	d)	e)

a) Complete the table. [5]
b) What feature of an ethene molecule tells you that it is unsaturated? [1] (AQA)

15. The diagram below represents a fractionating column used at an oil refinery. The points numbered 1 to 4 show the levels and temperatures at which different fractions are collected.

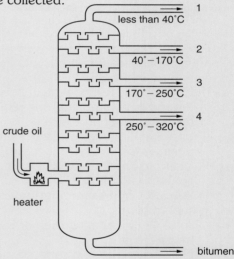

At which points would butane (boiling point 0°C) and octane (boiling point 126°C) be collected?

	Butane collected at point	Octane collected at point
A	1	2
B	2	1
C	2	3
D	3	2
E	1	4

[1] (OCR)

▶ Polymers

16. John reads this article in a magazine.

New Polymer Developed

Dr Philip Green is a chemist working for a major oil company. He has reported the development of a new bio-degradable polymer. This polymer has similar properties to poly(ethene) but, when it is left in the environment, it is quickly broken down by bacteria. Tests have shown that a supermarket bag made from the new polymer disappears when buried in soil for eight weeks.

Dr Green believes that all major supermarkets will use his polymer instead of poly(ethene) within the next few years. He said to our reporter, 'In the past, chemists have been blamed for developing materials such as poly(ethene) that cause major pollution problems. My team of chemists has spent 10 years trying to put this right. We think our new polymer will have a major impact on the litter problem in this country. The days of poly(ethene) bags spoiling the appearance of our countryside will soon be over. Everyone in my team is delighted that our efforts to make a non-polluting polymer have been successful'.

Use information from the article to help you answer these questions.

a) i) Explain why using the new polymer instead of poly(ethene) will reduce environmental problems. *[3]*

   ii) Supermarket bags made from the new polymer take much longer than eight weeks to break down when not in contact with the soil.
   Suggest why this is important for supermarkets. *[1]*

b) i) John shows the magazine article to his friend Holly.
   Holly says that oil companies are only interested in profit, and do not care for the environment.
   John tries to convince Holly that environmental issues are an important part of the job of chemists in the 21st century.
   What information from this article can he use to convince Holly? *[2]*

   ii) Holly tells John that the problem of litter from plastic rubbish could be solved by recycling all of the plastics that we use.
   John says that we are never likely to recycle all plastics.
   Explain why John is correct. *[2]* (OCR)

17. Ethene is used in the manufacture of the plastic poly(ethene). Ethene is heated under high pressure in the presence of a catalyst. Many ethene molecules join together to form a giant molecule of poly(ethene). The diagram below shows what happens in the reaction:

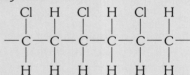

a) What is the name of this type of reaction? *[1]*

b) Describe how the ethene molecules join together to form poly(ethene). *[3]*

c) The poly(ethene) was heated and moulded into the shape of a bucket. The design of the bucket was changed. All the buckets were heated and then remoulded. What type of plastic is poly(ethene)? Explain your answer. *[2]* (EDEXCEL)

18. This diagram shows part of a molecule of the polymer PVC.

$$\text{—C—C—C—C—C—C—}$$

(Cl H Cl H Cl H above; H H H H H H below)

a) What is the structural formula of the monomer from which PVC is made? *[3]*

b) PVC is used to make 'artificial leather' for use in clothing, furniture, handbags etc. PVC is not a biodegradable material.
   i) Explain the meaning of the term **biodegradable**. *[1]*
   ii) Describe the problems that might be caused by the disposal of the PVC in these products. *[3]*

19. Plastic bottles are sometimes made of the addition polymer poly(propene).
a) What is meant by **addition polymer**? *[2]*
b) Draw the structure of a propene molecule showing all covalent bonds. *[2]*
c) Show how two propene molecules are joined as part of a poly(propene) molecule. Show all covalent bonds. *[2]* (EDEXCEL)

▶ Carbon chemistry

20. The structural formulae of several organic compounds are shown below.

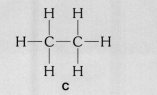

a) Which of these compounds are:
  i) alkanes,
  ii) alkenes,
  iii) alcohols?   *[4]*

b) Which is the only compound that can undergo addition polymerisation?   *[1]*

c) Which of the other four compounds can be dehydrated to make compound D? *[1]*

21. Ethanol can be made by fermentation or by the reaction between ethene and steam.

a) Write a balanced equation for each method.   *[4]*

b) What are the ideal conditions needed for production of ethanol by fermentation? *[3]*

c) Give an advantage and a disadvantage of each method.   *[4]*

d) Describe two uses of ethanol.   *[2]*

▶ Plant oils

22.   **Sunflower** → **Sunflower** → margarine
            seeds   **oil**

a) Suggest how sunflower oil can be extracted from sunflower seeds.   *[1]*

b) Sunflower oil contains unsaturated molecules.
  i) What type of chemical bond is present in an unsaturated molecule?   *[1]*
  ii) Describe how you could show that sunflower oil contains unsaturated molecules. *[2]*

c) Describe how sunflower oil can be hardened to form margarine.   *[3]*

23. Biodiesel can be made from vegetable oil obtained from plants such as oil-seed rape. Some scientists think that biodiesel may have advantages over diesel made from crude oil.
Biodiesel is a renewable source of energy, is biodegradable and contains no sulfur.

a) Describe how vegetable oil can be extracted from seeds.   *[1]*

b) Explain why:
  i) Biodiesel is a renewable source of energy.   *[1]*
  ii) A spillage of biodiesel into the sea may cause less harm than a spillage of diesel made from crude oil. *[1]*
  iii) A fuel which contains sulfur may harm the environment.   *[1]*

c) To provide enough vegetable oil to make a large amount of biodiesel would mean planting large areas of countryside with plants such as oil-seed rape.
Suggest why some people may object to this use of the countryside.   *[2]*

24. Some foods need to be cooked before they are eaten, others do not.

a) Name two foods that do not have to be cooked before they are eaten.   *[2]*

b) Give three reasons why some food should be cooked before it is eaten.   *[3]*

c) When meat is cooked proteins undergo chemical change.
Describe how the proteins change.   *[3]*

chapter 15

Energy Changes

We all need the energy we get from chemical reactions.
Energy heats our homes and powers industry.
Energy from the chemical reactions inside our bodies
keeps us all alive.

Whenever chemicals react, we get a transfer of energy.
The energy usually ends up as heat. Sometimes we also
get light energy or sound energy as well. Can you think
of a reaction which gives out heat, light and sound?

*The chemical energy in
these fireworks is transferred
to heat, light and sound energy*

▷ Fuels

Fuels store chemical energy.
When we burn a fuel, it reacts with oxygen and its
chemical energy is released.
We can use this energy to give us heat or light.

Food is the fuel for our bodies. We 'burn' sugar by
reacting it with oxygen in our cells. Luckily for us, in
respiration the reactions are carefully controlled.
We don't burst into flames! But fuels often do.

Choosing a fuel

When choosing a fuel you should consider its:
- energy content
- availability/cost
- toxicity
- pollution effects (see pages 147 to 149)
- ease of use/storage.

> **You need 3 things to make a fire:**
> - **fuel**
> - **heat**
> - **oxygen.**

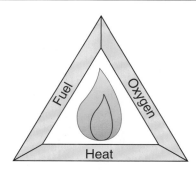

The fire triangle

*Whenever you come across **heat**,
And **fuel** and **oxygen** meet.
You're sure to get fire,
The flames will lick higher –
The **fire triangle's** complete!*

> *Investigation 15.1   Comparing Bunsen flames*
> Design an investigation to find if a yellow or a blue Bunsen
> flame provides more energy.
> Make sure your plan is as safe, reliable and fair as possible.
> Check your plan with your teacher before you start your tests.
> - Which type of flame is better? Why?

Notice the soot produced by the yellow Bunsen flame.
This is a sign of **incomplete combustion**. (See page 147.)
The gas we burn in a Bunsen burner is usually methane, CH_4.

Complete combustion: $CH_4(g) + 2O_2(g) \longrightarrow CO_2(g) + 2H_2O(g)$ This gives out most heat.

Incomplete combustion: $2CH_4(g) + 3O_2(g) \longrightarrow 2CO(g) + 4H_2O(g)$

and/or $CH_4(g) + O_2(g) \longrightarrow C(s) + 2H_2O(g)$
                                                                     carbon
                                                                     (soot)

▷ Energy content of fuels

One of the main factors to think about when choosing a fuel is its energy content. To compare this across different fuels we should look at the energy released per gram of fuel burned.

To get this information by experiment, you need to carry out a fair test. The only variable that should change each time we collect data is the type of fuel.

To make the investigation valid, we want the energy released as the fuel burns to heat up the water. (See Investigation 15.2 below.) Therefore we use a copper beaker (called a **calorimeter**).
- Why is the beaker made of copper?
- Why do you think that the copper beaker is called a 'calorimeter'?

You will also need to weigh how much fuel is used in each test.

The energy content of food is sometimes expressed in calories per 100 g. 1 calorie = 4.2 joules. Fats, oils and carbohydrates are 'high energy' foods and too much in your diet can cause obesity. That's why we can now buy 'low fat' crisps.

Then you can work out how much energy (in joules) is transferred to the water using this equation:

> **Energy = mass of water × specific heat capacity of water × rise in temperature**
> **(J)   (g)   (J/g/°C)   (°C)**

To convert this energy into the amount of energy per gram, we use this equation:

> **Energy per gram = $\dfrac{\text{energy released (J)}}{\text{mass of fuel used (g)}}$**

Example
0.2 g of fuel raised the temperature of 100 g of water by 25 °C.
How much energy does the fuel give out per gram?
(The specific heat capacity of water is 4.2 J/g/°C – this means that it takes 4.2 J of energy to raise the temperature of 1 g of water by 1 °C.)

Energy = 100 × 4.2 × 25 = 10 500 J
Energy per g = 10 500 ÷ 0.2 = 52 500 J/g (or 52.5 kJ/g)

Investigation 15.2   Comparing fuels
You can compare the heat content of different alcohols using the apparatus opposite.
- Design a fair test.
  Remember to weigh your spirit burner and its fuel before and after heating the water.
  Work out the energy content of each alcohol in joules per gram.
- Can you see any pattern in your data? (Look at the formula of each alcohol.) Use secondary data if necessary.
- What are the main sources of error in your investigation?
- How could you improve the reliability of your data?

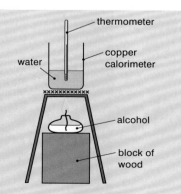

Apparatus for comparing fuels

▷ Exothermic and endothermic reactions

Exothermic reactions

We looked at the combustion of a fuel on page 147.
Think about what happens when we burn a fuel:

- Is energy given out or taken in as the fuel burns?

Reactions which give out heat are called **exothermic**.
Combustion, plus many oxidation and neutralisation
reactions are exothermic.

The charcoal (carbon) reacts with oxygen in this barbecue. Heat from the exothermic reaction cooks the food.

Experiment 15.3   Feeling hot!

Collect 25 cm^3 of copper sulfate solution in a small beaker.
Add a spatula of zinc powder. Stir it with a temperature sensor.
Carefully touch the outside of the beaker.

- What do you see happen?
- Does the beaker feel hot or cold?
- Does the reaction give out heat?
- Can you explain the shape of the graph.
- Can you write a word or symbol equation for this reaction?

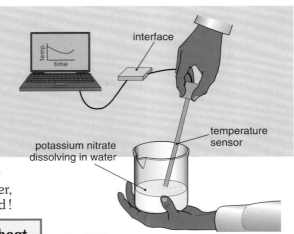

interface

temperature
sensor

zinc powder

copper sulfate
solution

The displacement reaction between zinc and
copper sulfate gives out heat.
It is an exothermic reaction.

> **Exothermic reactions give out energy, often as heat.**

Endothermic reactions

Some reactions take in heat from their surroundings.
These are called **endothermic** reactions.
Thermal decomposition is an example of an endothermic reaction.
It needs continuous heating.

Experiment 15.4   Feeling cold!

Add 3 spatulas of potassium nitrate to 25 cm^3 of water
in a small beaker. Stir it with a temperature sensor.
Hold the beaker in the palm of your hand.

- What do you feel?
- Is heat energy given out to your hand? Or is energy
  being taken from your hand?
- Can you explain the shape of the graph?

interface

temperature
sensor

potassium nitrate
dissolving in water

When potassium nitrate dissolves it takes in heat energy
from its surroundings. Its surroundings include the beaker,
the glass rod, the water, the air around it, and your hand!

> **Endothermic reactions take in energy, often as heat.**

▷ What happens to the temperature?

You have seen how some reactions give out heat,
and others take in heat.
We know that when fuels burn, heat is given out.
What happens to the temperature near the burning fuel?
Does the temperature rise or fall in an exothermic reaction?

Think about the experiments on the previous page:
What do you think happens to the temperature in each one?
What can you use to check your ideas?
(Try this out if you have time.)

We find that:

- in **exothermic** reactions, the **temperature goes up**,
- in **endothermic** reactions, the **temperature goes down**.

Now let us look at some reactions to see if
they give out heat or take in heat:

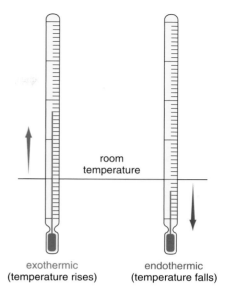

room
temperature

exothermic
(temperature rises)

endothermic
(temperature falls)

Experiment 15.5   Exothermic or Endothermic?

Use the apparatus as shown:
Work with each pair of substances in turn.
Record your results in a table.

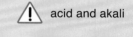 acid and akali

Reaction	temp. before mixing (°C)	temp. after mixing (°C)	Exothermic or Endothermic?
sodium hydroxide solution + dilute hydrochloric acid			
sodium hydrogencarbonate solution + citric acid			
copper sulfate solution + magnesium powder			
dilute sulfuric acid + magnesium ribbon			

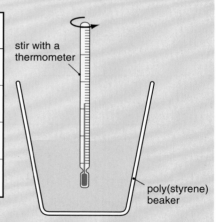

stir with a
thermometer

poly(styrene)
beaker

Look at your results:
- Did you find more exothermic or endothermic reactions?
- The first reaction is between an acid and an alkali.
  What do we call this **type** of reaction? (See page 125.)
- Which gas is given off in the second reaction in the table?
  (See page 136.)
- Which **type** of reaction takes place between copper sulfate
  and magnesium? (See page 71.)
- Which gas is given off in the last reaction in the table?
  (See page 129.)
- Would a temperature sensor and data logger provide you
  with better data? If you have time, test this out.

*Record the maximum or minimum
temperature during each reaction.*

*You can work out the actual energy
involved by assuming that the
solutions have the same specific
heat capacity as water (i.e. it takes
4.2 J of energy to change the
temperature of 1 g (or 1 cm³) of
solution by 1 °C).*

- *What measurements would you
  need to make?*

▷ Energy level diagrams

You have now seen some examples of exothermic and endothermic reactions.
Can you recall which gives out heat and which takes in heat? Look at the picture opposite:
It might help you to remember.

We can show the energy transfers in reactions on an **energy level diagram**.

These show us the energy stored in the reactants compared to the energy stored in the products.
Look at the examples below:
(Remember that reactants are the substances we start with and products are the new substances made in the reaction.)

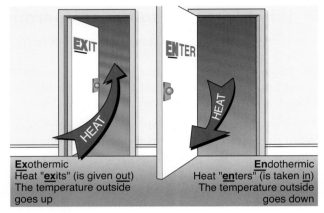

Exothermic
Heat "**ex**its" (is given <u>out</u>)
The temperature outside goes up

Endothermic
Heat "**en**ters" (is taken <u>in</u>)
The temperature outside goes down

Exothermic energy level diagram

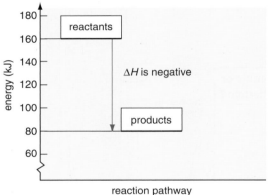

ΔH (we say 'delta H') is the symbol for the 'change in energy' in a reaction.
Look at the energy level diagram opposite:
It is for an exothermic reaction.
Notice that:
the products have **less energy** than the reactants.

> ΔH **is negative for an exothermic reaction.**

The difference in energy is given out as heat.
Therefore, the temperature rises.

For example,

$$HCl + NaOH \longrightarrow NaCl + H_2O \qquad \Delta H = -58\,kJ/mol$$

● Draw an energy level diagram to show this change.

Endothermic energy level diagram

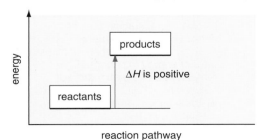

In endothermic reactions, the products have more energy than the reactants.

> ΔH **is positive for an endothermic reaction.**

The extra energy needed to form the products is taken in from the surroundings.
Therefore, the temperature falls.

▷ Chemistry at work : Energy matters

Heating up

We use the heat released in exothermic reactions whenever we burn a fuel. However we can also use exothermic changes in different ways. There are hand-warmers which use the energy given out when crystals form in a super-saturated solution. Self-heating cans often use the exothermic reaction of calcium oxide and water to release the energy to warm up their contents.

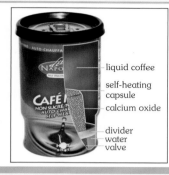

liquid coffee
self-heating capsule
calcium oxide
divider
water
valve

A self-heating can of coffee. Pressing a button in the bottom of the can releases water onto calcium oxide.

Cooling down

Sports injuries can be treated with cold packs to reduce swelling and numb pain. These packs can contain ammonium nitrate and water. The two are kept separate until squeezed. Then the ammonium nitrate dissolves in water taking in energy. The instant cold packs work for about 20 minutes.

Putting out fires

Can you remember the fire triangle? (See page 192.)
To put out a fire you need to remove one part of the triangle.
Have you ever had a fire in your kitchen?
People sometimes heat oils or fats above
their flash points. The flash point of a substance
is the temperature at which it ignites.
This happens in chip-pan fires.
If water is thrown on to burning oil or fat,
it makes the fire worse. The water turns to steam,
and sends burning oil flying into the air.
Look at the photo:
● How does the damp tea-towel put out the fire?
● Why must you leave the pan covered by
the tea-towel for some time after the fire?

Some fire extinguishers give off carbon dioxide gas.
They are good at putting out electrical fires.
They prevent oxygen getting to the fire.
● Why don't we use water on electrical fires?

Can you think of a problem with using
a carbon dioxide extinguisher outdoors?
Look at the photo opposite:
Fire crews at a plane crash spray foam onto the fire.
Carbon dioxide gas is trapped in the foam so it stays on the fire.

▷ Making and breaking bonds

We have already seen how atoms 'swap partners'
in chemical reactions.
This means that the bonds which join atoms
to each other must be broken.
New bonds must be made as the products form.

Do you think that energy is needed to break bonds?
Think of it as pulling apart 2 strong magnets.
You have to put in energy to separate them.
Therefore:

> **Breaking bonds requires energy. It is endothermic.**

What about making new bonds? That's like
the 2 magnets leaping across a gap because of their
attraction for each other.

> **Making new bonds gives out energy. It is exothermic.**

We can show these changes on an energy level diagram.
Let's look at the reaction between hydrogen (H_2) and
chlorine (Cl_2). They make hydrogen chloride (HCl).
The reaction needs energy to start it off:

$$H_2(g) + Cl_2(g) \longrightarrow 2\,HCl(g)$$

We must supply energy to break bonds!

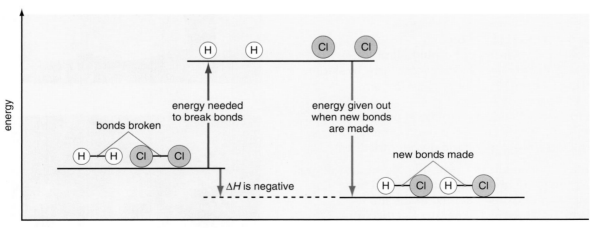

reaction pathway

- Which involves more energy –
  breaking the original bonds (endothermic) or making the new bonds (exothermic)?
- Is the overall reaction exothermic or endothermic?

▷ Bond energy

We can work out the energy needed to break different chemical bonds. This energy is called the **bond energy**.

Look at the table of bond energies opposite: Some bonds are stronger than others. Which is the strongest bond shown in the table?

Notice that the units are **kJ/mol**. ΔH is also measured in kJ/mol. This stands for kilojoules per mole. It is the energy needed to break a set number of bonds (a number which chemists call 'the mole').
You can read about the quantity called the mole on page 340.

Bond	Bond energy (kJ/mol)
H–H	436
Cl–Cl	242
H–Cl	431
C–H	413
C–C	347
C–O	335

Calculating ΔH

We can use the energy level diagrams and bond energies to work out the energy change (ΔH) for a reaction.

Let's look at the reaction between hydrogen and chlorine again. Look at its energy level diagram on the previous page:
You can use this and the bond energies at the top of this page to calculate ΔH:

How many H—H and Cl—Cl bonds are broken? We can think of this as the energy needed to start off the reaction.

Add up the bond energies:

$$+ [1 \times (H\text{--}H)] + [1 \times (Cl\text{--}Cl)]$$
$$= +(436 + 242)$$
$$= +678 \, \text{kJ/mol}$$

How many new HCl bonds are made?
Remember that making bonds is exothermic. Energy is given out. This is given a negative sign.

$$- (2 \times H\text{--}Cl)$$
$$= -(2 \times 431)$$
$$= -862 \, \text{kJ/mol}$$

Now we can work out the overall energy change (ΔH):

$$(+678) + (-862)$$
$$= -184 \, \text{kJ/mol}$$

So the reaction is exothermic.
More energy is given out making new bonds than is taken in breaking the original bonds.

The calculation can be shown on an energy level diagram

Bond energies are average values for the bonds in different molecules. Therefore these calculations only give us a rough value for ΔH.

▷ Chemistry at work: Hydrogen as a fuel

Can you remember the test for hydrogen gas?
Hydrogen reacts explosively with the oxygen in air.
When a lighted splint pops in hydrogen,
it reacts with oxygen to make water (steam).

hydrogen + oxygen ⟶ water
$$2H_2(g) + O_2(g) \longrightarrow 2H_2O(g)$$

One of the first uses of hydrogen was in airships.
Hydrogen is the lightest of all gases. However, its
violent reaction with oxygen led to disasters.
Which safer gas is used in airships today? (See page 58.)

Scientists are very interested in using hydrogen
as a fuel. Look at the equation above:

- Do you think hydrogen makes any pollution
  when it burns?
- What problems do you think hydrogen-powered
  cars have?

Chemists are looking for ways to store the gas
on the surface of transition metals. This would be safer
and take up less room in the car.

Hydrogen has played a big part in space travel.
It can be used as rocket fuel. The liquid hydrogen
is stored in tanks, ready to react with liquid oxygen.

It has also been used inside space-craft to power
fuel cells. A fuel cell is an efficient way of reacting
hydrogen with oxygen. It transfers the energy from
the reaction directly into electrical energy.
(See page 107.)

This hydrogen-powered car stores its fuel in the boot

Hydrogen is one of the fuels used in rockets. It is also used in fuel cells to provide electricity for the astronauts in space.

Investigation 15.6 The ideal fuel?

List the things that you think are important
for an ideal fuel. Think about:
how much energy is given out, transport, storage, safety,
pollution, how easy it is to light, plus your own ideas.
Your teacher will tell you which fuels you can test.
Plan an investigation to see which fuel is best.
- How will you make it a fair test?
- How will you make it safe?

⚠ Make sure your teacher checks your plan before
you start!

Summary

- Three things are needed for a fire – fuel, heat, and oxygen.
  If one part of this 'fire triangle' is removed, the fire goes out.
- The reaction of a fuel with oxygen is called **combustion**.
- When a hydrocarbon fuel burns in a plentiful supply of oxygen,
  it forms carbon dioxide and water.
- If a hydrocarbon fuel burns in a limited supply of oxygen,
  it makes carbon monoxide gas and carbon (soot).
  This is called **incomplete combustion** and releases less
  energy then complete combustion.
- **Exothermic** reactions give out heat. The temperature rises.
- ΔH is given a negative sign in exothermic reactions.
- **Endothermic** reactions take in heat. The temperature falls.
- ΔH is given a positive sign in endothermic reactions.
- We can work out how much energy is given out or taken in
  during a reaction by using bond energies.

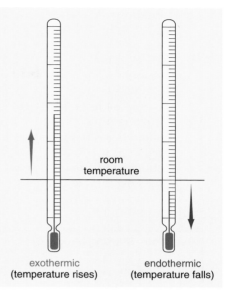

room
temperature

exothermic
(temperature rises)

endothermic
(temperature falls)

▷ Questions

1. Copy and complete:
   a) The 3 things we need for a fire are fuel,
      heat and . . . . . .
   b) If a hydrocarbon burns in plenty of oxygen,
      it produces carbon dioxide and . . . . . .
      The reaction is called complete . . . . . .
      In a poor supply of oxygen, we also get
      toxic carbon . . . . gas and possibly . . . . .
      We call this reaction . . . . combustion.
   c) Exothermic reactions give out . . . . , and
      the temperature . . . . .
      . . . . reactions take in heat, and the
      temperature . . . . .

2. Look at the data below:
   $$C + O_2 \longrightarrow CO_2 \qquad \Delta H = -394 \, kJ/mol$$
   $$2\,C + O_2 \longrightarrow 2\,CO \qquad \Delta H = -221 \, kJ/mol$$
   a) Which reaction gives out more energy per
      gram of carbon burned?
   b) Explain why this is important when
      designers make plants for a new industrial
      heater run on coke (C).

3. Tom weighed out 20 g of each of two weak
   acids, A and B.
   He added each powder to a solution of
   sodium hydrogencarbonate in an expanded
   polystyrene cup. Then he used a thermometer
   to find the change in temperature. He
   decided to check his results by repeating his
   experiment 3 times. Here are his results:

Weak acid	Temperature fell by (°C)		
	First test	**Second test**	**Third test**
A	4	9	8
B	2	2	3

   a) Are the reactions exothermic or
      endothermic?
   b) Work out the mean (average) result for A
      and B.
   c) What was the range of the temperature
      change for A and B? What does this tell
      you about the reliability of Tom's results?
      What advice would you give to Tom to
      improve this aspect of his investigation.
   d) Tom used 25 cm^3 of solution in each test.
      What was the mean (average) energy
      change in the reaction with A? (Assume
      that 4.2 J of energy changes the
      temperature of 1 cm^3 water by 1 °C).
   e) Was Tom justified in concluding that the
      reaction with weak acid A involved the
      largest energy change? Explain your
      answer.

4. a) Petrol burns in a car engine. Explain how carbon **monoxide** gas is formed.
   b) Carbon monoxide bonds to the haemoglobin in your blood. Explain how this makes it toxic. (See page 147.)
   c) Several holiday-makers die from carbon monoxide poisoning each year. Poorly ventilated gas heaters are often to blame. Imagine that you are a travel agent. Write a letter to the owner of some holiday homes explaining the dangers.

5. Look at this table:

Reaction	Starting temp.(°C)	Final temp.(°C)
A + B	19	27
C + D	20	25
E + F	19	17

   a) Decide whether each reaction is exothermic or endothermic? How can you tell?
   b) The volume of solution was the same in each reaction. Which had the largest energy change?

6. In an experiment a fuel raised the temperature of 500 g of water by 4 °C.

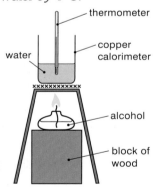

water

thermometer

copper calorimeter

alcohol

block of wood

   a) Work out the energy released in the experiment.
   (It takes 4.2 J of energy to raise the temperature of 1 g of water by 1 °C.)
   b) 0.25 g of fuel was burned in the experiment. Calculate the energy given out per gram of fuel.
   c) Evaluate the method used for this experiment.

7. When hydrochloric acid and sodium hydroxide react in a beaker, the temperature rises:

$$HCl + NaOH \longrightarrow NaCl + H_2O$$
$$\Delta H = -58\,kJ/mol$$

   a) Is the reaction exothermic or endothermic? Give 2 reasons for your answer from the information above.
   b) Copy and complete this energy level diagram:

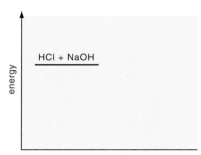

energy

HCl + NaOH

reaction pathway

8. Hydrogen burns in air to form steam.

$$2\,H_2 + O_2 \longrightarrow 2\,H_2O$$

   a) Copy the equation above and put in the state symbols.
   b) What is this type of reaction called?
   c) Draw an energy level diagram, like the one on page 198, which shows the breaking and making of bonds when hydrogen burns.
   (H_2 has an H–H bond, O_2 contains a double bond, O=O, H_2O contains two O–H bonds.)
   d) Use the bond energies in this table to calculate ΔH for the reaction:

Bond	Bond energy (kJ/mol)
H–H	436
O=O	498
O–H	464

   e) Why do you think that the bond energy of the O=O bond is higher than the other 2 values in the table above?

9. Give the advantages and drawbacks of using hydrogen as a fuel for cars.

Further questions on pages 242 and 243.

Rates of Reaction

chapter 16

Some reactions are fast, and others are slow.
Can you think of a reaction which happens
very quickly?
Fast reactions, like dynamite exploding,
start and finish within a fraction of a second.
Slow reactions, like concrete setting, may take
days, weeks, or even years to finish.
Can you think of another slow reaction?

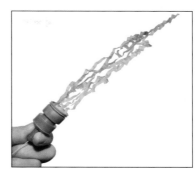

The chemicals in the base of a 'party-popper' react in a fraction of a second

This combustion reaction lasts a few seconds

The copper on this roof takes years to react in the air

What is 'rate of reaction'?

The rate of a reaction tells us **how quickly**
a chemical reaction happens.

It is important for people in industry
to know how fast a reaction goes. They have to
know exactly how much of their product
they can make each hour, day or week.
In a shampoo factory, the rate might be
100 bottles per minute.

We can't work out the rate of a reaction from its chemical
equation. Equations can only tell us how much product
it is possible to get. They don't say how quickly it is made.

We can only find the rate by actually doing experiments.

During a reaction, we can measure how
much reactant is used up *in a certain time*.
On the other hand, we might choose to measure
how much product is formed in a certain time.
Look at the equation opposite:

You can see how we can measure the rate of reactions on the next two pages

$$\text{Rate} = \frac{\text{amount of reactant used or product formed}}{} \div \text{time}$$

▷ Measuring rates of reaction

Let's look at a reaction that we have met before –
calcium carbonate and acid :

calcium carbonate	+	hydrochloric acid	⟶	calcium chloride	+	water	+ carbon dioxide

$$CaCO_3(s) \quad + \quad 2\,HCl(aq) \quad \longrightarrow \quad CaCl_2(aq) \quad + \quad H_2O(l) \quad + \quad CO_2(g)$$

reactants   products

How can we measure the rate of this reaction ?
We can measure how quickly one of the reactants
is being used up. However, it is not *easy* to
measure the amount of calcium carbonate or
hydrochloric acid as the reaction is happening.

Let's think about measuring one of the products.
Look at the equation above :
Which one of the products is a gas ?
It is much easier to measure **how much gas**
is being made as the reaction goes along.

Try one of the next experiments to measure
the rate of this reaction :

Experiment 16.1   Volumes of gas given off (1)

Set up the apparatus as shown :
Measure the volume of gas collected every 30 seconds.
Start your timing as soon as you put
the bung into the flask.   ⚠ acid
Put your results into a table like this :

Time (s)	Volume of gas (cm³)
0	0
30	

measuring cylinder

50 cm³ of
dilute
hydrochloric
acid

marble
chips

Plot a line graph of your results. Put 'time' along the horizontal axis
and 'volume of carbon dioxide gas' on the vertical axis.

Experiment 16.2   Volumes of gas given off (2)

Set up the apparatus as shown :
                        ⚠ acid

Repeat the method above.
Take the bung out of the flask when
you have collected 100 cm³ of gas.

dilute
hydrochloric
acid

syringe

Record your results and plot a graph as above.

• Explain why you can use a line graph to display
  your results. (See page 12.)

marble
chips

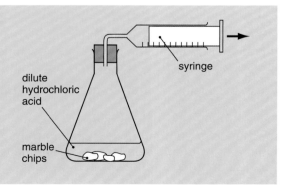

204

Experiment 16.3   Measuring the mass of gas given off

Set up the apparatus as shown :

Measure the mass every 30 seconds.
Record your results in a table like this :

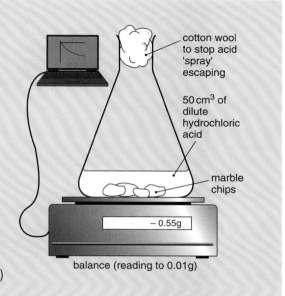

acid

cotton wool
to stop acid
'spray'
escaping

50 cm³ of
dilute
hydrochloric
acid

marble
chips

− 0.55g

balance (reading to 0.01g)

Time (s)	Mass (g)	Loss in mass (g)
0		0
30		

Some balances have a **TARE** button. You can
press this as you start timing. The balance will now
give you the loss in mass directly.

Plot a line graph of your results.
(Put loss in mass up the side and time along the bottom.)
The loss in mass is the mass of gas given off.

Alternatively, you may be able to link
your balance to a computer.
This will plot your results directly on to
a graph for you.

Graphs and rates of reaction

Are your graphs shaped like the one shown here ?
We can use graphs to measure the rate of a reaction
at any given time.
The slope (or **gradient**) of the line tells us how quickly
the reaction was going at that time.

> **The steeper the slope, the faster the reaction.**

Look at the graph :

● When is the reaction fastest ?

● How can you tell that the reaction is slowing down
  as time passes ?

● How do you know when the reaction has stopped ?

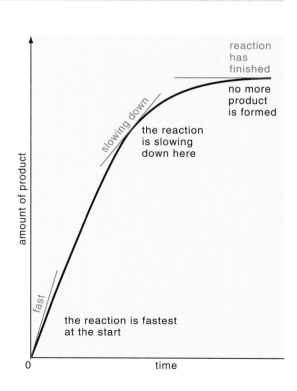

reaction
has
finished

no more
product
is formed

slowing down

the reaction
is slowing
down here

amount of product

fast

the reaction is fastest
at the start

0   time

▷ Effect of surface area

Have you ever tried to light a bonfire
or a camp fire? Which burns more quickly –
a block of wood or a pile of wood shavings?

We find that small pieces of solids, especially powders,
react faster than large pieces.

It's like frying two pans of chips.
One has potato cut into small, thin chips.
The other pan has bigger, thicker chips.
Which chips will be cooked first?
Which chips have the larger surface area?

Surface area is a measure of how much
surface is exposed. So for the same mass of potato,
small chips have a larger surface area than big chips.

*Lumps of coal burn slowly on a coal fire. However, coal
dust in the air down a mine can cause an explosion!*

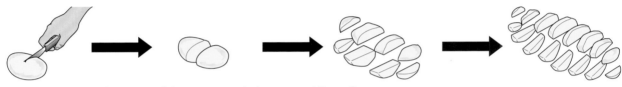

You increase the surface area of the potato each time you cut it smaller

Let's look at the effect of increasing surface area on
the rate of a reaction:

Experiment 16.4   The effect of surface area on rate of reaction
Set up the apparatus as shown:

Use the same method as Experiment 16.3.
Repeat your experiment twice.
The only thing you should change is the *size* of the
marble chips. Do it once with 5 g of small marble
chips, then again with 5 g of large marble chips.

⚠ acid

cotton wool

marble
chips

dilute
hydrochloric acid

- Which have the larger surface area – the small or the large chips?
- Why is it important to keep all other variables constant?

Print or plot both sets of results on the same graph.
Label the axes as in Experiment 16.3.

- Which line on your graph rises more steeply?
- Which size of marble chip reacts faster?
- What happens to the rate of reaction as we increase
  the surface area?

– 0.00g

balance (reading to 0.01g)
(connected to computer if possible)

As we increase the surface area, the rate of reaction increases.

The collision theory

As you know, all substances are made up of particles.
The particles might be atoms, molecules or ions.
Before we can get a chemical reaction,
particles must crash together. They must collide with
enough energy to cause a reaction.

This is called the **collision theory**.

Think about the rate of a reaction.
What do you think happens to the number of collisions
between particles if you speed up a reaction?

> **The more collisions between particles in a given time, the faster the reaction.**

*Particles must collide hard enough before they can react! The minimum amount of energy required to result in a reaction is called the **activation energy**. (See page 213.)*

Explaining the effect of surface area

Iron reacts with oxygen when you heat it in air.
What do you think it forms?

> ### Experiment 16.5   Sparklers!
> Compare what happens when you heat:
> 1. an iron nail
> 2. iron wool
> 3. iron filings.
> Hold the iron nail and iron wool in tongs as you heat them.
> Gently sprinkle a few iron filings into a Bunsen flame
> from the end of a spatula.
>
> • Put the 3 types of iron in order of increasing surface area.
> • What effect does increasing the surface area of the iron
>   have on the rate of its reaction?

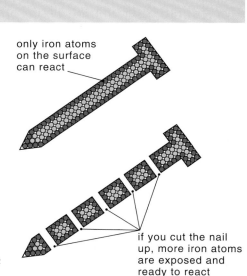

only iron atoms
on the surface
can react

if you cut the nail
up, more iron atoms
are exposed and
ready to react

Think about burning iron filings.
Powders have a very large surface area.
There are lots of iron atoms exposed at its many surfaces.
The oxygen molecules in the air can attack
any of these iron atoms.
With iron filings, there are lots of collisions
in a given time. The reaction is very fast.
Compare this with heating the iron nail.
The iron nail has a small surface area. It only reacts slowly.

In a lump, most particles are locked up inside,
In order to react, they have to collide!
But powders have lots of particles exposed,
If cut fine enough, they might even explode!

Explosions sometimes happen in flour mills or down mines. The
fine powders of flour or coal dust react rapidly if exposed to a
spark.

▷ Effect of concentration

Look at these instructions from a washing powder :

The washing powder dissolves into the water in your washing machine, making a solution.
The washing powder contains substances which react with stains to remove them.

What happens to the concentration of the solution in your machine as you add more cups of powder ?

If your washing is not badly stained, you can save money by adding less powder.
But what do you think happens if you use just 1 cup of powder on washing with difficult stains ?
You will probably have to wash the clothes several times.
It is better to follow the instructions given on the packet and just wash them once !

As you add more cups of powder, you are increasing the concentration of the solution.

So what happens to the rate at which the washing powder reacts with stains, as you increase the concentration ?

Which solution of washing powder is most concentrated ?

Experiment 16.6   Effect of concentration on rate of reaction

In this experiment you will **vary the concentration** of the acid each time.
• What other variables must be kept constant ?

Set up the apparatus as shown :
Time how long it takes to collect 20 cm³ of gas.
Record your results in a table like this :

Acid (cm³)	Water (cm³)	Time to collect 20 cm³ of gas (s)
10	40	
20	30	
30	20	
40	10	
50	0	

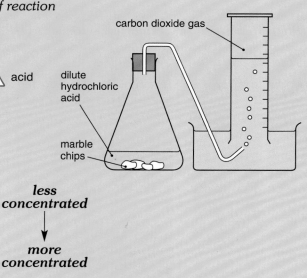

less concentrated
↓
more concentrated

• Which test gave the fastest reaction ?
• What happens to the rate of reaction as you increase the concentration of acid ?

As we increase the concentration, the rate of reaction increases.

Explaining the effect of concentration

You already know about the collision theory from page 207.
We can use it to explain why more concentrated solutions react more quickly.
Let's think about the reaction in the last experiment:

The acid particles can only react with the marble chips when they collide.
Look at these diagrams:

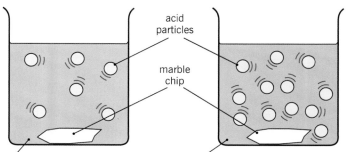

acid particles

marble chip

If this is a IM solution of acid . . . . . . .this is a 2M solution. There are twice as many acid particles **in the same volume of solution**

(We often use the units mol/dm^3 instead of M for concentration – see page 344.)

- *Which beaker has the faster reaction?*
- *Why?*

The acid particles move randomly through the water.

As you increase the concentration of the acid, there are
more acid particles in the same volume of solution.
Therefore there is a greater chance of acid particles colliding, and reacting, with particles on the surface of the marble. You increase the rate of the reaction.

It's a bit like dancing at a club. When a popular record is played, lots more people crowd on to the dance floor. The concentration of people on the dance floor increases. There is now a lot more chance of bumping into another dancer!

Gas reactions

Look at the syringes opposite:
If the end is sealed, how can you increase the pressure of the gases inside the syringe?
By pressing the plunger in, you now have the same number of gas particles in a smaller volume. In other words, you have increased the concentration of the gas.

Therefore:

> **In reactions between gases, increasing the pressure, increases the rate of reaction.**

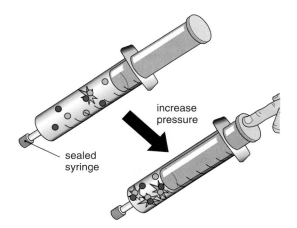

increase pressure

sealed syringe

There are more collisions in a given time when you increase the pressure of a gas

▷ Effect of temperature

Why do we keep our food in a fridge?
Some of the substances in food react with
oxygen in the air.

Have you ever tasted milk that has gone off?
If you have, you will know the sour taste of acids!
Oils and fats in many foods turn 'rancid'
when left in air. They react and turn into acids.

The **low temperature** in your fridge **slows down**
the reactions that make food go off.

Let's look at an experiment to measure the effect of
temperature on rate of reaction:

*What happens to the rate of chemical reactions
inside a fridge?*

Experiment 16.7   The effect of temperature on rate of reaction

Mix equal volumes of sodium thiosulfate solution and
dilute hydrochloric acid in a flask.
* What do you see?

The solution goes cloudy because we get
a precipitate of sulfur as the solutions react.
We can time how quickly the solution gets cloudy
to measure the rate of reaction.
The diagram opposite shows you how to do this:

Warm 50 cm³ of sodium thiosulfate solution to
one of the temperatures in the table below.
Then place the flask on your cross.
Add 5 cm³ of hydrochloric acid and swirl.
Time how long it takes for the cross to disappear.

⚠️ acid
sulfur dioxide

add dilute acid
and start timing

sodium
thiosulfate
solution

a cross
drawn
on paper

*Time how long it takes for the cross to
disappear*

Record your results in a table like this:

Temp. (°C)	Time for cross to disappear (s)
20	
30	
40	
50	

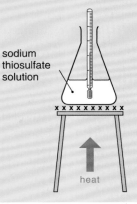

sodium
thiosulfate
solution

heat

* How did you make this a fair test?
* At which temperature did the cross disappear most quickly?
* What is the pattern in your results?

From this experiment we find that:

> **As we increase the temperature, the rate of reaction increases.**

Explaining the effect of temperature

What happens to the way particles move
when you heat them up?
The particles have more energy. They move around
more quickly.
As they travel faster, there are more collisions
in a certain time. Therefore, reactions get faster
as we raise the temperature.

But there is another reason why the rate increases.

Some colliding particles just bounce off each other.
They don't bang together hard enough to
start a reaction. They don't have enough energy.

However, at higher temperatures, the particles
are moving faster. They crash together harder.
Therefore, more collisions produce a reaction.

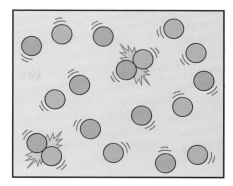
Reaction at 30 °C

> **So, raising the temperature:**
> **1. makes particles collide more often in a certain time, and**
> **2. makes it more likely that collisions result in a reaction.**

Because there are *more, effective collisions*
temperature has a large effect on rates of reaction.
If you raise the temperature by 10 °C, you
roughly double the rate of many reactions!

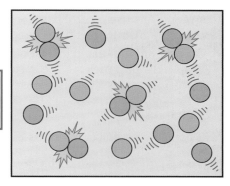
Reaction at 40 °C

Example

Here are some typical results:
Everything was kept the same in each experiment,
except the temperature.

Look at the graph:

Notice that you don't get any more of the product
at a higher temperature.
You get the **same amount**, but **quicker**.

- Is the reaction faster at 30 °C or 40 °C?

- How can you tell which reaction is faster?

- If you did the same experiment at 50 °C,
  what would the line on the graph look like?

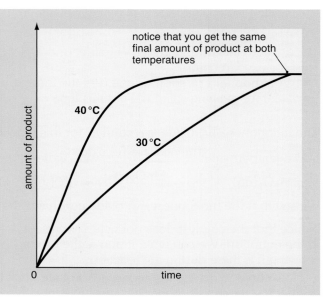

▷ Catalysts

You have already met catalysts on pages 47 and 145.
Can you remember which type of metals (and their compounds) make good catalysts? We use catalysts to reduce costs in the chemical industry.

> **A catalyst is a substance which speeds up a chemical reaction. At the end of the reaction, the catalyst is chemically unchanged.**

So we can get more product in a certain time using a suitable catalyst.
Have you ever seen a damaged car that has been repaired with filler?
Car repair kits use a catalyst to harden the filler quickly.
Look at the instructions from a repair kit:

- Do you need to use a lot of catalyst?
- Why must you prepare the surface **before** you mix the filler and catalyst?
- Why should you only mix the amount that you plan to use?

You can see a catalyst in action in the next experiment:

The filler sets in a few minutes once the catalyst is mixed in

INSTRUCTIONS FOR USE:
1. Prepare the surface to be repaired. Make sure it is clean and dry.
2. Mix the filler and catayst as shown:

filler
catalyst

3. Apply the filler immediately.

Experiment 16.8   Breaking down hydrogen peroxide

Pour some hydrogen peroxide solution into a boiling tube.
- What do you see around the inside of the tube?
- Are the bubbles forming quickly?

The solution gives off oxygen gas as it breaks down.
Try testing for oxygen gas with a glowing splint.
What happens?

Now add a little manganese(IV) oxide, and test for oxygen again.
- What happens as soon as you add the manganese(IV) oxide?
- Did your glowing splint re-light this time?

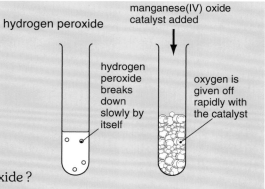

hydrogen peroxide

manganese(IV) oxide catalyst added

hydrogen peroxide breaks down slowly by itself

oxygen is given off rapidly with the catalyst

The manganese(IV) oxide is a catalyst for this reaction:

hydrogen peroxide $\xrightarrow{\text{manganese(IV) oxide}}$ water + oxygen

$$2\,H_2O_2(aq) \xrightarrow{\quad MnO_2(s)\quad} 2\,H_2O(l) + O_2(g)$$

We say that the manganese(IV) oxide *catalyses* the reaction.

Notice that the catalyst does not actually appear in the equation. We can write it above the arrow, showing it is there during the reaction.

- How can you get the manganese(IV) oxide back to use again after the reaction? (**Hint**: It is insoluble in water.)

212

How catalysts work

Do you remember the energy level diagrams from page 196? They show us the energy change when substances react together.

But before reactants can turn into products, they need enough energy to start off the reaction. On page 211 we said that sometimes particles with low energy can collide, but not react.

It's like lighting a gas cooker. We need to supply energy (from a spark or a match) before the gas starts to burn.

> **The minimum energy needed to start a reaction is called its activation energy.**

Look at the energy level diagram:
Once an exothermic reaction starts, it provides the energy itself to keep the reaction going.

> **A catalyst lowers the activation energy.**

Catalysts make it easier for particles to react. You can think of it like a high-jump competition. If you lower the bar, a lot more people can jump over it.
With a catalyst, a lot more particles have enough energy to react.

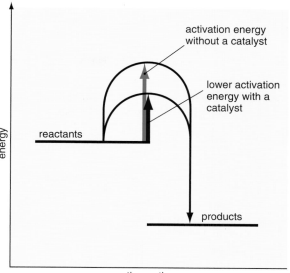

reaction pathway

*Just use a twist
Of your catalyst
And reactions will shoot
Through an easier route!*

Investigation 16.9   Catalysts

Here are two problems to investigate:

1. Which metal oxides catalyse the breakdown of hydrogen peroxide?
2. Does the amount of catalyst affect the rate at which hydrogen peroxide breaks down?

- What do you predict will happen?
- How will you make it a fair and safe test?
- How will you judge how quickly the hydrogen peroxide breaks down? Check your plan with your teacher before you start.
- Was your prediction correct?

▷ Enzymes

Have you ever seen adverts on TV for
'biological' washing powders? (See page 216.)
These contain **enzymes** which break down stains.

Enzymes are *biological catalysts*.
An enzyme is a large protein molecule.

Enzymes help reactions take place at the
quite low temperatures inside living things.

All plants and animals, including ourselves, depend on
enzymes to stay alive. Each enzyme catalyses
a particular chemical reaction.
For example, an enzyme called amylase is in our saliva.
It starts to break down starchy foods in our mouths.

On page 212 we saw the breakdown of hydrogen peroxide
using a catalyst. Hydrogen peroxide is a poison.
It can build up inside living things.
However, we have an enzyme in many of our cells
which can break down the poison.

You can watch the reaction in the next experiment:

Enzymes help to break down our food

Experiment 16.10   Breaking down hydrogen peroxide using an enzyme

Collect 25 cm³ hydrogen peroxide solution in a conical flask.
Add a small piece of fresh liver.

- What happens?

Test the gas given off with a glowing splint. ⚠ hydrogen peroxide

- What happens? Which gas is given off?

Some plants can also break down hydrogen peroxide.

Chop up a piece of potato or celery into small bits.
Add them to 25 cm³ hydrogen peroxide solution.

- What happens?
- Is the hydrogen peroxide broken down as quickly this time?

glowing splint

liver

hydrogen peroxide

An enzyme in the liver can break down hydrogen peroxide
very quickly. Like all enzymes, it is very efficient.
Each enzyme molecule can deal with thousands of
reacting molecules every second!

$$\text{hydrogen peroxide} \xrightarrow[\text{called catalase}]{\text{enzyme in liver}} \text{water} + \text{oxygen}$$

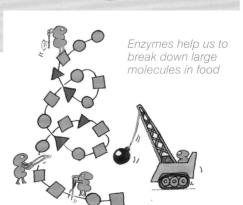

*Enzymes help us to
break down large
molecules in food*

How enzymes work

Enzymes are large protein molecules.
Each different enzyme has its own *special shape*.
The reactants fit into the enzyme, like a lock and key.
The reactant slots into the enzyme at its **active site**.
Look at the diagram below :

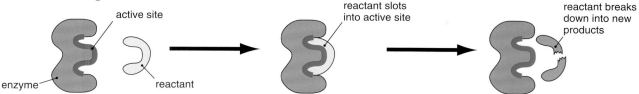

Other enzymes can build up big molecules from small ones.

Effect of temperature on enzymes

Look at this box of biological washing powder :
At which temperature does it work well ?
The enzymes in the powder don't work if you
use water that is too hot.
Try this experiment to see how temperature affects enzymes :

Experiment 16.11   Heating enzymes
You can compare how well the enzymes in ⚠ hydrogen peroxide
fresh liver and boiled liver work.
See how quickly they break down hydrogen peroxide.
Repeat Experiment 16.10. Then try it again using
liver you have boiled in water for 5 minutes.
● How did you make it a fair test ?
● What difference do you notice between fresh liver and boiled liver ?

Enzymes don't work at high temperatures.

As the temperature rises, what do you think happens
to the way an enzyme molecule shakes about ?

If the enzyme gets too hot, it changes shape.
So the reactants will no longer fit snugly into
the enzyme's active site. It can't catalyse the reaction.
We say that the enzyme is **denatured**.

Look at the graph opposite :
● What do we call the temperature at which an enzyme
  works best ?
● What is your body temperature ? Why do you think that
  most enzymes work best at around 40 °C ?

You can find out how we use enzymes on the next 2 pages.

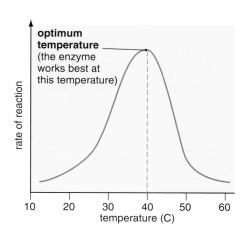

Changes in pH also denature enzymes

▷ Chemistry at work : Uses of enzymes

Yoghurt

For thousands of years, people have made **cheese** and **yoghurt** from milk. However, they didn't know about enzymes and how they helped!

Milk contains a sugar called *lactose*.
Natural bacteria in the milk turn this into *lactic acid*.
It is the first step in making cheese or yoghurt.
The lactic acid makes the milk taste sour and it helps to preserve it.

Nowadays cheese and yoghurt makers use pasteurised milk. This has been heated to kill any bacteria. So they add their own bacteria to start off the process.

Enzymes in the bacteria break down lactose into two simpler sugars. Then other enzymes convert the sugars into lactic acid.

It is thought that yoghurt was first made in the Middle East

Cheese

After the milk has gone sour, cheese makers add rennet. We get this from the stomachs of calves. The enzymes in it clot their mothers' milk. But now you can buy 'vegetarian' cheese. The enzymes used for clotting this cheese come from fungi (yeast).

The soggy lumps formed contain **curds** (the solid) and **whey** (the liquid). The curds are pressed into blocks and left to mature. In this final stage, other enzymes work on proteins and fats to give the cheese its taste and texture.
Cheeses, like cheddar, are left up to 9 months to mature. However, new enzymes can do the job in half the time.

The holes in this cheese were made by bacteria giving off carbon dioxide

Biological washing powders

Biological washing powders contain enzymes to break down stains. You can save energy by operating your washing machine at lower temperatures. They use **proteases** to breakdown proteins and **lipases** to attack fats. They also have **cellulases** to soften fabrics by breaking down 'bobbly' bits that can form. These washing powders also contain :
- detergents (see page 178)
- water softeners (see pages 292–3)
- bleach (to remove coloured stains)
- optical brighteners (to give that 'whiter-than-white' look).

▷ Chemistry at work : Uses of enzymes

More uses of enzymes

We have already seen how enzymes are used
in the brewing, baking and dairy industries. (See pages
166–7, and 216.) But enzymes also have many other uses.

An early example is in **making leather** from animal hides.
In olden times, hides were softened by smearing them
with dog and pigeon faeces (poo!). Fortunately,
enzymes came to the rescue at the start of the last century.
A German chemist, Otto Rohm, found out
why dog faeces worked. It was enzymes called **proteases**
that were breaking down proteins. These could be extracted
from the pancreas of a cow or pig. A lot less smelly !

It was from the 1950s that enzyme technology really started.
There are great benefits in using enzymes as catalysts.
They are over 10 000 times more efficient than other
catalysts used in industry. One enzyme molecule can
catalyse 10 million reactions in a single second !
They also work at lower temperatures, saving energy.

Do you like those chocolates with the soft, gooey centres ?
Have you ever wondered how they are made ?
The chocolate is poured over a solid mixture which contains
sucrose (the sugar we add to tea or coffee) and an enzyme.
The chocolate sets and then the enzyme breaks down
the sucrose into glucose and fructose.
These smaller sugars are much more soluble than sucrose
and dissolve in the small amount of water in the original
mixture.
Look at the table opposite :

Industry	Enzymes used to ...
medical	treat cancer / make penicillin
confectionary/ sweets	break down starch syrup into glucose syrup (**carbohydrases** are used) / change glucose into fructose which is sweeter so less is needed and is used in slimming foods (**isomerase**) / break down sucrose into glucose and fructose / make artificial sweeteners
meat	make the meat tender
baby food	start off digestion of food (**proteases** and **lipases**)

We looked at the good points about using enzymes,
but scientists have had to overcome some problems.
For example, if you want to catalyse one particular reaction
you need a **pure enzyme** (not the mixture found in cells).
This can be tricky, which makes pure enzymes expensive.

You also need to use expensive enzymes over and over again.
But it is hard to remove enzymes from liquid products.
So the enzymes can be stuck to plastic beads.
We say that they are **immobilised**.
They do not get washed out with the product,
allowing a continuous process.

Biotechnologists have also found other ways
to immobilise enzymes. For example, they can be trapped
in tiny pores inside inert structures or in beads of
calcium alginate.

Investigation 16.12 Magnesium and acid

Magnesium ribbon reacts with dilute acid,
giving off hydrogen gas. ⚠ acid

Use your ideas from this chapter to investigate
the factors which affect the rate of this reaction.

- Choose which variable you will investigate.
- Predict and explain what you think will happen.
- Plan a fair and safe test to collect reliable data.
Show your plan to your teacher before you start.

Summary

Chemical reactions are speeded up by increasing :

- **surface area**
- **concentration (or pressure if gases are reacting)**
- **temperature**.

The **collision theory** explains why these factors
affect the rate or speed of a reaction. When particles
collide more often in a certain time, reactions speed up.
With higher temperatures, the collisions are also harder.
This means that more collisions produce a reaction.

Some reactions are also speeded up by a **catalyst**.
The catalyst itself is not chemically changed
at the end of the reaction.

- Enzymes are biological catalysts. They are protein molecules.

To make my potions faster – I grind up
snail shells, add more toad legs to
the same amount of water, add
a touch of my secret catalyst –
and boil the brew!

POTIONS
WHILE
U-WAIT

WITCHES'
BREW

▷ Questions

1. Copy and complete :
   a) Small pieces of solid, especially powders,
      have a . . . . surface area.
      The . . . . the surface area, the faster the
      reaction.
   b) The more concentrated a solution is, the . . . .
      it reacts. This is because there are more
      particles in the same . . . . . Therefore, the
      particles . . . . more often in a certain time.
   c) The . . . . the temperature, the faster the
      reaction. Hot particles have more energy,
      so they move around more . . . . . This
      means that they collide . . . . often. The
      collisions are also . . . . and more effective.
   d) A . . . . is a substance which speeds up a
      reaction, but is chemically . . . . itself at the
      end of the reaction.
   e) An . . . . is a catalyst found in living things.

2. *Explain* how the witch in the cartoon above
   makes her potions so quickly !

3. Imagine that the pupils in a school
   playground are reacting particles. A reaction
   happens each time the pupils bump into each
   other hard enough to say 'Ow !'.
   a) What happens to the pupils as the
      temperature goes up ?
   b) Explain how a) affects the rate of their
      reaction.
   c) Think of 2 ways to increase the
      concentration of pupils in the school
      playground.
   d) Explain how c) affects the rate of their
      reaction.
   e) Think up your own model, using pupils,
      that could help to explain the effect of
      surface area on rate of reaction.

4. Hassan and Zoe want to measure the rate of reaction between zinc and dilute sulfuric acid. They set up their apparatus as shown below:

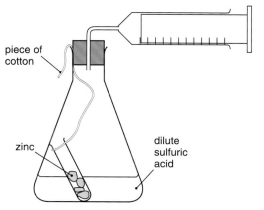

piece of cotton

zinc

dilute sulfuric acid

a) Why is the zinc in a test tube?
b) How could they start the reaction?

Zoe read the volume of gas collected every minute. Hassan recorded the results, as shown below:

Time (mins)	0	1	2	3	4	5	6	7	8
Volume of gas (cm³)	0	15	24	32	33	39	40	40	40

c) Plot Hassan's results on a graph. Put a circle around the point on your graph which seems to be a mistake. What do we call this type of result? (See page 9.) Now draw a 'line of best fit'.
d) During which minute was the reaction fastest?
e) How long did the reaction take to finish?
f) Name the gas given off in the experiment.
g) If you add a little copper sulfate, the gas is given off more quickly.
Draw a dotted line on your graph to show what would happen.

5. a) What is a **catalyst**?
b) Hydrogen peroxide decomposes to form water and oxygen gas. The reaction is catalysed by some transition metal oxides. You are given oxides of copper, manganese and nickel. Describe how you can test which is the best catalyst.
c) How can you get the metal oxides back to use again when the reaction has finished?

6. A group of students are studying the effect of surface area on rate of reaction.
They use marble chips (calcium carbonate) reacting with dilute hydrochloric acid.

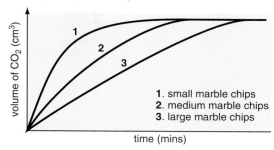

volume of CO_2 (cm³)

1
2
3

1. small marble chips
2. medium marble chips
3. large marble chips

time (mins)

a) What did the students investigating the reaction have to do to make it a **fair test**?
b) Which size of marble chips has the largest surface area (given the same mass of each)?
c) Which marble chips reacted fastest? How can you tell?
d) The students doing the experiments also tried reacting the same mass of **powdered** calcium carbonate with the acid. What would their results look like on the graph above?

7. The reaction in question 6 can also be followed by recording the mass of a flask containing the reactants.
a) Why is there a gradual loss in mass?
b) Draw the apparatus you would use to do this experiment. Why do you use cotton wool in the mouth of the flask?
Here are some results obtained at 20 °C:

Time (mins)	Mass of flask and its contents (g)
0	80.00
1	78.50
2	77.50
3	76.95
4	76.60
5	76.41
6	76.33
7	76.30
8	76.30

c) Draw a graph of mass against time.
d) Draw a dotted line on your graph showing the reaction at 30 °C.

Further questions on pages 244 and 245.

chapter 17
Reversible  Reactions

Have you ever seen a toothbrush that
changes colour as you use it?
The 'smart' plastic contains substances which change
colour as they get warm. (See page 334.)
What happens when the toothbrush cools down again?
Look at the picture opposite:
- Why do you think that the brush changes
  colour at both ends?

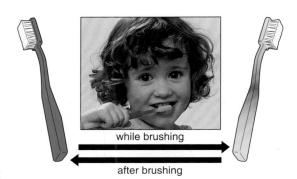

while brushing

after brushing

> **Changes which can go backwards or
> forwards are called reversible.**

Some chemical reactions are reversible.
The products can react to give us back the
reactants that we started with. Let's look at an example:

Experiment 17.1   *Changing copper sulfate*
Gently heat some blue (hydrated) copper sulfate crystals
as shown:
Stop when the blue crystals have changed.

- What happens? What do you notice at
  the mouth of the test-tube?

⚠️ copper sulfate

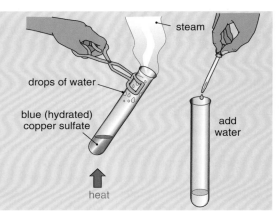

steam

drops of water

blue (hydrated)
copper sulfate

add
water

heat

Let the test-tube cool down.
Now add a few drops of water from a dropper.
- What happens?
- How can you get the white powder back again?

When you heat blue (hydrated) copper sulfate, it loses
the water which is bonded in its crystals.
This reaction is called **dehydration**. It is endothermic.
You get a white powder called anhydrous copper sulfate.
If you add water, the blue (hydrated) copper sulfate
is formed again. This is called **hydration**. It is exothermic.

Reversible reactions are shown by arrows pointing in both directions:

hydrated copper sulfate $\rightleftharpoons$ anhydrous copper sulfate + water
    (blue crystals)   (white powder)

$$CuSO_4.5\,H_2O \quad \rightleftharpoons \quad CuSO_4 \quad + \; 5\,H_2O$$

> **The energy absorbed in
> the forward reaction is
> the same as the energy
> given out in the reverse
> reaction.**

You can use the reverse reaction to test for water.
Look at the rule for energy changes in reversible reactions
in the box opposite:
If the forward reaction is exothermic, the reverse reaction will absorb
an equal amount of energy.

More reversible changes

Indicators changing colour are also examples of
reversible changes.
They are one colour in acid, and another colour in alkali.
Can you remember the colour of litmus
in acid and in alkali?

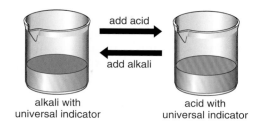

alkali with
universal indicator

acid with
universal indicator

Look back to Experiment 10.1 on page 125.
You were trying to neutralise an acidic solution
with alkali. Which indicator did you use?
Why don't you need to start all over again
if you add too much acid?

Experiment 17.2   'Changing colour'
Phenolphthalein indicator is colourless in acid,
and pink-purple in alkali.

Add some phenolphthalein to a beaker of water.
Rinse another beaker with sodium hydroxide solution.
The second beaker looks empty, but it now has
drops of alkali in it.
Pour the water from the first beaker into it.

• What happens? Why?

• A third beaker has a little dilute acid in it.
Pour your solution into this beaker.

• What happens? Why?

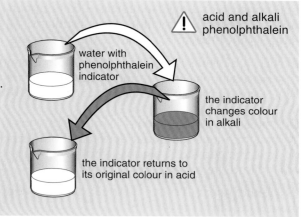

acid and alkali
phenolphthalein

water with
phenolphthalein
indicator

the indicator
changes colour
in alkali

the indicator returns to
its original colour in acid

Here is another example of a reversible reaction:
When ammonia and hydrogen chloride gases meet
they react. They form a white cloud of ammonium chloride.

You can reverse this reaction in the next experiment:

Experiment 17.3   Heating ammonium chloride – a reversible reaction
Gently warm some ammonium chloride in a boiling tube.

• What do you see on the cool part of the tube, near the top?
• How can you explain what happens in the tube?

loose plug of
mineral wool

ammonium
chloride

ammonium chloride   heat

Ammonium chloride solid is broken down by heat, forming
ammonia and hydrogen chloride gases.
At lower temperatures, the ammonia and hydrogen chloride
react to re-form the ammonium chloride.

ammonium chloride $\rightleftharpoons$ ammonia + hydrogen chloride

$NH_4Cl(s)$ $\rightleftharpoons$ $NH_3(g)$ + $HCl(g)$

▷ Dynamic equilibrium

Many reactions are 'one-way only'.
The starting materials react to form the products.
The products can't easily be turned back
into the starting materials again.
Imagine trying to turn a cake back into
flour, sugar, butter and eggs!

However, sometimes products can re-form reactants.
You have seen some reversible reactions
on the last 2 pages.

If you don't let any reactants or products escape,
both forward and reverse reactions can happen
at the same time. Reactants make products,
and, at the same time, products make reactants.

In such a 'closed system', eventually both the forward and
reverse reactions will be going **at the same rate**.
When this happens it is called **dynamic equilibrium**.

Do you know what the word 'dynamic' means?
You might have read about dynamic footballers,
or dynamic pop-stars. People who have lots of
energy and are always on the move!

'Dynamic' refers to movement. In dynamic equilibrium
there is constant movement.
Reactants are changing to products. Products are changing
to reactants. Eventually, we reach a point of equilibrium.

> **At equilibrium, the rate of the forward reaction
> and the reverse reaction is the same.**

- What do you think happens to the amount of reactants and
  products at that point?

When **equilibrium** is reached, the amount of each
substance *stays the same*. There *appears* to be no change
in the closed system.

Look at the cartoon opposite:
It helps to explain dynamic equilibrium.

Both back and forth, and to and fro,
The equilibrium does go.
Forwards and backwards, at the same rate,
The system reaches a steady state!

*If you look from the side, the man stays in the
same position. He is running up at the same rate
as the escalator moves down.*

The position of equilibrium

Dynamic equilibrium is not like a balancing see-saw.
There does not have to be equal amounts of
reactants and products in the equilibrium mixture.

Look at the cartoon opposite :

The man can appear to be still near the top or
near the bottom. He doesn't have to be half-way up
the escalator. As long as he is climbing at the same
rate as the steps are going down, he is in
dynamic equilibrium.

Therefore, the position of equilibrium can lie
in favour of the reactants or the products.
It can lie to the left or to the right, as you look at
the equation.

You have seen a reaction at equilibrium before on
page 93 :

$$\underset{\substack{\text{left-hand} \\ \text{side}}}{H_2O(l)} \rightleftharpoons \underset{\substack{\text{right-hand} \\ \text{side}}}{H^+(aq) + OH^-(aq)}$$

There are only a few H^+ and OH^- ions in water.

- Which side of the equation does the position of
  equilibrium lie ?

Look at the pictures below :
They might help you to understand how the
position of equilibrium can lie to the left or
to the right.

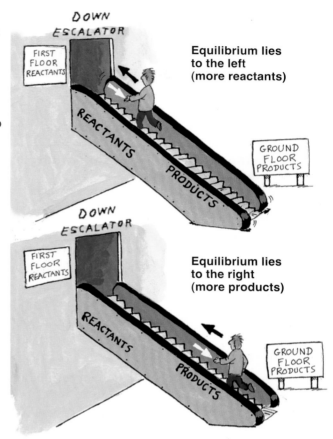

**Equilibrium lies
to the left
(more reactants)**

**Equilibrium lies
to the right
(more products)**

The man running up the 'down-escalator' can appear to be
still on any part of the staircase

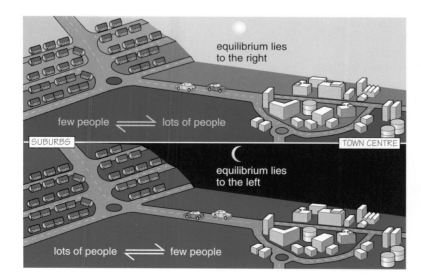

Just after rush hour there are more
people in the town centre than in the
suburbs. Some people go home early
but these are replaced in the town
centre by shoppers.

At night, most people have gone
home. Some night-workers arrive in the
town centre, but these are replaced
in the suburbs by people leaving
work late.

▷ Affecting the position of equilibrium

We can change the position of equilibrium
by altering conditions.
For example, we can change concentrations of reactants
or products. Temperature also affects the
equilibrium mixture.
If the reaction has gases in it, then changing the
pressure might affect the equilibrium as well.

> **The position of equilibrium shifts to try to cancel out any changes you introduce.**

The French chemist, Henri Louis Le Chatelier (1850–1936), did much of the early work on equilibrium mixtures

Changing concentrations

For example, you might increase the concentration
of a reactant. This will make the forward reaction
go faster for a while, until we reach equilibrium again.

In effect, we have added more to the left-hand side,
so the equilibrium shifts to the right,
as if to remove it.

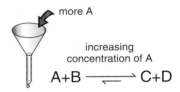

equilibrium moves to right to
reduce the concentration of A

Changing temperature

Can you think what happens if you ***increase*** the
temperature of an equilibrium mixture?
The position of equilibrium shifts to
reduce the temperature.

Look at the reaction below:
It is used in industry to make ammonia (NH_3) in
the Haber process. (See pages 231–3.)

$$N_2(g) \; + \; 3H_2(g) \; \rightleftharpoons \; 2NH_3(g) \quad \Delta H \; = \; -92 \text{ kJ/mol}$$

The forward reaction is exothermic, giving out heat.
The reverse reaction is endothermic. It takes in heat.
So if you increase the temperature, the reverse reaction
is favoured, as this will lower the temperature.
Therefore, at high temperatures, the equilibrium mixture
will have more N_2 and H_2 than before. (See page 233.)

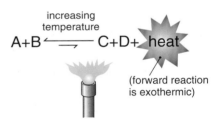

equilibrium moves to left to
get rid of the extra heat

Changing pressure

Changing the pressure can also affect a mixture at equilibrium.
However, there must be different numbers
of gas molecules on either side of the equation.
If you increase the pressure, the equilibrium shifts to
try to reduce it again.
Moving to the side with the least number of gas molecules
will lower the pressure.

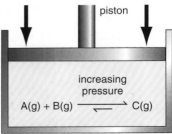

equilibrium moves to the side with
the least number of gas molecules
to reduce the pressure

You can change the position of equilibrium in
the next experiment:

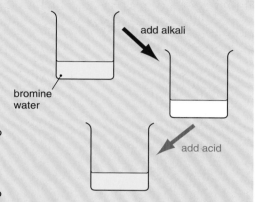

Experiment 17.4   Bromine water
Collect some bromine water as shown:
- What colour is it?
The strength of the yellow colour depends on
how many bromine molecules (Br_2) are in the solution.

Add a little sodium hydroxide solution (alkali).
- What happens to the colour?
- What has happened to the number of bromine molecules?

Now add dilute hydrochloric acid.
- What change do you see now?
- What has happened to the number of bromine molecules?

bromine water
acid and alkali

add alkali

bromine
water

add acid

Bromine water is an equilibrium mixture:

$$Br_2(aq) + H_2O(l) \rightleftharpoons Br^-(aq) + OBr^-(aq) + H^+(aq)$$
yellow   colourless

When you add alkali (OH^-), H^+ ions are removed.
(Remember neutralisation on page 130:
$H^+ + OH^- \longrightarrow H_2O$.)
The equilibrium moves to the right to try to
replace the H^+ ions. This means that more
Br_2 and H_2O must react, and the yellow colour fades.

When you add acid, H^+ ions are added to the mixture.
The equilibrium shifts to the left to get rid of them.
This makes more Br_2 molecules, and the yellow
colour gets deeper again.

"Which way will the balance lie?"
Le Chatelier did cry.
"It will move to cancel out
Whichever change you bring about!"

Now try this experiment:

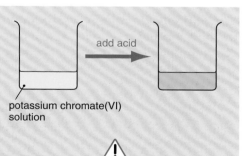

Experiment 17.5   Chromate in equilibrium with dichromate
    chromate ions + H^+ $\rightleftharpoons$ dichromate ions
        yellow   orange

Add a little dilute sulfuric acid to
some potassium chromate(VI) solution.
- Which colour change do you see?
- What has happened to the number of chromate and
  dichromate ions in the beaker?
- How could you get back more chromate ions?
Try adding some sodium hydroxide solution (alkali). What happens?
- Can you explain the changes in this experiment?

add acid

potassium chromate(VI)
solution

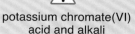

potassium chromate(VI)
acid and alkali

▷ Strong and weak acids

We have looked at acids before in Chapter 10.
You know that all acids contain hydrogen (page 127) and that
H^+(aq) ions are in all acidic solutions (page 130).

However, some acids are better at producing
H^+(aq) ions than others. These are called **strong acids**.

Imagine you start with the same number of acid molecules,
in the same volume of water. A strong acid will give more
H^+(aq) ions than a weak acid.

Hydrochloric acid is a strong acid.
Ethanoic acid is a weak acid.

You will see some differences between strong and weak acids
in the next experiment:

*A car battery contains
sulfuric acid – a strong acid*

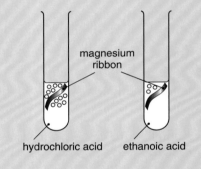

*Oranges
contain citric
acid – a weak
acid*

Experiment 17.6   Testing a strong and weak acid

Try the tests below using solutions of hydrochloric acid and
ethanoic acid.
- Why must both acids have the same concentration?

⚠ acid

Test 1   Add universal indicator solution to both acids.
- What are their pHs?

Test 2   Add the same amount of magnesium ribbon to both acids.
- What difference do you notice?

Test 3   Add a spatula of sodium carbonate to both acids.
- What difference do you notice?

Test 4   Test the acid's conductivity, as in the circuit on page 83.
- What difference do you notice in the brightness of the bulb?

magnesium
ribbon

hydrochloric acid   ethanoic acid

Strong acids split up (**dissociate**) almost completely in water.
For example:

$$HCl(g) \xrightarrow{\text{water}} \underbrace{H^+(aq) + Cl^-(aq)}_{\text{hydrochloric acid}}$$

However, weak acids reach **dynamic equilibrium**.
The acid molecules split up, just like a strong acid.
But at the same time, the H^+ ions and negative ions that are made
join up again. They form the original acid molecules.
For ethanoic acid:

$$CH_3COOH(aq) \rightleftharpoons CH_3COO^-(aq) + H^+(aq)$$

Equilibrium is reached when the molecules split up at the
same rate as the ions join back together again.

The weaker the acid, the further the position of equilibrium lies
to the left. There will be more molecules in the solution, and fewer
H^+(aq) ions – the ions that cause acidic properties.

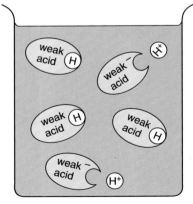

*Partial dissociation of a weak acid.
Solutions of weak acids contain
'undissociated' molecules in equilibrium
with H^+ ions and negative ions. With
fewer ions present, they don't conduct
electricity as well as strong acids.*

▷ Proton donors and acceptors

We can refer to $H^+(aq)$ ions as **protons**. Let's think why.
An atom of hydrogen is made up of one proton ($+$) and one electron ($-$) orbiting it.
When that electron is lost we are left with an H^+ ion — which is a single proton!

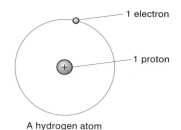
A hydrogen atom

We know that water must be present for an acid to form.
The H^+ ions in an acidic solution bond to neighbouring water molecules. The H^+ ions are said to be **hydrated**.
We show them as $H^+(aq)$ ions.

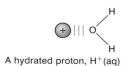
A proton or H^+ ion

> **We can define an acid as a proton donor.**

For example, nitric acid in water:

$$HNO_3(aq) \longrightarrow H^+(aq) + NO_3^-(aq)$$

A hydrated proton, $H^+(aq)$

We often think of bases as the opposites of acids.
So if acids give away protons, bases accept protons.
They can take the H^+ ion from a water molecule.
For example, ammonia is a weak alkali:

$$\underset{\text{ammonia}}{NH_3(aq)} + H_2O(l) \rightleftharpoons NH_4^+(aq) + \underset{\text{hydroxide ion}}{OH^-(aq)}$$

> **We can define a base as a proton acceptor.**

Alkalis are **soluble bases**.
When they dissolve in water, they produce hydroxide ions, $OH^-(aq)$.
We say that ammonia is a weak base or alkali, because in its solution there is an equilibrium mixture. There are more ammonia molecules in its solution than hydroxide ions.
This is like the weak acid we looked at on the previous page.

Acids give protons and bases accept them. This busker is probably a bass guitarist!

Strong bases or alkalis will **dissociate** (split up into ions) almost completely.
Examples of strong alkalis are sodium hydroxide (NaOH) and potassium hydroxide (KOH).

Choosing indicators

We can measure the volumes of acidic and alkaline solutions that react together using **titration**. (See pages 236 and 354–5.)
To signal the end of the reaction, we use an acid–base indicator.
When it changes colour we have just the right amount of acid and alkali for a complete reaction.

For a strong acid reacting with a strong alkali, any indicator will do.
However, not all indicators work well in other combinations.

> For a strong acid plus a weak alkali, use **methyl orange** indicator.
>
> For a weak acid and a strong alkali, use **phenolphthalein indicator**.

▷ Chemistry at work : Developing ideas about acids

Svante Arrhenius was born in Sweden on 19th February, 1859. He was always bright and taught himself to read at the age of three!
He went on to study at the University of Uppsala.
Svante had always done well in his studies, so he was very disappointed when he barely passed his project.

The tutors at Uppsala did not think much of his ideas on the conductivity of solutions.
They just could not believe his idea that molecules could split up in water and form ions (charged particles). Also, he was so young to be suggesting the work of his elders was incorrect!

However, other, younger scientists, could appreciate the logic in Svante's theory. The more scientists he met and chatted to, the more refined his theory became. Then, in the 1890s, the discovery of charged particles within atoms added weight to Svante's arguments.

Svante Arrhenius (1859–1927). His ideas about acids and ions took a long time to be accepted.

He continued to develop his ideas and proposed that acid molecules split up (dissociate) in water. He said they produced hydrogen ions (H^+) and that alkalis gave hydroxide ions (OH^-).
In 1903 his acceptance by the scientific community was sealed when he was awarded the Nobel Prize for Chemistry.

The Arrhenius theory of acids and bases was further developed in 1923. In a weird coincidence, two scientists came up with the same idea in different countries at the same time.

Johannus Brønsted from Denmark, and Thomas Lowry from England, both suggested a revised version of Svante's ideas.

The existing theory worked well most of the time. However, there were some troubling examples that could not be explained. For example, ammonia (NH_3) forms an alkaline solution, but where do the hydroxide ions (OH^-) come from? Ammonia certainly doesn't split up to form them.

Thomas Lowry (1874–1936)

The Brønsted and Lowry definition could explain this. It said that acids split up giving away hydrogen ions and that bases accept hydrogen ions.
So in the case of ammonia:

$$NH_3(aq) + H_2O(l) \rightleftharpoons NH_4^+(aq) + OH^-(aq)$$

This refinement of the original theory was easier for other scientists to accept than a completely new theory. So it soon became part of the body of chemical knowledge.

● Explain the different time scales for the acceptance of the two theories of acids and bases outlined above.

Johannes Brønsted (1879–1947)

Summary

- Many reactions are 'one-way' only.
  Reactants form products. The products formed cannot turn back into the reactants again.
- Some reactions are **reversible**.
  Reactants form products. However, in different conditions, the products can also react together to form the reactants again.
- If nothing is allowed to escape, a reversible reaction can reach a position of **equilibrium**.
  The amount of reactants and products stays the same.
- At this point, the reaction is in **dynamic** equilibrium.
  The forward reaction and the reverse reaction are going at the same rate.
- The position of equilibrium can be altered by changing the conditions. The position of equilibrium always shifts to try to cancel out the change introduced. (Adding a **catalyst** does not affect the composition of an equilibrium mixture. However, it does speed up the rate at which equilibrium is reached. So a catalyst cannot increase yield – you get the same amount of product, but faster!)
- Strong acids and alkalis **dissociate** (split up) into ions almost completely in water.

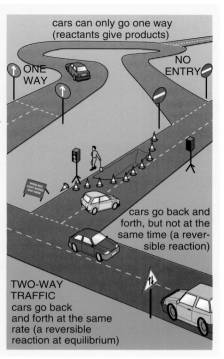

cars can only go one way (reactants give products)

ONE WAY

NO ENTRY

cars go back and forth, but not at the same time (a reversible reaction)

TWO-WAY TRAFFIC
cars go back and forth at the same rate (a reversible reaction at equilibrium)

▷ Questions

1. Copy and complete :
   Reactions which can go forwards or backwards are called . . . . .
   An . . . . mixture contains constant amounts of reactants and products in a closed system.
   When the rate of the forward reaction equals the rate of the reverse reaction, we call it . . . . equilibrium.
   We can change the position of equilibrium by altering concentrations or . . . . . (Pressure affects equilibria if there is an imbalance of . . . . molecules in the equation.)
   A . . . . does not affect the position of equilibrium, but equilibrium is reached more quickly.

2. Hydrated copper sulfate is made up of blue crystals.
   a) Explain what happens if you heat the blue crystals.
   b) How can you get hydrated copper sulfate back again ?
   c) Write an equation showing this reversible reaction.

3. Here is an example of dynamic equilibrium :
$$ICl(l) \; + \; Cl_2(g) \; \rightleftharpoons \; ICl_3(s)$$
brown   yellow/   yellow
liquid   green gas   solid

The equilibrium can be set up in a glass U-tube :

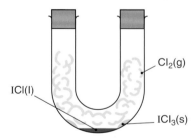

$Cl_2(g)$

$ICl(l)$

$ICl_3(s)$

   a) The equilibrium mixture contains the 3 states of matter. What does this statement mean ?
   b) The pressure inside the U-tube is increased. What happens to the equilibrium mixture? Explain your answer.
   c) What would happen to the equilibrium mixture if the chlorine gas was allowed to escape ? Explain your answer.

Further questions on page 246.

chapter 18

Have you ever seen a farmer or a gardener speading fertiliser on their soil?
Farmers add millions of tonnes of fertilisers to the soil each year.

Most fertilisers, such as ammonium nitrate, contain **nitrogen**. Its chemical formula is NH_4NO_3.
Plants need nitrogen to grow. They use it to make proteins.

Look at the way nitrogen is re-cycled in nature:

nitrogen in the air

decay by de-nitrifying bacteria

a few plants can use nitrogen directly from the air e.g. clover, beans and peas

nitrogen is turned into nitric acid in thunderstorms

The nitrogen cycle

RIP

bacteria

nitrates and ammonium compounds in the soil

taken in through roots

the plants use nitrogen to make protein

plants die

animals excrete animals die

yum

animals eat plants

animals use the nitrogen to make protein

We disturb the **nitrogen cycle** by farming crops.
The crops take nitrogen from the soil as they grow.
Then the farmer harvests the crop. The plants are cut down and taken away. Therefore, most parts are never allowed to rot back into the soil.
Nitrogen is removed, but not replaced.

The natural cycle can replace some nitrogen.
Look at the cycle above:
• How can nitrogen get back into the soil?
These sources cannot supply enough nitrogen for next year's crop. So farmers need fertilisers.

Almost 80% of the air is nitrogen gas. So you might think that plants have plenty of nitrogen available.
However, only a few types of plant can use nitrogen directly from the air. These include peas, beans and clover.

Most plants need nitrogen in a soluble form in the soil.
Then they can absorb it through their roots.

Crops are harvested. They don't have a chance to replace the nitrogen they have taken from the soil by decaying naturally

▷ Fixing nitrogen

We have a very cheap supply of nitrogen – the air.
But how can we turn it into a form that plants can use?

Turning nitrogen from the air into nitrogen compounds
that plants can use is called **'fixing'** nitrogen.
Chemists have found a way to 'fix' nitrogen.

The first step is to change *nitrogen gas into ammonia*.
Look at the next section to find out how this is done.

The Haber process

A German chemist called Fritz Haber discovered
how to make ammonia from nitrogen gas in 1909.

nitrogen + hydrogen   ammonia

$$N_2(g) \; + \; 3\,H_2(g) \qquad 2\,NH_3(g)$$

The same reaction is used today to make
millions of tonnes of ammonia.
You can read more about the Haber process on page 233.
Look at the flow diagram of the process below:

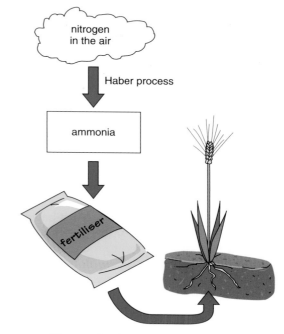

Nitrogen fixation

The Haber process

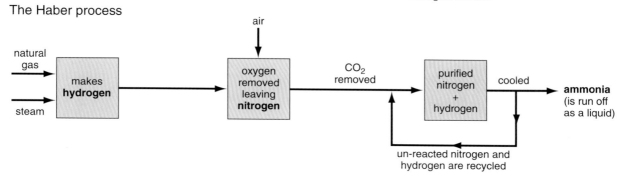

Notice from the equation above that the reaction is reversible.
At the same time as ammonia is made,
it splits up into nitrogen and hydrogen again.
However, any un-reacted nitrogen and hydrogen are
recycled and used again.

Energy from the process is also re-cycled to save money.
Many reactions in the process are exothermic. The energy
they give out is used to heat reaction vessels and
to make steam. As well as making hydrogen,
the steam powers compressors. These produce
the high pressures needed in the process.

As well as saving money, using less energy is also better
for our environment. We use less of the Earth's limited
resources and release less pollution.

Facts about the Haber process
Raw materials
Air (for nitrogen)
Natural gas (to make hydrogen)
Steam (to make hydrogen and to
generate high pressures)
Conditions
Temperature : about 450 °C
Pressure : about 200 atmospheres
Catalyst : mainly iron

▷ Chemistry at work : The Haber process

The Haber process is used in industry to make ammonia (NH_3).

Ammonia is a very important chemical.
It is used to make fertilisers, such as ammonium nitrate. (See page 237.)
Ammonia is also used to make nitric acid which can produce explosives. (See page 235.)

These uses meant that it was important to discover how to make ammonia on a large scale.

The story starts in Germany just before the First World War.
In preparation for war, the Germans realised that they would need to make their own fertilisers and explosives. Both were made from nitrogen compounds imported from South America.
The supplies of nitrogen compounds – from bird droppings in Peru and sodium nitrate from Chile – had nearly run out. When war started, they wouldn't be able to ship them back to Germany anyway.
So scientists raced to find a way to turn nitrogen gas in the air into useful nitrogen compounds.

Fritz Haber was a lecturer in a technical college in Germany.
In 1909 he managed to make an equilibrium mixture containing nitrogen, hydrogen and ammonia.

$$N_2(g) + 3H_2(g) \quad 2NH_3(g)$$

Haber's apparatus is shown opposite :
He had shown that it was possible to turn nitrogen gas into ammonia. However, with this equipment he could only make about 100 g of ammonia!

The German chemical company BASF then invested over a million pounds to 'scale up' the process.
The process needed high pressures. The first reaction vessel blew up under the strain!

It was a chemical engineer named Carl Bosch who eventually solved the problem.
He devised a double-walled steel vessel which could work safely at 300 times atmospheric pressure.

Over 6500 experiments were carried out to find the best catalyst. An **iron catalyst** worked well.
It was improved further by adding traces of other metal oxides.

Fritz Haber (1868–1934) got the Nobel Prize for Chemistry in 1918 for making ammonia from nitrogen and hydrogen. He was forced to leave Germany in 1933 when Hitler came to power. He died soon afterwards a disillusioned man.

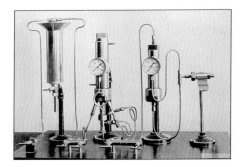

Haber's apparatus

Carl Bosch (1874–1940) was awarded the Nobel Prize for Chemistry in 1931 for his work on high-pressure reactions

▷ Chemistry at work : The Haber process

In 1913, the first ammonia factory was making
about 30 tonnes a day.
Without this supply, the First World War
might have ended much sooner.

However, the Haber process is still used today to make
millions of tonnes of ammonia. About 85 % of this ammonia
goes on to make fertilisers to help feed the world!

This is an example of both the good and bad effects of
the chemical industry.

Conditions chosen for the Haber process

Look at the reaction below :

$$N_2(g) \; + \; 3\,H_2(g) \quad\overset{\text{iron}}{\underset{\text{catalyst}}{\rightleftharpoons}}\quad 2\,NH_3(g) \qquad \Delta H \; = \; -92\,kJ/mol$$

**The catalyst speeds up the rate
at which we reach equilibrium.
However, it does not increase
the yield of ammonia.**

People running the process must choose
the conditions carefully. They want to make
as much ammonia as possible, as quickly as possible.

Look at the graphs opposite :

- What happens to the amount of ammonia in the
  equilibrium mixture when you
  – raise the pressure?
  – raise the temperature?

You can get high yields of ammonia by using high
pressure and a low temperature.

The conditions chosen for the Haber process are

- **a pressure between 150 and 300 atmospheres**
- **a temperature between 400 and 450 °C**.

- Why do you think that we don't use even higher
  pressures?
  Think about the safety problems that Bosch faced.
  What about the costs involved?

- Why aren't lower temperatures used?
  Think about your work on rates of reaction.
  It's no use getting a very high yield of ammonia
  if you have to wait days to get it!

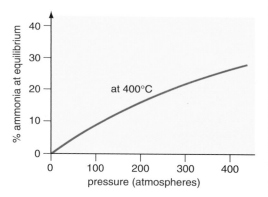

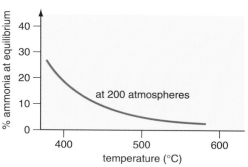

The equilibrium mixture contains only about 15 % ammonia.
It is constantly drawn off by cooling it down into a liquid.

The unused gases are recycled.
You can see a flow chart for the Haber process on page 231.

▷ Ammonia

Remember that we make ammonia in the Haber process.
The ammonia can be used as a fertiliser itself.
However, most is changed into compounds of ammonia.
These compounds have advantages over ammonia
as fertilisers. (See page 236.)

You might have come across ammonia before.
Have you ever smelt a bottle of smelling salts?
If you have, you will certainly remember
the sharp, unpleasant smell of ammonia gas. You might
also have come across it in ammonia-based cleaning
fluids.

In the next two experiments you can make and test
some ammonia gas.

*This farmer is injecting ammonia into a field. Look at
the properties of ammonia on the next page.*

● *What are the disadvantages of using ammonia
directly as a fertiliser?*

Experiment 18.1   Making ammonia in the lab

Set up the apparatus as shown in a fume-cupboard:

Heat the mixture and collect a test tube of
ammonia gas.

You can tell when the tubes are full by holding
a damp piece of red litmus paper near
the mouth of the test tube. The red litmus
changes blue when the test tube is full.

● Which properties of ammonia does this
method of collection depend on?

Put a bung in the test tube when it is full
of ammonia.

● How could you collect a gas that is more dense than air?

You can now test some properties of ammonia gas:

Experiment 18.2   Does ammonia dissolve in water?

Hold a stoppered test tube of ammonia under
some water in a beaker.
Take out the bung, with the mouth of the tube
under the surface.
Shake it gently from side to side.
Compare this to a test tube full of air.

● What happens to the level of the water in the test tube
full of ammonia?
● What does this test tell you about how soluble ammonia
is in water?

Your teacher might show you the Fountain experiment
which shows how soluble ammonia is.

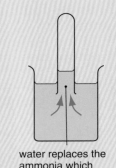

water replaces the
ammonia which
dissolves in the
water

Properties of ammonia

From the last two experiments we can work out
a lot about ammonia gas.
Has it got any colour?
Look at the way we collected it:
Do you think it is more dense or less dense than air?
We used litmus paper to see when the test-tube
was full of ammonia. How could we tell?
What does this tell you about ammonia gas?
Litmus is used as the test for ammonia gas.

> **Ammonia is the only common *alkaline gas*.
> It turns damp red litmus paper blue.**

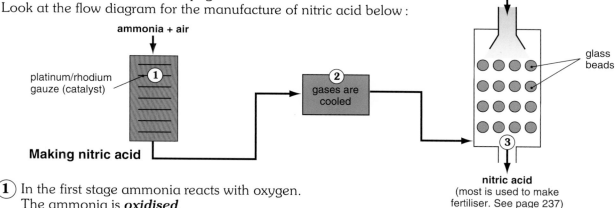

*Can you remember how ammonia reacts with
hydrogen chloride gas? (See page 221.)*

Ammonia into nitric acid

About 10% of the ammonia made in the Haber process
is turned into nitric acid.
You will see how the nitric acid can react with
ammonia to make fertiliser on page 237.
Look at the flow diagram for the manufacture of nitric acid below:

Making nitric acid

ammonia + air

platinum/rhodium gauze (catalyst) — ①

② gases are cooled

water

glass beads

③

nitric acid
(most is used to make
fertiliser. See page 237)

① In the first stage ammonia reacts with oxygen.
The ammonia is *oxidised*.
This reaction will not happen without a catalyst.
Platinum, mixed with rhodium, is used at 900 °C:

$$\text{ammonia} + \text{oxygen} \xrightarrow{\text{platinum/rhodium}} \text{nitrogen monoxide} + \text{water}$$

② Then the nitrogen monoxide is mixed with air.
It gets oxidised to nitrogen dioxide:

nitrogen monoxide + oxygen ⟶ nitrogen dioxide

③ Finally, the nitrogen dioxide and more oxygen react
with water to make **nitric acid, HNO_3**.

nitrogen dioxide + oxygen + water ⟶ nitric acid

• Write the symbol equations for each stage in the process.

▷ Making fertilisers

As you know from page 234, ammonia can be used as a liquid fertiliser itself. However, it has some disadvantages. As a liquid, it has to be injected into the soil. It is much easier to spread solid pellets of fertiliser on a field. Ammonia makes the soil alkaline. You also lose some nitrogen as ammonia gas evaporates from the soil.

We can make solid fertilisers by reacting ammonia (an alkali) with **an acid**. Remember that a salt forms when an acid and an alkali react together. (See page 126.) Can you remember what this type of reaction is called?

You can make your own fertiliser in the next experiment:

Solid fertilisers are made into pellets so that we can spread them easily

---

Experiment 18.3   Making ammonium sulfate fertiliser

Collect $25\,cm^3$ of ammonia solution in a small conical flask.
Use a pipette and filler to measure this accurately.
Add dilute sulfuric acid, $1\,cm^3$ at a time, from a burette.
After adding each cm^3 of acid, swirl your flask.
Dip a glass rod into the solution. Then test a drop of the solution on a piece of blue litmus paper.

⚠️ acid and alkali

Keep adding acid until the litmus just turns pink.
● How much sulfuric acid did you need to neutralise the ammonia solution?

Then pour the solution into an evaporating dish.
Heat it on a water bath until about half of the solution has evaporated off. (Don't let it boil dry.)
Leave the rest of the solution to evaporate off slowly.

● What do your crystals of fertiliser look like?
● What is the fertiliser's chemical name?

burette

dilute sulfuric acid

ammonia solution

Titration using a burette

The sulfuric acid neutralises the ammonia solution.
The salt that we make is called **ammonium sulfate**:

ammonia + sulfuric acid ⟶ ammonium sulfate
$2\,NH_3(aq) + H_2SO_4(aq) \longrightarrow (NH_4)_2SO_4(aq)$

Notice that compounds of ammon**ia** are called ammon**ium** salts.

Investigation 18.4   Testing your own fertiliser

Plan an investigation to see if your ammonium sulfate affects the growth of a seedling.
Ask your teacher to check your plan before you start.

Ammonium nitrate fertiliser

Ammonium nitrate is a very important fertiliser.
Do you know which acid reacts with ammonia
to make ammonium nitrate?

On page 235, we saw how ammonia is turned into
nitric acid. Most of this nitric acid is used to
make ammonium nitrate fertiliser.
Look at the diagram below:

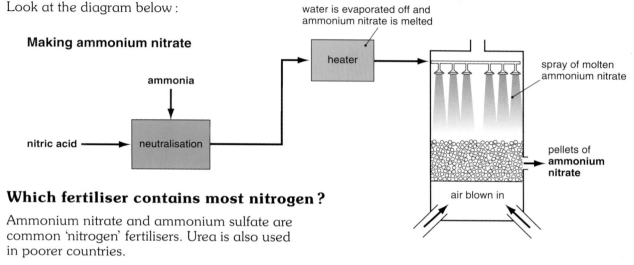

Making ammonium nitrate

Which fertiliser contains most nitrogen?

Ammonium nitrate and ammonium sulfate are
common 'nitrogen' fertilisers. Urea is also used
in poorer countries.

But which of these contains most nitrogen?
Can you remember how to calculate
relative formula masses. (See page 37.)
Look at the examples below:

Ammonium sulfate, $(NH_4)_2SO_4$	Ammonium nitrate, NH_4NO_3	Urea, $CO(NH_2)_2$
$2 \times N = 2 \times 14 = \quad 28$ $8 \times H = 8 \times 1 = \quad 8$ $1 \times S = 1 \times 32 = \quad 32$ $4 \times O = 4 \times 16 = +\ 64$ Formula mass $= \quad 132$	$2 \times N = 2 \times 14 = \quad 28$ $4 \times H = 4 \times 1 = \quad 4$ $3 \times O = 3 \times 16 = +\ 48$ Formula mass $= \quad 80$	$1 \times C = 1 \times 12 = \quad 12$ $1 \times O = 1 \times 16 = \quad 16$ $2 \times N = 2 \times 14 = \quad 28$ $4 \times H = 4 \times 1 = +\ 4$ Formula mass $= \quad 60$
Fraction of N $= \dfrac{28}{132}$	Fraction of N $= \dfrac{28}{80}$	Fraction of N $= \dfrac{28}{60}$
% of N $= \dfrac{28}{132} \times 100 = 21.2\%$	% of N $= \dfrac{28}{80} \times 100 = 35\%$	% of N $= \dfrac{28}{60} \times 100 = 46.7\%$

So urea has the highest percentage of nitrogen.
Urea is found in manure. It is not as soluble as
ammonium fertilisers so it takes longer to release
its nitrogen. On the other hand, it is not
washed out of the soil (**leached**) so easily.
The leaching of fertiliser causes pollution problems. (See page 240.)

▷ Nitrogen, phosphorus and potassium – N, P and K

Plants also need other elements, besides nitrogen, for healthy growth.
Phosphorus (P) and **potassium (K)**, like nitrogen, are often added to the soil in fertilisers.

Look at the diagram of the plants opposite :
Each one has missed out on one of the 3 essential elements.

- What happens if a plant does not have enough :
  – nitrogen ?
  – phosphorus ?
  – potassium ?

Farmers can buy fertilisers which provide nitrogen, phosphorus and potassium. Nobody has ever made a compound of these 3 elements that plants can use. Therefore, the bags of fertiliser contain mixtures of compounds.

You can find out about your soil with a special soil-testing kit. It tells you which element(s) your soil needs.
You can then buy a fertiliser to suit your soil.

Look at the bags of fertiliser below :
The percentage of nitrogen is the first number, followed by phosphorus, then potassium.
These numbers are called the **N : P : K** value of the fertiliser.

nitrogen is needed for the proteins in leaves and stalks

phosphorus speeds up the growth of roots, and the ripening of fruit

potassium protects plants against disease and frost damage. It promotes seed growth

There was a young man from Leeds
Who swallowed a packet of seeds
With N, P and K
And pesticide spray
He's now fully covered in weeds !

A   B   C

- Which fertiliser would you use :
  – if your soil needed topping up with all 3 elements ?
  – if your soil was very low in phosphorus ?

Investigation 18.5   Fertilisers

Plan an investigation to answer **one** of these problems :
- How does the amount of fertiliser affect plant growth ?
- What effect do nitrogen, phosphorus and potassium have on plant growth ?
- Which mixture of nitrogen, phosphorus and potassium is best ?

Make sure that you plan a fair test. Is it safe ?
Check your plan with your teacher before you start.

Here are some compounds that you could use as fertilisers :
- ammonium nitrate, NH_4NO_3
- calcium phosphate, $Ca_3(PO_4)_2$
- potassium chloride, KCl.

Test if any of these affect the pH of the soil.
Will this affect your test ?

▷ A fertiliser factory

There is a large fertiliser factory at Billingham on Teesside. On this page we can see why this is a good place to have the factory.

The factory can make its own ammonia and nitric acid.
These react with each other to make **ammonium nitrate**, NH_4NO_3.
- Which essential element does this give to plants?

There is also a sulfuric acid plant on the site.
The sulfuric acid is used to make phosphoric acid from phosphate rock.
The phosphoric acid reacts with more ammonia to make **ammonium phosphate** fertiliser, $(NH_4)_3PO_4$.

You can see how large the site at Billingham is

Look at the formula of ammonium phosphate above:
- Which *two* essential elements does it provide?

Potassium is put into fertilisers as **potassium chloride**, KCl, or **potassium nitrate**, KNO_3.
- Which potassium compound provides *two* essential elements?

Look at the map:

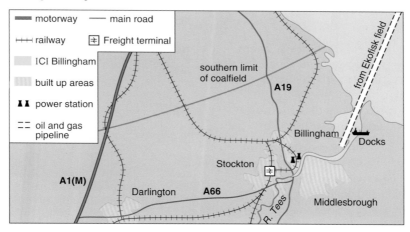

So why build the factory at Billingham? There are large towns nearby that can supply the work-force.
It is near the North Sea. This is important for a supply of natural gas. Remember that natural gas is a raw material for the Haber process. (See page 231.)
- Which river is nearby? Plenty of water is needed to make steam. What is steam used for when making ammonia?

- Is the factory near a port to bring in phosphate rock and sulfur (to make sulfuric acid) from other countries?
The company can also export their fertilisers from a port.
- Why are good road and rail links important?
- Discuss the issues that **automation** (replacing workers with machines) would raise for a company and the local community.
- What would be your concerns if a large fertiliser factory was built near your home?

You need good transport systems to distribute your product

▷ The trouble with fertilisers . . .

As you know from page 230,
most plants take up nutrients through
their roots. So fertilisers have to be
soluble in water.

Sometimes too much fertiliser is added to the soil.
Sometimes it is added at the wrong time
of year. (Why do you think that spring
is the best time?)
This can cause pollution in rivers.

Some fertiliser is washed down through
the soil by rain. It is *leached* out of the soil.
The dissolved fertiliser drains from the fields
into rivers.

Tiny plants, called algae, thrive on the fertiliser.
They start to cover the surface of the water.
This cuts off light to other living things
in the river.

When the algae die, bacteria decompose them.
The bacteria multiply quickly with so much food.
They use up much of the oxygen dissolved
in the water.
This means that fish and other water animals
cannot get enough oxygen. Soon they die.

This chain of events is called **eutrophication**.

Nitrate fertilisers are very soluble. They are
finding their way into our drinking water.
People are starting to worry about the
possible health risks.

In this country, as many as 5 million people
are at times drinking more than the
recommended amount of nitrate.

There is concern about stomach cancer
and 'blue baby' disease (when a new-born baby's
blood is starved of oxygen).

However, others argue that there is no evidence.
Links between the levels of nitrate in our water
and disease have not been proved.
But most people agree that it is wise to limit
the amount of nitrate we drink.

- Why is it difficult to prove beyond doubt that
  nitrates in drinking water cause certain diseases?

Fertiliser is 'leached from the soil'

Algae thrive

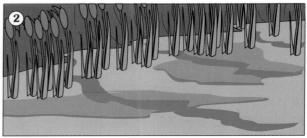

*Bacteria decompose the algae and use up the
dissolved oxygen. Fish die without oxygen.*

*Plants in this polluted river are thriving on fertilisers
leached from fields*

Summary

- Farmers add fertiliser to their soil to provide nutrients. Nitrogen, phosphorus and potassium are all needed for healthy plant growth.

- Nitrogen fertilisers are made from ammonia (NH_3). Ammonia is made in industry by the Haber process :

  nitrogen + hydrogen   ammonia
  $$N_2(g) + 3H_2(g) \quad 2NH_3(g)$$

- Ammonia is an alkali. It reacts with acids, making the salts we use as fertilisers. For example,

  ammonia + nitric acid $\longrightarrow$ ammonium nitrate
  $$NH_3(aq) + HNO_3(aq) \longrightarrow NH_4NO_3(aq)$$

- Fertilisers can be washed out (leached) from the soil. They get into rivers and lakes causing eutrophication. (See page 240.)

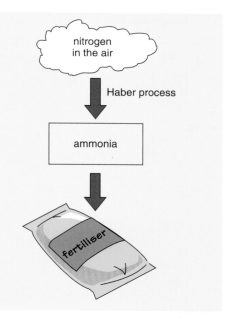

nitrogen in the air

Haber process

ammonia

fertiliser

▷ Questions

1. Copy and complete :
   Farmers add 3 essential nutrients to their soil :
   .... (N), ... (P) and .... (K).
   Ammonia is made by the .... process.
   Nitrogen and .... are reacted together to make the ammonia.
   Ammonia reacts with .... to make salts.
   For example,
   ammonia + ... acid ⟶ ammonium nitrate
   Fertilisers can be washed out, or l_____,
   from the soil. Then they can cause
   e_____ in rivers and lakes.

2. a) Draw a flow diagram of the nitrogen cycle.
   b) Farmers add nitrogen fertilisers to their soil. Explain why they have to do this when there is so much nitrogen in the air.
   c) In olden days farmers used to 'rotate' their crops, as do some organic farmers.
      One of their fields was left without a crop every few years. Clover was grown in the field. This was ploughed into the soil before seeds were planted the following year. Why was this a good idea?

3. Ammonia is made in industry by the Haber process.
   a) Write the word and symbol equation for making ammonia.
   b) What are the raw materials for the nitrogen and hydrogen used in the process?
   c) What is the catalyst used?
   d) What temperature and pressure are chosen for the process?
   e) Look back to page 233.
      Explain why the conditions in d) are used.

4. Describe how we make nitric acid in industry. Include a flow diagram of the process.

5. What factors would you look for in siting a fertiliser factory?

6. Rivers and lakes can be polluted by fertilisers. Design a leaflet for farmers explaining the problem.
   Suggest how they might help reduce the problem.

Further questions on page 247.

▷ Energy changes

1. The apparatus was used to measure the temperature change of 100 g water that resulted from burning fuel.

Results of this experiment are shown in the table.

Mass of burner and fuel/g	before	233.75
	after	232.25
	fuel burned	1.5
Temperature of 100 g water/°C	after	64
	before	31
	temperature rise	33

a) Calculate the energy released by burning the fuel.
(Specific heat capacity of water is 4.20 J/g/°C.) *[3]*
b) Calculate the energy released per gram of fuel. *[2]*
c) This value is less than the theoretical value calculated using bond energies.
Suggest why. *[1]*

2. a) What piece of apparatus is needed to show that a reaction is exothermic? *[1]*
b) How could you show that the reaction between barium hydroxide solution and dilute hydrochloric acid is exothermic? *[4]* (EDEXCEL)

3. Calor gas is a convenient energy source for boating, camping and caravaning.
It is available as butane Calor gas or propane Calor gas.
The following label has been taken from a cylinder containing butane Calor gas, C_4H_{10}.

a) Suggest why these cylinders should **not** be:
  i) stored or used in cellars or basements, *[1]*
  ii) changed near 'naked lights'. *[1]*
b) From your knowledge of the properties of hydrocarbons suggest why propane Calor gas, C_3H_8, is preferred as a fuel in winter rather than butane Calor gas even though it is more expensive and gives out less heat per gram than butane Calor gas. *[2]*
c) The following equation represents the complete combustion of the formula mass in grams of butane in air. The structural formulas of the chemicals involved are shown.

$$2\left(\begin{array}{c} H\ \ H\ \ H\ \ H \\ |\ \ \ |\ \ \ |\ \ \ | \\ H-C-C-C-C-H \\ |\ \ \ |\ \ \ |\ \ \ | \\ H\ \ H\ \ H\ \ H \end{array}\right) + 13(O{=}O)$$

$$\longrightarrow 8(O{=}C{=}O) + 10(H\overset{O}{}H)$$

Bond	Energy (kJ/mol)
C—H	413
O=O	498
C=O	743
O—H	464

Calculate:
  i) the energy needed to break all the bonds in the reactants, *[3]*
  ii) the energy released as new bonds are formed in the products, *[2]*
  iii) the energy change. *[1]* (AQA)

4. The symbol equation below shows the reaction when methane burns in oxygen.

$$CH_4 + 2O_2 \longrightarrow CO_2 + 2H_2O$$

An energy level diagram for this reaction is shown below.

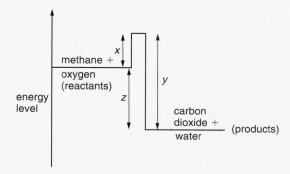

a) Which chemical bonds are broken and which are formed during this reaction? [4]

b) Explain the significance of x, y and z on the energy level diagram in terms of the energy transfers which occur when these chemical bonds are broken and formed. [6]   (AQA)

5. Hydrogen (H_2) and chlorine (Cl_2) react together to make hydrogen chloride (HCl). The reaction is highly exothermic.

The equation for the reaction can be written as:

$$H—H + Cl—Cl \longrightarrow H—Cl + H—Cl$$

a) During the reaction some bonds are made and others are broken.

i) Copy the equation below and put a ring around the bond(s) which is(are) broken. [1]

$$H—H + Cl—Cl \longrightarrow H—Cl + H—Cl$$

ii) Copy the equation below and put a ring around the bond(s) which is(are) made. [1]

$$H—H + Cl—Cl \longrightarrow H—Cl + H—Cl$$

iii) What energy change (transfer) takes place when bonds are broken? [1]

iv) What energy change (transfer) takes place when bonds are made? [1]

v) How does the difference in the energy involved in breaking and making bonds explain why the reaction between hydrogen and chlorine is exothermic? [1]   (OCR)

6.

HYDROGEN
FUEL OF THE FUTURE

It has been suggested that hydrogen could be used as a fuel instead of the fossil fuels that are used at present. The equation below shows how hydrogen burns in air.

$$2H_2 + O_2 \longrightarrow 2H_2O + heat$$

The hydrogen would be made from water using energy obtained from renewable sources such as wind or solar power. The water splitting reaction requires a lot of energy.

a) Hydrogen was successfully used as a fuel for a Soviet airliner in 1988. Why would hydrogen be a good fuel for use in an aeroplane? [2]

b) The water splitting reaction is shown in the equation below.

$$2H_2O \longrightarrow 2H_2 + O_2$$

Use information from the table in question 8 on page 202 to help you answer this question.

i) Calculate the energy needed to split the water molecules in the equation into H and O atoms.

$$2H_2O \longrightarrow 4H + 2O \qquad [2]$$

ii) Calculate the energy change when the H and O atoms join to form H_2 and O_2 molecules.

$$4H + 2O \longrightarrow 2H_2 + O_2 \qquad [2]$$

iii) Is the overall reaction:

$$2H_2O \longrightarrow 2H_2 + O_2$$

exothermic or endothermic? Use your answers to i) and ii) to explain your choice. [4]   (AQA)

▷ Rates of reaction

7. Calcium carbonate reacts with dilute hydrochloric acid as shown in the equation below.

$$CaCO_3(s) + 2HCl(aq) \longrightarrow CaCl_2(aq) + H_2O(l) + CO_2(g)$$

The rate at which this reaction takes place can be studied by measuring the amount of carbon dioxide gas produced.
The graphs below show the results of four experiments, 1 to 4. In each experiment the amount of calcium carbonate, the volume of acid and the concentration of the acid were kept the same but the temperature of the acid was changed each time. The calcium carbonate was in the form of small lumps of marble.

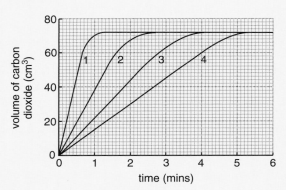

a) Apart from altering the temperature, suggest **two** ways in which the reaction of calcium carbonate and hydrochloric acid could be speeded up. [2]

b) Which graph, 1 to 4, shows the results of the experiment in which the acid had the highest temperature? Explain fully how you know. [2]

c) i) In experiment 2, how does the rate of reaction after one minute compare with the rate of reaction after two minutes? [1]

   ii) Explain, as fully as you can, why the reaction rate changes during experiment 2. [2] (AQA)

8. A student used the following apparatus to carry out an investigation of the reaction between magnesium and dilute hydrochloric acid.

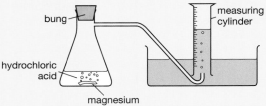

The student put $50\,cm^3$ of dilute hydrochloric acid into the conical flask, added a 3 cm length of magnesium ribbon and inserted the bung. The student recorded the time to produce $50\,cm^3$ of hydrogen gas.
The student repeated the experiment four more times using hydrochloric acid solutions of different concentrations made by using different volumes of the dilute hydrochloric acid and water.

Experiment	Volume of dilute hydrochloric acid (cm³)	Volume of water (cm³)	Time to collect 50 cm³ of gas (s)
1	50	0	20
2	40	10	66
3	30	20	45
4	20	30	60
5	10	40	145

a) One time appears to be inconsistent.
   i) Which time appears to be inconsistent? [1]
   ii) Suggest a time that is consistent with the other results. [1]

b) i) Use the results to state how the rate of reaction changes with decreasing concentration of hydrochloric acid. [2]
   ii) Explain, in terms of particles, how decreasing the concentration of the hydrochloric acid affects the rate of reaction. [3]

c) Suggest another way of measuring the volume of gas produced. [1]

d) Suggest another way to change the rate of a reaction. [1]

e) Copy and balance the equation below, and fill in the missing state symbols.

$$Mg(\quad) + HCl(aq) \longrightarrow MgCl_2(\quad) + H_2(g)$$
[2]   (EDEXCEL)

9. Some students have studied the reaction between some metals and dilute hydrochloric acid. The metals react with the acid until they cannot be seen.
Bubbles of gas are produced during **all** the reactions.
The students know that:

a metal + an acid $\longrightarrow$ a salt + hydrogen

a) Write a word equation for the reaction of zinc with dilute hydrochloric acid. [2]
b) Magnesium reacts with dilute hydrochloric acid to produce magnesium chloride and hydrogen.
Write a balanced symbol equation for this reaction. [2]
c) They carried out an investigation to discover whether the concentration of acid used had any effect on the time taken for the magnesium to react completely. Equal volumes of acid and equal lengths of magnesium ribbon were used in each test. The table shows a set of results.

Concentration of acid/ moles per dm³	Time taken for magnesium to react completely/seconds
2.0	13
1.5	22
1.2	30
0.8	70
0.6	145
0.5	250

i) Plot the points on a graph to show the time taken for the reaction to finish when different concentrations of acid are used (reaction time up the side, concentration of acid along the bottom). [2]
ii) Finish the graph by drawing the best curve. [1]
iii) Use your graph to help you to describe how increasing the concentration of acid affects the rate of reaction. [1]
iv) Use your graph to predict the time taken for the reaction if the concentration of acid was 1.0 moles per dm³. [1]

v) Write down **two** other factors which could change the rate of this reaction. In each case state how the rate would change. [2]
vi) Use the idea of particles to explain the effect of changing concentration on the rate of reaction. [2]
d) One student asked if the class could repeat their experiment using potassium instead of magnesium. The teacher said 'Definitely not!'
Use your knowledge of the reactivity series to explain why the teacher was very wise to say this. [2] (OCR)

10. The table below gives the results from the reaction of excess marble chips (calcium carbonate) with (A) 20 cm³ of dilute hydrochloric acid, (B) 10 cm³ of the dilute hydrochloric acid + 10 cm³ of water.

	Total mass of carbon dioxide produced in grams	
Time in minutes	(A) 20 cm³ dilute HCl	(B) 10 cm³ of dilute HCl + 10 cm³ of H₂O
0	0.00	0.00
1	0.54	0.27
2	0.71	0.35
3	0.78	0.38
4	0.80	0.40
5	0.80	0.40

a) Plot the two curves on a graph (time along the bottom). Label the first curve (A) and the second curve (B). [3]
b) **Sketch, on the same grid**, the curve you might have expected for reaction (A) if it had been carried out at a higher temperature. Label this curve (C). [1]
c) **Sketch, on the same grid**, the curve you might have expected for reaction (B) if the marble chips had been ground to a powder. Label this curve (D). [1]
d) Explain your answer to part c). [1]
(WJEC)

▷ Reversible reactions

11. Ethanol (C_2H_6O) is manufactured from ethene (C_2H_4) and steam (H_2O).
The equation for the reaction is:
$$C_2H_4 + H_2O \rightleftharpoons C_2H_6O$$

a) The sign $\rightleftharpoons$ indicates that this is a reversible reaction.
What is meant by a reversible reaction? [2]

b) The graph shows how the conversion of ethene into ethanol, at three different temperatures, will change as the pressure changes.

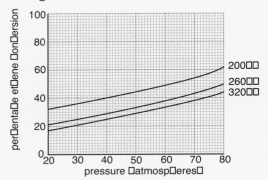

i) How does the conversion of ethene into ethanol change as the pressure increases? [1]

ii) How does the conversion of ethene into ethanol change as the temperature increases? [1]

iii) What is the percentage conversion of ethene at 260°C and 55 atmospheres pressure? [1]

iv) A catalyst is used in this reaction. Why is a catalyst used? [1]
What effect does a catalyst have on the percentage conversion of ethene?
[1]   (OCR)

12. In the manufacture of nitric acid, ammonia is oxidised to nitrogen monoxide.
$$4NH_3(g) + 5O_2(g) \rightleftharpoons 4NO(g) + 6H_2O(g)$$
The reaction is exothermic.
In industry a mixture of dry ammonia and air at about 7 atmospheres pressure is passed over a platinum catalyst at 900°C. Explain how changing the conditions used could affect the **yield** of nitrogen monoxide.
[6]   (AQA)

13. In the Contact process for the manufacture of sulfuric acid, sulfur dioxide is oxidised to produce sulfur trioxide.
$$2SO_2 + O_2 \rightleftharpoons 2SO_3$$
This equilibrium reaction is exothermic from left to right.
State and explain the effect of each of the following conditions on the rate and yield of this reaction.
a) increase in temperature   [3]
b) increase in pressure   [2]
c) use of a catalyst.   [3]

14. The monomer chloroethene is made from ethene in a two-stage process.
The first stage is to convert ethene to 1,2-dichloroethane.
$$2C_2H_4(g) + 4HCl(g) + O_2(g) \rightleftharpoons$$
ethene
$$2C_2H_4Cl_2(g) + 2H_2O(g)$$
1,2-dichloroethane
State and explain the effect of increasing the pressure on:
a) the yield of 1,2-dichloroethane,   [2]
b) the rate of reaction.   [2]   (AQA)

15. A student heated some blue copper sulfate crystals. The crystals turned into white copper sulfate.

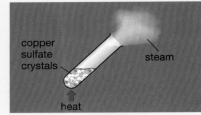

a) The blue copper sulfate had to be heated continuously to change it into white copper sulfate. State whether the reaction was exothermic or endothermic. Explain your answer.   [1]

b) The word equation for this reaction is shown below:

hydrated [+ heat energy] $\rightleftharpoons$ anhydrous + water
copper sulfate   copper sulfate
(blue)   (white)

i) What does the symbol $\rightleftharpoons$ tell you about this reaction?   [1]

ii) How could the student turn the white powder back to blue?   [1]   (AQA)

▷ Products from air

16. Ammonia is manufactured by the Haber process, in which nitrogen and hydrogen react according to the equation shown below.

$$N_2 + 3H_2 \rightleftharpoons 2NH_3$$

a) What is meant by the symbol $\rightleftharpoons$? *[1]*

b) Name **one** compound manufactured on a large scale from ammonia. *[1]*

c) Why is iron used in the Haber process and what effect does it have ? *[2]*

d) The table below shows the percentage of ammonia present at equilibrium when nitrogen and hydrogen react at different temperatures and different pressures.

Pressure (atmosphere)	Ammonia present at equilibrium (%)				
	Temperature (°C)				
	100	200	300	400	500
10	88.2	50.7	14.7	3.9	1.2
25	91.7	63.6	27.4	8.7	2.9
50	94.5	74.0	39.5	15.3	5.6
100	96.7	81.7	52.5	25.2	10.6
200	98.4	89.0	66.7	38.8	18.3
400	99.4	94.6	79.7	55.4	31.9
1000	99.9	98.3	92.6	79.8	57.5

i) What is meant by the statement that a reaction has reached equilibrium ? *[1]*

ii) What will be present in the equilibrium mixture with the ammonia? *[1]*

iii) Use the table to say how the **percentage of ammonia** present at equilibrium is affected by : increasing the temperature ; increasing the pressure. *[2]*

e) How will the **rate of reaction** of nitrogen with hydrogen be affected by increasing the temperature ? *[1]*

f) The maximum yield of ammonia shown in the table occurs at 1000 atmospheres pressure and 100°C. Why are these conditions **not** used commercially ? *[2]*

(EDEXCEL)

17. Ammonia is made from nitrogen and hydrogen in the Haber process.

$$N_2(g) + 3H_2(g) \rightleftharpoons 2NH_3(g)(+ heat)$$

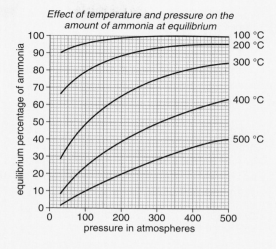

Effect of temperature and pressure on the amount of ammonia at equilibrium

The Haber process uses a temperature of 450°C and a pressure of 200 atmospheres. Explain why these conditions are chosen.

[3]   (AQA)

18. Read this article about coffee production in Costa Rica, a country in Central America. Answer the questions that follow.

Coffee is one of the most valuable crops of Costa Rica. Coffee production accounts for nearly a fifth of the country's exports. The coffee plantations are found mainly in the Central Valley of Costa Rica. Large amounts of nitrate fertilisers are used throughout the year in the production of the coffee.

Scientists have shown that coffee plants take up about 40% of the nitrate fertiliser during the main growing season. Towards the end of the growing season the plants only take up about 6% of the nitrate fertiliser.

Underneath the Central Valley, there are natural underground reservoirs in the rock which hold large volumes of fresh water. These reservoirs supply water to more than one million people, about a quarter of the population.

Scientists have found that nitrates have contaminated the underground water. They have also shown that a large amount of this nitrate contamination has come from fertilisers used in coffee production.

a) Why do farmers use nitrate fertilisers ? *[1]*

b) How does the use of nitrate fertilisers help the people of Costa Rica ? *[1]*

c) Suggest how the nitrates got into the underground water. *[1]*

d) Why are scientists concerned about pollution of the underground water ? *[1]*

e) Suggest how the problem of nitrate pollution could be reduced in Costa Rica.

[2]   (AQA)

chapter 19

Ionic compounds

This section explains the properties of materials.
To understand the ideas, you need to know about
the structure of atoms. (See Chapter 3.)

When we look at the Periodic Table, we find a group
of elements with almost no reactions at all.
- Can you remember the name of this group?
  What can you say about the arrangement of electrons
  in their atoms?

We have seen on page 56 that the atoms of these
very stable noble gases all have **full outer shells of electrons**.
Other atoms would also become stable with full outer shells.

In this chapter we will see how metals and non-metals
bond together in compounds.

Metals, like sodium, react with non-metals, like chlorine. The compound formed is made of ions.

▷ Ionic bonding

Let's look at sodium chloride as an example:

Ionic bonds form between *metals and non-metals*.

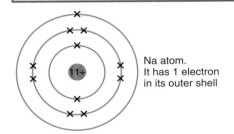

Na atom.
It has 1 electron
in its outer shell

A sodium atom has 11 electrons (2,8,1)

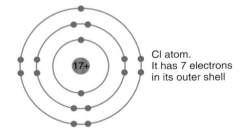

Cl atom.
It has 7 electrons
in its outer shell

A chlorine atom has 17 electrons (2,8,7)

Both atoms would be more stable if they had full outer shells.
They will always take the easiest way to
fill their outer shell with electrons.

Notice that sodium has just 1 electron in its outer shell.
- How do you think it could get a full outer shell?
  Would it be easier for sodium to lose just 1 electron,
  or to gain 7 electrons?

Look at the chlorine atom above:
It has 7 electrons in its outer shell.
- How do you think it could get a full outer shell?
  Would it be easier for chlorine to lose 7 electrons,
  or to gain just 1 more electron?

Metal atoms give electrons to non-metal atoms

When sodium reacts, it **loses** its 1 outer electron.
This leaves a full shell.
Where do you think sodium's electron goes to?

Chlorine accepts the electron from sodium.
It **gains** the 1 electron it needs to fill its
outer shell.
Look at the diagram below:

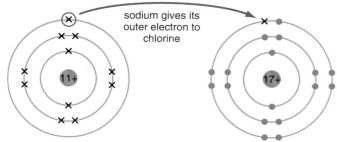

sodium gives its
outer electron to
chlorine

Remember that all atoms are neutral.
They have an equal number of positive protons
and negative electrons. So the charges cancel out.

However, after sodium has given an electron to chlorine,
the electrons and protons in both atoms no longer balance.

Let's add up the charges:

sodium	10 electrons	$=$	$10-$	*chlorine*	18 electrons	$=$	$18-$
	11 protons	$=$	$11+$		17 protons	$=$	$17+$
			$1+$				$1-$

The atoms, which we now call **ions**, become charged: Na^+ and Cl^-.
You can show the ions like this:

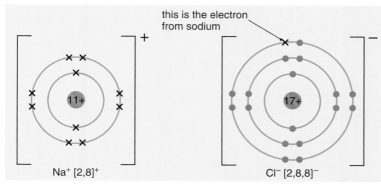

this is the electron
from sodium

Na^+ $[2,8]^+$ Cl^- $[2,8,8]^-$

As you know, opposite charges attract.
Therefore, the Na^+ ions and Cl^- ions are strongly attracted
to each other. This electrostatic attraction sticks the ions together,
and is called an **ionic bond**.
Millions of ions bond together to form crystals as the attraction acts
in all directions. You can see a diagram of this on page 253.

*Opposites attract!
But why will Na^+ and Cl^-
never be alone?*

▷ More ionic compounds

We have looked at sodium chloride as an example of an ionic compound.
When sodium and chlorine react together, each sodium atom gives 1 electron to each chlorine atom.
The atoms react together – one-to-one.
So the formula of sodium chloride is NaCl.

Magnesium is in Group 2 of the Periodic Table.
Therefore, we know that it has 2 electrons in its outer shell. (See page 60.)
How could it get a full outer shell?
What do you think happens if we react magnesium, instead of sodium, with chlorine?
Look at the diagram below:

When metal meets non-metal, the sparks they do fly.
"Give me electrons!" non-metals do cry.
"You've got a deal!" the metal responds,
Then ions attract to form their bonds.

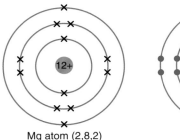

Mg atom (2,8,2)

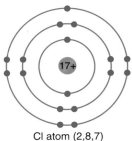

Cl atom (2,8,7)

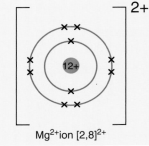

Cl⁻ ion [2,8,8]⁻   Mg²⁺ion [2,8]²⁺   Cl⁻ ion [2,8,8]⁻

The magnesium atom loses 2 electrons when it reacts.
Therefore, its ion has a 2+ charge, Mg^{2+}.

You can see that each magnesium atom can give an electron to each of 2 chlorine atoms.
So the formula of magnesium chloride is **$MgCl_2$**.

Now let's look at sodium reacting with oxygen:
An oxygen atom has 6 electrons in its outer shell.
How can it get a full outer shell?
Look at the diagram below:

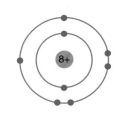

Na atom (2,8,1)   O atom (2,6)

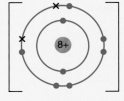
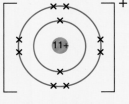

Na⁺ ion [2,8]⁺   O²⁻ ion [2,8]²⁻   Na⁺ ion [2,8]⁺

The oxygen atom gains 2 electrons, each with a negative charge.
Therefore its ion has a 2− charge, O^{2-}.
It takes 2 sodium atoms to fill the outer shell of 1 oxygen atom.
So the formula of sodium oxide is **Na_2O**.

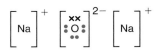

'Dot and cross' diagram for Na_2O.
It shows only outer shell electrons.

Working out the formula

You have now seen how 3 ionic compounds are formed.
Look at the ions which make up each compound:

Ionic compound	Formula	Ratio of ions present
sodium chloride	NaCl	1 Na$^+$: 1 Cl$^-$
magnesium chloride	MgCl$_2$	1 Mg^{2+} : 2 Cl$^-$
sodium oxide	Na$_2$O	2 Na$^+$: 1 O^{2-}

Now add up the charges on the ions in each compound.
● What do you notice? Do they balance out?

Ionic compounds are **neutral**.
The charges on their ions cancel each other out.
Knowing this, and the charge on the ions,
we can work out the formula of any ionic compound.

Example

Magnesium oxide
Magnesium ions have a 2+ charge, Mg^{2+}.
Oxide ions have a 2− charge, O^{2-}.
The charge on 1 magnesium ion balances out
the charge on 1 oxide ion.
$(2+) + (2-) = 0$
Therefore, they bond in the ratio one Mg^{2+} to one O^{2-}.
The formula is **MgO**.

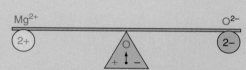

The formula of magnesium oxide is MgO

Example

Aluminium oxide is a little more difficult.
Aluminium ions have a 3+ charge, Al^{3+}.
Oxide ions have a 2− charge, O^{2-}.
So how many aluminium ions and oxide ions
combine to balance each other out?
2 Al^{3+} ions will cancel out 3 O^{2-} ions.
$2 \times (3+) = 6+$ and $3 \times (2-) = 6-$
$(6+) + (6-) = 0$
Therefore the formula is **Al$_2$O$_3$**.

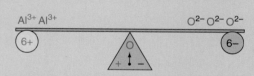

The formula of aluminium oxide is Al$_2$O$_3$

Notice that the metal always comes in front
of the non-metal in the formula.
● Copy and complete the table below:

	chloride, Cl$^-$	bromide, Br$^-$	oxide, O^{2-}
sodium, Na$^+$	NaCl		Na$_2$O
magnesium, Mg^{2+}	MgCl$_2$		
aluminium, Al^{3+}			Al$_2$O$_3$

▷ Properties of ionic compounds

You have now seen how metals bond with non-metals.
Metals give electrons to non-metals. This means that:

> **Metals always form positive ions, and
> non-metals form negative ions.**

Look at the photo of two ionic compounds:
Can you see any similarities?
You can test an ionic compound
in the next experiment:

Here are crystals of two ionic compounds

Experiment 19.1   Testing ionic compounds

1. Spread a few grains of sodium chloride on a microscope slide.
   Focus your microscope on the grains.

   - What shape are the grains?
   - Are the angles at the surface of the grains similar?
     We call solids with regular angles **crystals**.

2. Heat some sodium chloride crystals strongly in a test tube.

   - What happens? Does sodium chloride have a **high melting point**?

3. Add 2 spatulas of sodium chloride to half a beaker of water.
   Stir it with a glass rod.

   - What happens? Is sodium chloride **soluble in water**?

4. Set up the circuit as shown:
   Dip the electrodes into some solid sodium chloride.

   - Does the solid conduct?

   Now half-fill the beaker with water, and stir.

   - Does the bulb light up now?
   - Does the **solution conduct electricity**?

   Switch off your power pack as soon as the test is completed.

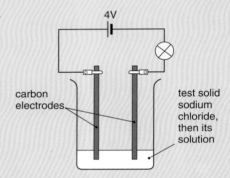

carbon electrodes

chlorine gas

4V

test solid sodium chloride, then its solution

Here is a list of the properties of ionic compounds:

Ionic compounds:

- are made of crystals (which can be split along certain angles)

- have high melting points

- are often soluble in water

- conduct electricity when molten or dissolved in water,
  but not when solid. (Page 83 explains why.)

*We transferred electrons and
lost our hearts,
'Til electrolysis do us part.*

*(Let's hope it doesn't rain – or the
marriage could soon be dissolved!)*

▷ Giant ionic structure

Scientists need to know how the ions are arranged in ionic compounds. This helps to explain how ionic compounds behave.
We can get information by firing X-rays at a crystal. The X-rays make a pattern as they pass through the crystal. The pattern gives us clues about the arrangement of the ions.

Dorothy Hodgkin used X-rays to help show the structure of materials

Scientists have found that the ions form **giant structures**. Millions of positive and negative ions are fixed in position.

Look at the diagram below:

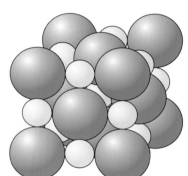

Cl⁻ ion

Na⁺ ion

*Part of the giant ionic structure of sodium chloride. It is called a giant ionic **lattice***

Can you see that the ions are arranged in a regular pattern? This explains the regular angles of ionic crystals.

Why do you think that ionic compounds have high melting points?
Remember that opposite charges attract, forming strong ionic bonds. These forces of attraction act in all directions.
Imagine trying to separate all the ions in the giant structure above! It takes a lot of energy to overcome all that electrostatic attraction.

Many ionic compounds dissolve in water.
To explain this we need to look at a water molecule in more detail:
The electrons in H_2O are not evenly spread. One end of the molecule is slightly negative compared to the other end.

Look at the diagram opposite:

The water molecules are attracted to the ions and pull them from the giant structure (lattice). The compound dissolves. Its ions are then free to move around. (The electrolysis of ionic compounds is explained in Chapter 7.)

the oxygen end of a water molecule is slightly negative compared to the hydrogen end

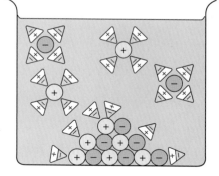

An ionic compound dissolving in water

▷ Chemistry at work : Ionic compounds – halides

We have already met lots of important ionic compounds
in this book.
For example, Chapter 8 is about salt (sodium chloride).
Look back at some of its uses.
Here are some other useful ionic compounds of the
halogens – the **halides**.

Sodium fluoride

If you eat sweet, sticky food, it can rot your teeth.
Bacteria in your mouth feed on the sugar,
making acid as they do so. This acid causes
tooth decay.
Fluoride ions help to prevent tooth decay.
As children grow, the fluoride forms part of the
calcium compounds in their teeth.
This makes it more difficult for acid to attack teeth.

Some water authorities add fluoride to their water supplies.
This greatly reduces the number of fillings children have.

However, too much fluoride can actually harm teeth.
In fact, higher doses are toxic (poisonous) !
This has led to many arguments about fluoride in water.

- What do you think about adding fluoride to our water supplies ?
- How would you set up a survey to test if fluoride is effective
  and safe ?

Sugary foods cause tooth decay

Silver halides

These are used on photographic film and paper.
Silver halides are affected by light.
Try the experiment below :

Experiment 19.2   Effect of light on silver halides
Add silver nitrate solution to two test tubes,
each containing sodium chloride. What happens ?
The white precipitate is silver chloride.
Do the same with two test tubes of sodium bromide,
then sodium iodide.
Place a test tube of each silver halide in a dark
cupboard for 5 minutes.
Leave the other 3 silver halides in the light.
- Describe the differences you see.

The silver halides are reduced to silver in sunlight.
This explains their use in photography.

Developing a photograph

Summary

- When metals react with non-metals they form **ionic** compounds.
  Metal atoms give electrons to non-metal atoms.
  This makes **positive metal** ions, and **negative non-metal** ions.
- The oppositely charged ions are attracted to each other
  by strong electrostatic forces. This is called **ionic bonding**.
- The ions are arranged in huge structures, called **giant ionic lattices**.
- Ionic compounds :
  - are made of crystals
  - have high melting points
  - are often soluble in water
  - conduct electricity when molten or dissolved in water
    (when free ions are present), but not when solid.

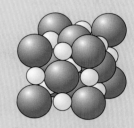

The strong electrostatic forces of attraction operate in all directions in a giant ionic lattice.

▷ Questions

1. Copy and complete :
   .... compounds are made from metals and non-metals.
   Metals always form .... charged ions, and non-metals form .... charged ions.
   These ions are arranged in .... ionic structures. Their crystals have .... melting points and many will .... in water.
   They conduct .... when molten or when in solution, but not when they are .....

2. Which of these substances have ionic bonds ?

ammonia, NH_3	water, H_2O
copper oxide, CuO	zinc chloride, $ZnCl_2$
lead bromide, $PbBr_2$	potassium fluoride, KF
methane, CH_4	

   How did you decide ?

3. Explain, using diagrams to help, how the atoms in a) and b) below, transfer electrons to form ions.
   (Remember that the 1st shell can hold 2 electrons, the 2nd shell 8 electrons, and the 3rd shell also holds 8 electrons.)

   a) Lithium, Li (which has 3 electrons) and fluorine, F (which has 9 electrons).

   b) Potassium, K (which has 19 electrons) and chlorine, Cl (which has 17 electrons).

4. Here are 3 metal ions :
   lithium, Li^+ calcium, Ca^{2+},   iron(III), Fe^{3+}
   Here are 3 non-metal ions :
   fluoride, F^- iodide, I^-,   sulfide, S^{2-}
   Draw a table, like the one on page 251, to show the formulas of the compounds between these metals and non-metals.

5. Explain, using diagrams to help, how these atoms transfer electrons to form ions :
   a) Magnesium, Mg (which has 12 electrons) and fluorine, F (which has 9 electrons).
   b) Magnesium, Mg (which has 12 electrons) and oxygen, O (which has 8 electrons).

6. Look at the Periodic Table on page 392 :
   a) What is the link between the group a *metal* is in and the charge on its ions ? You can use sodium, magnesium and aluminium as examples.
   b) Try to find a simple mathematical equation that tells you the link between the charge on a *non-metal* ion and its group number. Use oxygen and fluorine as examples.
   c) Carbon (which has 6 electrons) never forms ions. Why not ?
   d) Hydrogen (which has 1 electron) can form both H^+ ions and H^- ions. Explain why ?
   e) Why do some people say that hydrogen should be above lithium, Li, in the Periodic Table, and others argue that it should be above fluorine, F ?

Further questions on pages 276 to 278.

chapter 20

Covalent bonding

In the last chapter we saw how metals bond with non-metals.
Remember that metal atoms need to lose electrons, and non-metals need to gain electrons.

But think of all those hydrocarbons in Chapter 11!
Like many compounds, they are formed from *just non-metals*.
Which non-metals do they contain?

Both carbon and hydrogen atoms need to gain electrons.
So how do their atoms bond to each other?
How can they *both* gain electrons to get full outer shells?
It can't be like ionic bonding because there is no metal atom to give away electrons.

Two non-metal atoms can *both gain electrons by sharing!*
If their outer shells overlap, they share pairs of electrons.
In this way both atoms can gain electrons.

Methane is an important fuel.
Its formula is CH_4.
The carbon and hydrogen atoms are joined to each other by covalent bonds.

> **Shared electrons form covalent bonds between non-metal atoms.**

Let's look at the smallest hydrocarbon, methane.
It's formula is CH_4.

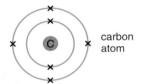

carbon atom

Carbon has 4 electrons in its outer shell.
It needs to gain 4 electrons to get a full outer shell.
(Remember that the 2nd shell can hold 8 electrons.)

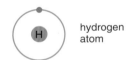

hydrogen atom

Hydrogen has just 1 electron.
If it can gain 1 more electron, it will fill its shell.
(The 1st shell is filled by just 2 electrons.)

1 carbon atom and 4 hydrogen atoms overlap like this:

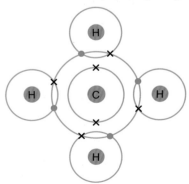

Count the electrons in the outer shell of each carbon and hydrogen atom in methane.

- Is each shell complete?
  Can you see that the carbon now has 8 electrons in its outer shell?
  How many electrons does each hydrogen have?
  Is hydrogen stable with this arrangement of electrons?

It is electrostatic attraction between the pairs of electrons and the C and H nuclei that hold the atoms together in the molecules. This attraction (which is the covalent bond) only operates between the pairs of nuclei.

In Chapter 11 we showed covalent bonds like this:

= carbon
= hydrogen

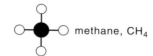 methane, CH$_4$

This 'ball and stick' diagram shows the 4 single covalent bonds between C and H atoms. Each covalent bond is a pair of electrons.

Let's look at another example of covalent bonding.
Chlorine gas does not exist as single atoms.
Remember that a chlorine atom has 7 electrons in
its outer shell. It needs 1 more electron to fill it.
How can it manage to do this?

Like many gases, it is di-atomic. Two chlorine atoms
bond together to make a Cl$_2$ molecule.
Look at the diagrams below of the Cl$_2$ molecule:

Chlorine atoms share a pair of electrons in a covalent bond

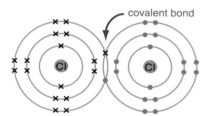

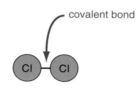

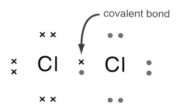

Three ways to show the covalent bond in Cl$_2$. The right-hand one is a 'dot and cross' diagram (showing the outer electrons only).

Count the electrons in the outer shell of each chlorine atom.
Do they both have 8 electrons? Are they both stable compared
to their atoms?

Double bonds

Do you recall the molecule ethene from Chapter 11?
Its formula is C$_2$H$_4$. We said that its carbon atoms
are joined by a 'double bond'.
Can you remember how it forms poly(ethene)? (See page 154.)

Carbon dioxide is another molecule which contains
a double bond. Look at the diagram below:

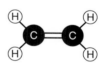

Ethene has a double covalent bond between its carbon atoms

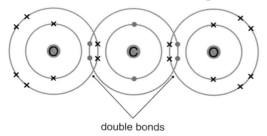

double bonds

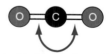

a double bond between the
carbon and each oxygen atom

double bonds
(each double bond contains 2 pairs
of electrons, i.e. 4 electrons)

'Dot and cross' diagram for CO$_2$

● How many electrons are in each area of overlap?
  How many electrons are in the single covalent bonds
  in CH$_4$ and Cl$_2$?
Each **pair of electrons** in the overlapping shells is a covalent bond.
So a double bond is made up of 4 electrons (2 pairs).

▷ Giant covalent structures

The great variety of life on Earth depends on
carbon's ability to form covalent bonds with itself.
As the element, carbon atoms can bond to
millions of other carbon atoms in both
diamond and graphite.

Carbon in the form of diamond

Do you know the hardest substance on Earth?
Look at the photo opposite:
Diamond's hardness makes it a very useful material.
You can read more about its uses on page 260.

Diamond is made from only carbon atoms.
Think back to the previous page.
How many covalent bonds can each carbon atom form?
Look at the diagram below:

Some drills are tipped with diamonds. This one
cuts through rock when drilling for oil.

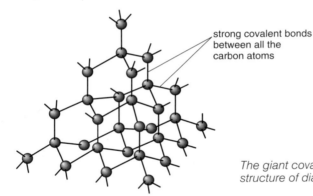

strong covalent bonds
between all the
carbon atoms

The giant covalent
structure of diamond

Each atom forms 4 strong covalent bonds with
its neighbours.
The atoms are arranged in a **giant covalent structure**. Some
people call this a giant molecular or macromolecular structure.

● Does diamond have a high or a low melting point?
Explain why.

Another substance with a giant covalent structure
is silica. Its chemical name is silicon dioxide (SiO_2).
It melts at over 1500 °C.

Substances with giant covalent structures
are **not soluble in water**.
Their particles are not charged (unlike ionic compounds).
So water molecules are not attracted to them.

They **don't conduct electricity** in any state.
There are no free ions or electrons to carry the charge.
(Graphite is an exception; see the next page.)

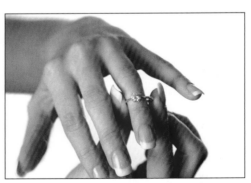
Although diamond is very hard, it can be
split in certain directions (cleaved) to make
jewellery

Carbon in the form of graphite

Another form of carbon is graphite.
Diamond and graphite are **allotropes** of carbon.
Allotropes are different forms of the same element
(in the same state).
The carbon atoms in graphite are also held together
in a giant covalent structure.

However, some of its properties are very different from
a typical giant covalent substance, such as diamond or silica.

Graphite (mixed with clay) is used in pencil 'leads'

If you touch a lump of graphite, it feels smooth and slippery.
Your pencil contains graphite. As you move it across
your paper it flakes off, leaving a trail of carbon atoms.
Look at the diagram below :

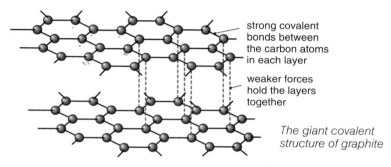

strong covalent
bonds between
the carbon atoms
in each layer

weaker forces
hold the layers
together

*The giant covalent
structure of graphite*

The layers can slide over each other easily.

● How does this explain graphite's use in pencils?

Do you know the only non-metal element that conducts
electricity well? Think of the electrodes we use in
electrolysis. (See Chapter 7.)

● How many neighbouring carbon atoms are bonded to each
other by strong covalent bonds in graphite?
Why is this strange?

The 4th electron from each carbon atom is found
above and below the plane of the layers.
These electrons are only held loosely to the carbon atoms.
They can drift along the layers in graphite.
We call them 'delocalised' electrons. These electrons make
graphite a good conductor of electricity. Look at the
diagram opposite:

● Do you think that graphite conducts better along its
layers or across them?

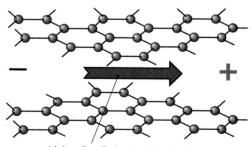

*'delocalised' electrons can move
along the layers in graphite*

Graphite does not melt when you heat it.
At over 3000 °C it turns straight into a gas.
We say it **sublimes**. This is called **sublimation**.

● Why does it take a lot of energy for graphite to sublime?

Uses of diamond

Jewellery

We all know about the use of diamonds in jewellery.
If you have ever looked in a jeweller's shop,
you will know how expensive they are.

Diamonds have ideal properties for gemstones.
A well-cut diamond reflects light better than
any other gem (it has 'brilliance').
It also disperses white light into
the colours of the spectrum (it has 'fire').

A diamond brooch

Cutting tools

Diamond is the hardest of all substances.
It is used on the edges of drills on oil wells.
(Look back to the top photo on page 258.)
It also lines circular saws used to cut metal,
stone and other hard materials.

A new process can now fuse together tiny
diamonds. So instead of metal coated in diamonds,
you can now use a solid diamond tool on a lathe.
This can cut and shape the hardest materials.

Diamond coated surgical instruments are used
for delicate operations, like those on the eye.

An engraver using a diamond-tipped 'pen' to produce a design on glass

Another use

Which types of materials are the best conductors of heat?
Most people would say that metals are.
However, diamond is a better thermal conductor!
On the other hand, it is a poor electrical conductor.
This leads to its use in electronics to get rid
of heat produced in circuits.

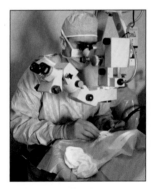

This surgeon is operating on an eye using diamond coated instruments

Diamond mines

The diamond industry is very big business in South Africa.
Diamonds are often mined from some of the deepest mines in the
world. The diamonds are formed under huge pressure and high
temperatures way beneath the surface.
The working conditions are tough for the miners and many in
South Africa live in poor townships. The dust from the mining
waste is a constant problem.

● Discuss the issues raised by the diamond industry.

▷ Chemistry at work : Carbon

Uses of graphite

Pencils

We have already talked about graphite's use in pencils on page 259. It is mixed with clay to make it harder. You know that there are different grades of pencil, such as H, HB, or 2B.

- Which grade of pencil makes the darkest lines on paper?
- How do you think the amount of clay varies in each type of pencil?
- Which type do you think needs sharpening most often?

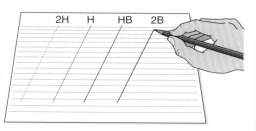

Lubricant

Graphite feels very slippery. It can be used as a powder to lubricate metal parts on machinery.

- When would you need to use solid graphite rather than oil? Remember that graphite only turns to a gas at over 3000 °C!

In some lubricants, graphite is added to oil to improve its properties.

The layers of carbon atoms slide over each other easily, making graphite a good lubricant

Graphite into diamond!

The diamonds used in cutting and grinding tools are called **industrial diamonds**.
They are not mined from the ground, like gemstone diamonds.
They are made in factories from graphite.

Graphite is squeezed in a press and heated.
A metal is used as a solvent and a catalyst.
The temperature is about 1400 °C in the press, and the pressure is 60 000 times normal air pressure!
The diamonds are made in a few minutes.
Under these conditions, anything containing carbon, even wood, changes into diamonds!
The diamonds made are small and unattractive.
Manufacturers can vary the sizes to suit the use by changing conditions.

Industrial diamonds are small and unattractive. However, scientists in the USA have developed a new technique. They recently made a diamond over 25 cm long!

Bucky-balls

In 1985, a new form of carbon was discovered.
Its molecules are made from 60 carbon atoms
joined together. The atoms fold around and
make a ball-shaped molecule.

Look at the photo opposite :
The new molecule looks just like a football !
The carbon atoms form pentagons and hexagons,
like the panels on a football.

Its full name is **buckminster-fullerene**.
Scientists named it after an architect, Buckminster Fuller.
In 1967, he designed a bucky-ball shaped building
in Montreal. Buckminster-fullerene is a black solid that
dissolves in petrol to make a deep red solution.

The new molecule has excited a lot of interest. Other similar
molecules have been discovered. One is shaped like a rugby ball
and others are shaped like tubes.

A model of buckminster-fullerene

The fullerene family

Look at some of the fullerene family below:

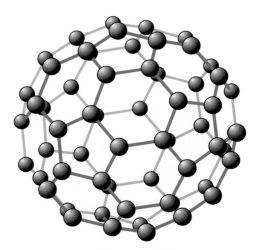

The original bucky-ball, C_{60}

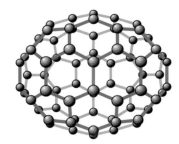

A molecule of C_{70}, shaped like a rugby ball

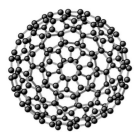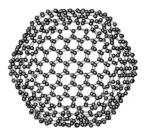

We can also make larger cage-like structures such as C_{240} and C_{540}

▷ Chemistry at work : Fullerenes

Uses of the fullerenes

Chemists have also made an exciting group of carbon allotropes called **'bucky-tubes'**. Look at the diagram opposite :
They make these by joining together fullerenes.

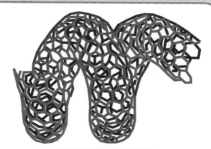

These tiny tubes are sometimes referred to as **nanotubes**.
The nanotubes are used as:

- semi-conductors in miniature electrical circuits
- industrial catalysts (the catalyst is attached to the nanotube producing a large surface area)
- reinforcing tennis rackets (they are very strong but light).

You can read more about nanotechnology on page 336.

- What are the advantages and potential drawbacks of developing new catalysts with incredibly large surface areas ?

Potential uses

Bucky-TV

Researchers in the USA have found a way of lining up 'bucky-tubes' on the surface of glass. The 'bucky-tubes' stick out at right angles from the glass which could be used as flat screens for TVs or computer monitors.
The tubes could each fire electrons from the end sticking out of the screen.
This would replace bulky cathode ray tubes used in most sets.
It promises to give us brighter pictures that you can view from a wider range of angles.

Sir Harry Kroto

Bucky-strength

Israeli and American scientists have shown that bucky-tubes can withstand pressures almost a million times atmospheric pressure. This makes them about 200 times stronger than any other fibre. They believe the bucky-tubes can be embedded in other materials like plastics to make incredibly strong materials. These might be used, for example, in making bullet-proof vests.

Bucky-mules

Bucky-balls and cages can trap atoms or groups of atoms inside their structures.
Scientists are developing ways of using these 'molecular mules' to carry drugs or radioactive atoms, used to treat cancer, to the sites in the body where they are needed.

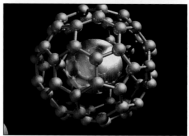

▷ Simple molecular structures

You already know that covalent bonds are strong.
So have you ever wondered why lots of substances
with covalent bonds are easy to melt or boil?

With 2 non-metals, each atom wants more
Both need electrons to make up the score.
A deal is struck, "Our electrons we'll share."
Covalent bonds formed by electron pairs.

Let's take methane, CH_4, as an example:
It has a very low boiling point.
It boils at -161°C!
(Compare this with the values given for the
giant covalent structures in the table opposite.)

By the time you get to room temperature
(about 20 °C), methane is already a gas.
And yet it has strong covalent bonds holding
its atoms together, just like diamond.
So why is it so easy to boil methane?

To answer this question, you must realise that
no covalent bonds are broken when methane boils.
When it boils, single CH_4 molecules move apart.
They separate from each other but they are still CH_4 molecules.

Compare this with the giant covalent structures of diamond,
silicon or sand. To boil these, we have to break apart
the whole giant structure. Millions of covalent bonds
do have to break in these examples.

Giant covalent structure		Simple molecular structure	
Substance	Boiling pt. (°C)	Substance	Boiling pt. (°C)
diamond	4830	methane	−164
silicon	2355	water	100
sand (silica)	2230	chlorine	−35

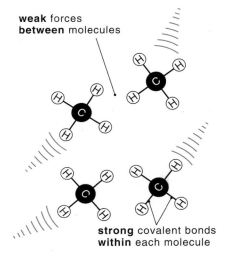

weak forces
between molecules

strong covalent bonds
within each molecule

Methane has a simple molecular structure

> **Substances, with low melting points and boiling
> points have simple molecular structures.**

Simple molecular substances have strong covalent bonds joining
their atoms within each molecule. However, they only have weak
forces between individual molecules.
We say that they have **weak inter-molecular forces**.

Experiment 20.1   Heating iodine

Use tweezers to put a few iodine crystals in a flask. Replace the bung.
Hold the base in your hands for a few minutes, as shown:
Look carefully through the flask, against a white background.

- What can you see?
- Do you think it is easy or hard to separate iodine molecules
  from each other? Do you think it has a giant structure?
- Iodine turns straight from a solid into a gas.
  What is this type of change called? (See the bottom of page 259.)

⚠ iodine

Iodine, I_2, has a simple molecular structure.
Weak forces hold its molecules in place in its crystals.
It does not conduct electricity in any state. That's because its
molecules carry no overall charge.

Experiment 20.2   Heating sulfur

Collect 2 spatulas of powdered sulfur in a test tube. ⚠ sulfur dioxide gas can be formed
Warm it slowly and gently over a Bunsen flame
in a fume-cupboard.
- What happens? Describe the changes you see as it melts.

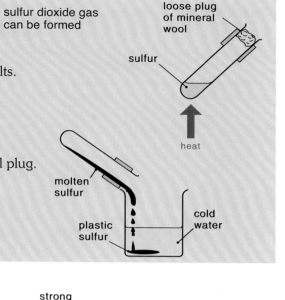

When it has melted, heat the sulfur more strongly.
Keep the flame away from the mouth of the test tube.
It could ignite any sulfur vapour escaping.
- How does the sulfur change?

When the liquid is about to boil, remove the mineral wool plug.
Pour it quickly into a beaker of cold water.
- What happens?

Get the solid formed out of the beaker.
Pull at it gently.
- What happens? Is the solid stretchy?

1. Sulfur exists as S_8 molecules at room temperature.
   These crown-shaped molecules are packed neatly
   together to form crystals.

2. It melts at 115 °C. This low melting point tells us
   that sulfur has a simple molecular structure.
   There are relatively weak forces **between** its molecules.

3. As the molten sulfur is heated more strongly,
   the rings of 8 sulfur atoms open up.
   They then join together to make long chains.

4. Near its boiling point the long chains start to
   break up. They escape from the liquid as a gas.

5. If you pour the molten sulfur into cold water,
   plastic sulfur is made. The sulfur does not
   have time to change back into its S_8 molecules.
   It is 'frozen' in a long chain structure, like rubber.
   Eventually the plastic sulfur goes hard.
   It changes back slowly into S_8 molecules again.

> **In general the larger the molecule, the
> stronger the intermolecular forces.**

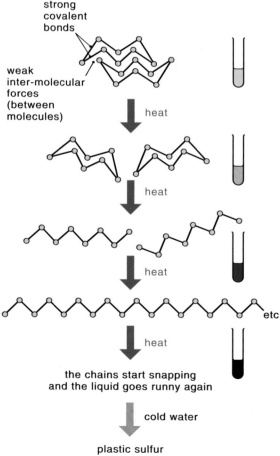

Summary

- Non-metal atoms are bonded to each other by **covalent bonds**.
- A covalent bond is a ***shared pair of electrons***.
- Covalent substances can have either:
  - a **giant covalent structure**, or
  - a **simple molecular structure**.
- Giant covalent structures are huge 3-dimensional networks of atoms. Millions of atoms are all joined by strong covalent bonds. This means that they have high melting points and boiling points.
- Simple molecular substances are made of small molecules. The atoms in each molecule are joined together by strong covalent bonds. However, there are only ***weak*** forces of attraction ***between*** molecules. In general, the larger the molecule, the stronger the intermolecular forces (so the higher the melting and boiling points).
- Substances with covalent bonding do ***not*** conduct electricity (except for graphite).

Chlorine atoms share a pair of electrons in a covalent bond

▷ Questions

1. Copy and complete:
   Non-metal atoms bond to each other by .... electrons. These are called .... bonds.
   Covalently bonded substances with high .... points and high boiling points, have .... structures.

   On the other hand, substances with low melting points and low boiling points have .... molecular structures. These have strong covalent bonds within each ...., but weak forces .... molecules.

   No covalently bonded substances conduct electricity, except .....

2. Look at the table below:
   Substances A, B, C and D all have covalent bonds.

Substance	Melting point (°C)	Boiling point (°C)
A	−125	−90
B	5	170
C	2200	3900
D	55	325

   a) Which substance has a giant covalent structure?
   b) Which type of structure do the other substances have?
   c) Which substance is a gas at room temperature (about 20 °C)?
   d) Which substance is a liquid at room temperature?

3. Fluorine forms di-atomic molecules, F_2.
   (Fluorine atoms have 9 electrons.)
   Draw a diagram to show the bonding in an F_2 molecule.

4. Water is a covalent compound.
   Its formula is H_2O.
   (Hydrogen atoms have 1 electron.)
   (Oxygen atoms have 8 electrons.)
   a) Draw a diagram to show the bonding in an H_2O molecule.
   b) How can you tell that water has a simple molecular structure?

5. Carbon exists as diamond and graphite.
   a) What do we call different forms of the same element?
   b) Diamond and graphite have very different properties and uses.
   Use their structures to explain these differences.

6. Alan heated some sulfur in a test-tube inside a fume-cupboard.
   He heated it until it was almost boiling.
   a) Describe the changes Alan would ***see*** in the test-tube.
   b) Explain these changes.
   c) Why did he do his experiment in a fume-cupboard?

7. a) Match each of these descriptions (A to E) to the correct substance (1 to 5) :

Descriptions:
  A. A black solid that dissolves in petrol to form a deep red solution.
  B. A gas that turns limewater milky.
  C. A transparent solid that is the hardest substance on Earth.
  D. A dark grey solid that is a non-metal but conducts electricity well.
  E. A gas that we can use as the fuel in a cooker or a Bunsen burner.

Substances:
  1. Carbon dioxide.
  2. Graphite.
  3. Methane.
  4. Buckminster-fullerene.
  5. Diamond.

  b) Which substances in the above list are compounds?
  c) Which of the substances is made up of molecules, each containing 5 atoms?
  d) i) Divide the substances into two groups: 'High boiling points' and 'Low boiling points'. Discuss any difficulties you have classifying the substances.
     ii) How do the structures of the substances explain you answer to part i)?
  e) Which of the substances have atoms joined to each other by strong covalent bonds? (Careful!)
  f) Draw a diagram to show the bonding in a carbon dioxide molecule.
     (You may wish to use a 'dot and cross' diagram.)
  g) Which of the substances is shown below?

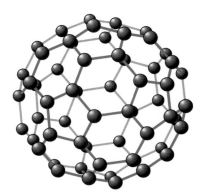

8. a) Oxygen atoms have 8 electrons. Explain why we can represent an oxygen molecule as:

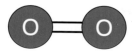

  b) What do we call a molecule that is made up of two atoms?
  c) Give the name and formula of another molecule with the formula X_2.
  d) Why is the boiling point of oxygen $-183\,°C$?
  e) Sulfur exists as rhombic and monoclinic sulfur. Both are crystalline solids. What name describes these two forms of sulfur?
  f) Sulfur atoms have 16 electrons. It does not exist as S_2 molecules at $20\,°C$, but as rings of 8 atoms in S_8 molecules:

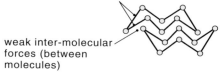

     Explain how each atom in an S_8 molecule has achieved a complete outer shell of electrons.
  g) What is the missing word in this sentence?

     Sulfur has a higher melting point and boiling point than oxygen because its molecules are. . . .

9. This question is about the fullerenes. Carry out some research on the fullerenes. Find out :
   a) how buckminster-fullerene was discovered,
   b) why the discovery can be used as an argument for funding 'pure' scientific research,
   c) the uses and potential uses of the fullerenes.

Further questions on pages 278 to 280.

chapter 21

Metals and structures

▷ Metals

Think of some of the things around your home
that are made of metal.
Did you include all the wiring, any radiators,
your hot-water tank, or your cutlery and pans?

- Do you know which metals these things are made from?
- Which properties make metals good for these uses?

In the last two chapters we have looked at
ionic and covalent bonding. But the atoms in a metal
are held together in a different way.

▷ Metallic bonding

Do you remember all the properties of metals?
Any ideas we have about the bonding and structure
of metals must be able to explain their properties.
Metals have giant structures. Most:

- have high melting and boiling points
- conduct electricity and heat
- are hard and dense
- can be hammered into shapes (they are malleable)
- can be drawn out into wires (they are ductile).

We believe that metal atoms (or positively charged ions)
are held together by a **'sea' of electrons**.
Look at the diagram of sodium metal below:

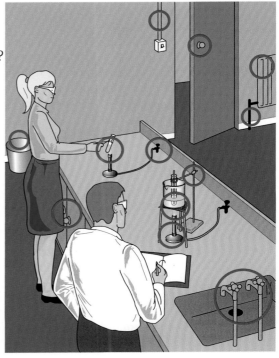

Metals are very useful materials

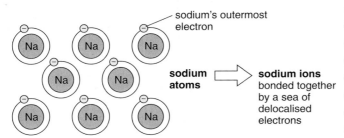

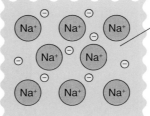

metallic bonding:
electrons can drift
anywhere in the shaded area.
There are strong
electrostatic forces of
attraction between the
delocalised electrons
and the positive metal ions,
bonding the giant lattice
together.

Each sodium atom gives up its electron from its outer
shell into the 'sea' or 'cloud' of electrons.
The electrons can drift about in the metal.
We call them **'delocalised' electrons**.

These free electrons explain how electricity
can pass through solid metals.

- What happens when one end of the metal is made
  positive and the other end negative?

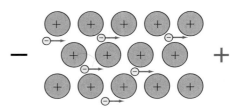

*Electrons move towards the positive charge. These free
electrons can also transfer heat through metals quickly.*

▷ Chemistry at work : Super-conductors

Imagine the energy savings we could make from a 'frictionless' train; a train that could hover above its track, making no contact. Well that is what researchers have developed using super-conductors.

The MAGLEV (magnetic – levitation) train has super-conducting magnets on board, which let it 'float' above the coils in its track.

A **super-conductor** is a substance whose electrical resistance drops to zero below a certain temperature. This was discovered in 1911, using mercury. The metal was cooled by liquid helium to –269 °C. You can't get much colder than that! The resistance that every metal has to electrons flowing through its structure just disappeared.

This test train has raced along its track at 361 m.p.h. However, one worry people have are the possible health hazards of the large magnetic field around the train.
● *What do you think about developing the high speed, low energy train?*

Since then, scientists have been striving to invent material that will super-conduct at higher temperatures. Now the record is up to –135 °C.

This might sound cold but it is well above the temperature of a common coolant, liquid nitrogen. The higher this temperature, the easier it will be to develop new applications.

The prize that the researchers really want is to transfer electricity along the National Grid with *no energy losses*. Normal copper wires heat up when electric current passes through it. If we could use new super-conducting materials, there would be no resistance at all.

So we wouldn't have to burn as much fossil fuel in power stations, reducing air pollution.

A super-conducting cable

Engineers have now built electricity generators that are 99% efficient.

Tests also show that super-conducting wires would take up 7000% less room than conventional cables.

In medicine, you have seen the MRI body scanner on page 58.

Now there are much smaller monitors that don't subject the patient to such large magnetic fields.

Researchers are also trying to develop new super-fast electronic circuits to use in computers.

Look at the projected growth of industries based on super-conductors:

● What could affect the projections on the graph? How might they grow even faster?

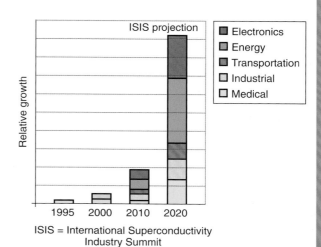

ISIS = International Superconductivity Industry Summit

269

▷ Alloys

Have you ever travelled in an aeroplane?
Did you know that the aeroplane was made mainly
from aluminium?

● Why do you think aluminium is chosen for this job?

Notice that aeroplanes are made **mainly** from aluminium.
If they were made from pure aluminium,
their wings would soon snap off!
Pure aluminium is not strong enough to cope with
the great stress put on wings during flight.

So how can we combine the low density of aluminium
with the strength an aeroplane needs?
Alloys give us the answer.

Aeroplanes are made from aluminium alloys

> **An alloy is a mixture of metals.**

If small amounts of another metal are added to aluminium,
it becomes a lot stronger.
The metals are mixed together when they are molten.
(Notice that metals never **react** with each other.
They form a mixture, not a new compound.)

You can see how alloying affects the structure
of a metal in the next experiment:

Experiment 21.1   Bubble bath!

Using the apparatus shown, make rows of
small bubbles in the dish. The plunger should be
pushed in slowly and steadily. This will make
sure that the bubbles are the same size.

The bubbles represent the atoms in a metal.
Fill the dish with bubbles.

● Do the bubbles line up in rows?
● What happens when a bubble bursts? Can you see
  how easily the rows of bubbles slide past each other?

Now inject a larger bubble into the middle of the dish.
This is like adding an atom of a different metal.
In other words, you have made an alloy!

● Can you see how this disturbs the regular pattern
  of bubbles?

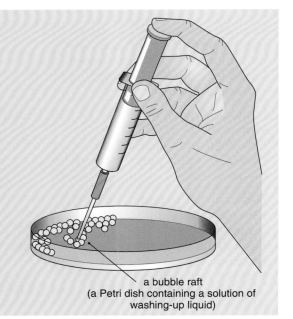

a bubble raft
(a Petri dish containing a solution of
washing-up liquid)

Explaining the properties of alloys

Your observations from the last experiment will help you to understand alloys.

Look at the diagram below:

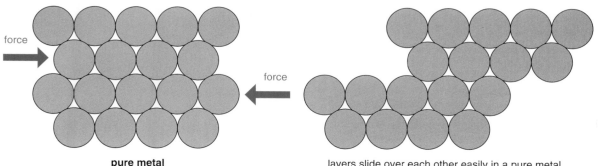

pure metal

layers slide over each other easily in a pure metal

In a pure metal, the atoms are all the same size.
Layers can slide over each other easily.
This happens when metals are hit with a hammer.
• What do we call this property?
It also happens when a metal is stretched.
• What do we call this property?

However, look what happens when we add atoms of a different size:

• *How do you think this copper blade was made?
Why didn't it smash?*

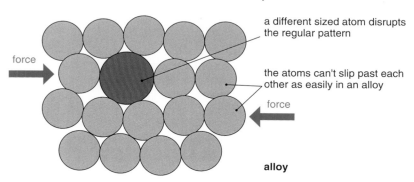

a different sized atom disrupts the regular pattern

the atoms can't slip past each other as easily in an alloy

force

alloy

The layers of atoms can't slide over each other as easily now.
They get 'jammed' in place.
The alloy is a lot harder and stronger than the original metal.

You can read about some uses of alloys on the next two pages.

▷ Chemistry at work : Alloys

Aluminium alloys

You know that alloys are mixtures of metals.
The blend of metals has different properties to the original metals. We saw how alloying aluminium greatly increases its strength so it can be used to make aeroplanes. The alloy is called duralumin.
It has about 4 % copper and a little magnesium added to the aluminium.
Aluminium alloys are expensive because we have to extract aluminium by **electrolysis**. (We can't use the cheaper reduction by carbon because aluminium is too reactive to be displaced by carbon.)
The process uses a lot of energy in melting the aluminium oxide and then passing electricity through it.

Titanium alloys

Titanium alloys are expensive. Although it is not a reactive metal, we can't extract it using cheap carbon. When we try to, titanium reacts with the carbon. This makes the metal brittle. So we have to use a more reactive metal, such as sodium or magnesium, to displace it. Also, of course, the reactive metal has had to be extracted by electrolysis! The extraction of titanium takes several steps, which also makes it an expensive process.

Memory metals

Some alloys are smart! They can 'remember' their original shape. If they get deformed, at a certain temperature they will return to this shape. The main '**shape memory metal**' is called **nitinol**. It is an alloy of nickel and titanium. Its uses include :
- braces for teeth
- under-wires for bras
- fire sprinklers
- antennae for mobile phones
- frames for spectacles and sun-glasses
- supports to open blocked arteries.

This type of Nitinol has a transition temperature, which is below room temperature, so it reverts to its original shape automatically. These sunglasses are expensive at the moment.

Artificial joints

The joints in our bodies take a lot of wear and tear.
The 'ball-and-socket' type joints in the hip and shoulder have to work especially hard. Sometimes the joints are attacked by arthritis. This is a very painful and crippling disease.

Look at the photo opposite :
Now people can be fitted with replacement joints.
These are made of plastic and titanium alloys.
- What properties must an alloy used inside your body have ?

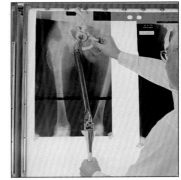

This hip joint is made from an alloy of titanium. Like aluminium, titanium is protected by a tough oxide layer on its surface.

▷ Chemistry at work : Alloys

Gold alloys

Pure gold, as well as being very expensive, is a soft metal.
Just think of the wear and tear on a gold wedding ring.
The metal is made more hard-wearing by alloying it with copper.

Have you got anything made from gold?
If you have, do you know what 'carat' gold it is?
This tells you how much copper is added to the gold.
The higher the carat, the less copper.
24-carat gold is pure gold. On the other hand, 9-carat
gold is about two-thirds copper and one-third gold.

Copper alloys

Have you ever heard a brass band or a pop group with a brass section?
Brass is a mixture of copper and zinc.
Musical instruments made of brass make a pleasing, sonorous sound.
It can also be stamped or pressed into the intricate shapes of the instruments.

Brass has other useful properties as well.
It is much stronger than copper or zinc.

Brass has a lower melting point than either metal.
This makes it easier to cast into shapes.

You've probably heard of the '**bronze** age'. People have been using the alloy of copper and tin for
thousands of years. Bronze made it possible to make more complex and effective tools and weapons.

Lead alloys

Have you ever used a soldering iron?
Solder is used to join up parts of electrical circuits.

This alloy is made from lead and tin. It has a low melting point for a metal.
Why is this important? The solder is made into thick wire. It melts when
held against a hot soldering iron.

Other alloys melt at even lower temperatures (about 70 °C).
These are used in automatic fire sprinklers.

Mercury alloys

Mercury is the only liquid metal at 20 °C. It is useful for alloying because it can 'dissolve' other
metals, without the need for melting as in other alloys. Its best known use is for fillings in teeth.
The dentist can mix the mercury **amalgam**, which is gooey so it can be pressed into the tooth
cavity. It then sets hard quite quickly. People are worried about the toxicity of mercury in dental
care, and new plastic alternatives are now available. The amalgam contains about 50% mercury
plus silver, copper, tin and zinc.

Summary

- Metal atoms are bonded together by a 'sea' of electrons. These **delocalised** electrons are free to drift between the atoms (which can be thought of as positive ions).
- Metal atoms are arranged in **giant structures**. This explains why most metals have high melting points. The atoms (or positive ions) are usually *packed closely* together in these giant structures (lattices). This is why most metals have a high density.
- Mixtures of metals are called **alloys**. An alloy is stronger than the metals used to make it.

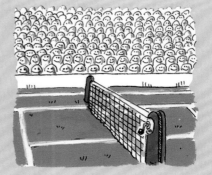

Metal atoms are usually 'close-packed'

Summary of structure and bonding

In this section, we have seen how atoms bond to each other.
We have also looked at how the atoms, ions or molecules are arranged in structures.
The bonding and structure of a substance explain its properties.

Here is a summary of the last 3 chapters so that you can compare the different types of structure and bonding :

Bonding	ionic (between metals and non-metals)	covalent (between non-metals)		metallic (between metals)
Structures	giant ionic	giant covalent	simple molecular	giant metallic
Melting point	high	high	low	high
Conduct electricity ?	not when solid, but they do when molten or dissolved in water (when ions are free)	no	no	yes (has free electrons)
Example	sodium chloride	diamond	water	zinc

▷ Questions

1. Copy and complete :
   A '. . . .' or 'cloud' of electrons join metal atoms (or . . . . charged ions) together in . . . . structures. Metals can . . . . electricity because these . . . . are free to drift through the structure.
   Most metals have . . . . melting points because it takes a lot of energy to break down their . . . . structures.

2. Copy and complete :
   There are 3 types of bonding – ionic, ... and . . . . .

   . . . . bonds form between metals and non-metals. The ions are arranged in . . . . structures, which have high melting points. Sodium chloride is an example. It does not conduct electricity when . . . . , as its ions are fixed in position. However, when molten or . . . . in water, the . . . . become free to move. It is electrolysed as it conducts.

   Covalent bonds form between non-. . . . atoms. These substances can have either a . . . . structure, like diamond (with a . . . . melting point), or be made up of simple . . . . . , like water (with a . . . . melting point). Neither type of structure conducts electricity. The only exception is . . . . .

   Substances with giant . . . . structures are the only ones that conduct electricity when solid.

3. a) Make a list of the general properties of metals.
   b) Write down any metals you know that don't have the usual metallic properties.
   c) Name a property that **all** metals have in common.

4. Make a summary table about alloys under these headings :

Alloy	Metals in alloy (include % if possible)	Uses

5. Anne has been given samples of aluminium, copper and an alloy of the two metals. She has rods, wires and blocks of the 3 metals.
   a) How are alloys of aluminium and copper made ?
   b) Design a series of tests so that Anne can find out which of the 3 metals :
   i)   is hardest
   ii) is most difficult to snap (has the greatest tensile strength)
   iii) resists corrosion best.
   Say how you made each one a **fair test** and how you would collect **reliable data**.
   c) What results would you expect for the hardness and tensile strength tests ? Explain why.

6. Look at this table :

Substance	Melting point (°C)	Conducts electricity	
		when solid	when molten
A	1500	✓	✓
B	115	✗	✗
C	−0.5	✗	✗
D	660	✗	✓

   Explain your answers to each question below :
   a) Which substance is made up of ions arranged in a giant structure ?
   b) Which is a metal ?
   c) Which substances are made up of relatively small individual molecules ?
   d) Which is a non-metal **solid** at 20 °C with a simple molecular structure ?
   e) Which substance is broken down as it conducts electricity ?

7. Sodium, magnesium and aluminium are in Groups 1, 2 and 3 of the Periodic Table. Sodium has 11 electrons, magnesium has 12 and aluminium has 13 electrons.
   a) Draw each atom showing the arrangement of electrons.
   b) The metal elements have metallic bonding. How many electrons do you think the atoms of each metal in a) donate into the 'sea' of electrons ?
   c) Aluminium is the best conductor of electricity. Sodium is the worst. Can you explain this ?

Further questions on pages 281 to 283.

▷ Ionic compounds

1. Look at this information about sodium and chlorine:

a) i) How many protons does an atom of sodium have? [1]
ii) How many electrons does an atom of chlorine have? [1]
iii) As well as electrons and protons, most atoms contain a third kind of particle. What is its name? [1]
b) When sodium forms compounds it usually does so as a positive ion with one unit of charge.
i) What change in electron structure occurs when a sodium atom becomes a sodium ion? [1]
ii) What change, if any, occurs in the nucleus when the ion is formed? [1]
c) When chlorine forms an ionic compound it gains one electron. What symbol is used to represent the chloride ion formed in this way? [1]
d) Explain why the formula for the compound formed when sodium and chlorine react is NaCl and **not** $NaCl_2$. [2]
e) What does the information given below tell you about the **structure** of each substance?
Chlorine melts at −101°C.
Sodium chloride melts at 801°C. [4]
(AQA)

2. When lithium reacts with fluorine, lithium fluoride, LiF, is formed. It is made up of positive and negative ions.
a) How are the positive lithium ions formed from lithium atoms? [1]
b) How are the negative fluoride ions formed from fluorine atoms? [1]
c) **Explain** how the ions are held together in lithium fluoride. [1]
d) Name this type of bonding. [1]
e) Explain why the bonding in lithium fluoride, LiF, produces a high melting point solid. [2] (WJEC)

3. Atoms can form ions with a single negative charge. To do this the atom must
A gain a proton
B gain an electron
C lose a proton
D lose an electron [1] (AQA)

4. The elements of Group 1 form ions with a single positive charge e.g. Na^+. The elements from Group 2 form ions with a double positive charge e.g. Ca^{2+}. Explain this in terms of their electronic structures. [2] (OCR)

5. Use the Periodic Table on page 392 to help you answer this question.
a) An atom of the element with atomic number 9 has a mass number 19.
i) State the number of neutrons and protons in the nucleus of the atom.
ii) State the number of electrons in an atom of the element.
iii) Describe, with the aid of diagrams, the formation of ions in the reaction of the element with sodium.
iv) Give the chemical name of the compound formed when the element reacts with sodium. [7]
b) Explain why solid sodium chloride does not conduct electricity, but aqueous sodium chloride does. [3] (EDEXCEL)

6. The diagrams below show the electronic structure of an atom of calcium and an atom of oxygen.

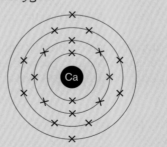 

a) Describe, as fully as you can, the ions that are formed when atoms of these elements react. [2]
b) Calcium oxide is an insoluble ionic compound. Why do ionic compounds have high melting points? [2]

c) What must be done to the solid calcium oxide to make it conduct electricity? Suggest a reason for your answer. *[2]*
(AQA)

7. The Periodic Table on page 392 may help you answer parts a), i) and ii) of this question. The diagram below shows the structure of sodium chloride.

○ sodium ion

● chloride ion

a) i) How does the structure of a sodium ion differ from the structure of a sodium atom? *[1]*
   ii) How does the structure of a chloride ion differ from the structure of a chlorine atom? *[1]*
   iii) What type of chemical bond is present in sodium chloride? *[1]*

b) Use the diagram of sodium chloride to help you to explain why:
   i) sodium chloride crystals can be cube shaped, *[1]*
   ii) solid sodium chloride has a high melting point, *[2]*
   iii) solid sodium chloride is an electrical insulator, *[2]*
   iv) molten sodium chloride will undergo electrolysis. *[2]* (AQA)

8. Using the list of ions on page 391 write the formula of:
a) iron(II) sulfate,
b) iron(III) oxide. *[2]* (WJEC)

9. Sodium (Na) and chlorine (Cl_2) react together to form sodium chloride (NaCl).
a) Write a symbol equation for this reaction. *[1]*
b) Sodium chloride contains sodium ions (Na^+). Write an equation to show the formation of a sodium ion from a sodium atom. Use the symbol e^- to represent an electron. *[1]* (OCR)

10. Magnesium oxide is a compound made up of magnesium ions and oxide ions.

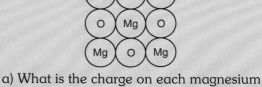

a) What is the charge on each magnesium ion? *[1]*
b) Explain how the magnesium ions get this charge. *[2]* (AQA)

11. The element magnesium (atomic number 12) reacts with chlorine (atomic number 17) to form the compound magnesium chloride, $MgCl_2$.
a) Give the meaning of each of the following words:
   i) element   ii) compound *[2]*
b) The diagrams below show the electron shells in a magnesium atom and in a chlorine atom.
   Copy and complete the diagrams to show the arrangement of electrons.

Magnesium atom (Mg)   Chlorine atom (Cl) *[2]*

c) What happens to these electron arrangements when magnesium reacts with chlorine to form magnesium chloride, $MgCl_2$? *[4]*
d) The compound magnesium chloride has *ionic bonding*. Explain what this means. *[2]* (AQA)

12. The table shows the properties of two substances.

Substance	Does it dissolve in water?	Melting point (°C)	Does it smell?	What colour is it?	Does it conduct electricity when …	
					solid?	melted?
A	yes	1000	no	white	no	yes
B	no	125	yes	yellow	no	no

a) Which of these substances has a giant ionic structure? *[1]*
b) Give **two** reasons for your choice? *[2]*
(AQA)

277

13. The diagrams below show the arrangements of electrons in the *inner* energy levels (shells) of potassium and chlorine atoms.

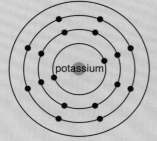

a) Use the Periodic Table on page 392 to help you answer this part of the question. Copy and complete both the diagrams by adding the correct number of electrons in the outer energy level of each atom.   [2]

b) Describe what happens when potassium atoms and chlorine atoms react to form potassium ions and chloride ions.   [2]

c) Use the table on page 391 to help you answer this part of the question.

  i) Write the symbols for the ions of potassium, magnesium and chloride.[2]

  ii) Use your answers to write the formulas of potassium chloride and magnesium chloride.   [2]   (AQA)

14. The diagram represents the arrangement of electrons in a magnesium atom.

a) Copy and complete the table.   [2]

	Number of			Electron arrangement
	protons	neutrons	electrons	
magnesium-24 oxygen-16		8	8	2,8,2

b) Magnesium oxide contains ionic bonding. Explain fully, in terms of transfer of electrons and the formation of ions, the changes which occur when magnesium oxide is formed from magnesium and oxygen atoms   [4]   (OCR)

▷ **Covalent bonding**

15. Use the Periodic Table on page 392 to help you answer this question.
Explain, with the aid of diagrams, how covalent bonds are formed between carbon and hydrogen in methane, CH_4. Indicate which noble gas electron configuration is achieved by the atoms.
             [3]   (EDEXCEL)

16. Chlorine will combine with the non-metal element, carbon, to form this molecular compound.

a) What is the type of bond in this molecule?
                    [1]

b) Explain how these bonds are formed. (You may use a diagram.)   *[2]*   (AQA)

17. Which of the substances A, B, C or D consists of small molecules?

Substance	Melting Point	Electrical conductivity of solid	Electrical conductivity of solution in water
A	high	nil	good
B	high	good	insoluble
C	low	good	insoluble
D	low	nil	nil

          [1]   (AQA)

18. The diagram represents an atom of fluorine. Only the outer electrons are shown.

The atoms in a fluorine molecule, F_2, are covalently bonded.
Draw a diagram to show the arrangement of electrons in a fluorine molecule.   *[2]*   (OCR)

19. One oxygen atom will bond to two hydrogen atoms to form a water molecule, H_2O. One nitrogen atom will bond to three hydrogen atoms to form ammonia, NH_3. Explain, with the help of diagrams showing the arrangement of electrons, why oxygen and nitrogen bond to different numbers of hydrogen atoms. *[4]* (OCR)

20. Hydrogen reacts with chlorine according to the equation:
$$H_2(g) + Cl_2(g) \longrightarrow 2HCl(g)$$
   a) Name the type of bond present in hydrogen chloride. *[1]*
   b) How is this type of bond formed? *[1]*
   c) State **two** physical properties of hydrogen chloride which result from this type of bonding. *[2]*
   d) Draw a diagram of the bonding in hydrogen chloride showing how all the electrons in each atom are arranged. *[3]* (EDEXCEL)

21. a) Draw a dot and cross diagram to show the electronic structure of a molecule of chlorine(I) oxide, Cl_2O. Show only the electrons in the outermost shell (highest energy level), and use DOTS for electrons from the oxygen atom and CROSSES for electrons from the chlorine atoms. *[2]*
   b) Explain, in terms of the forces present, why chlorine(I) oxide has a low melting point. *[2]* (EDEXCEL)

22. The hydrogen halides (hydrogen fluoride, hydrogen chloride, hydrogen bromide and hydrogen iodide) are important chemicals. The diagram below represents a molecule of hydrogen chloride.

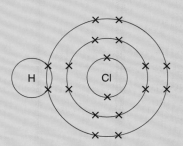

   a) What type of particles are represented by the crosses (X)? *[1]*
   b) What type of chemical bond holds the atoms in this molecule together? *[1]*
   c) Would you expect hydrogen chloride to be a gas, a liquid or a solid, at room temperature and pressure? Explain your answer. *[3]* (AQA)

23. a) Silicon and carbon are both in Group 4 of the Periodic Table. They form the oxides, silicon dioxide (SiO_2) and carbon dioxide (CO_2). The properties of these two oxides are shown in the table below:

Property	Carbon dioxide	Silicon dioxide
Melting point	sublimes at −78°C	1610°C
Boiling point	sublimes at −78°C	2230°C
Electrical conductivity	poor	poor

   Give the structures and bonding of carbon dioxide and silicon dioxide. Explain your answers. *[4]*
   b) Draw a diagram to show the arrangement of outer electrons in a molecule of silicon chloride. (Atomic numbers: silicon = 14, chlorine = 17.) *[4]* (EDEXCEL)

24. Explain, as fully as you can, why a water molecule contains two hydrogen atoms but a hydrogen chloride molecule contains only one.

$$H\diagdown \atop O \atop | \atop H \qquad H\!-\!Cl$$

   (You may use a diagram in your answer if you wish.) *[3]* (AQA)

25. Fluorine forms a **gaseous** compound with oxygen, with the formula F_2O.
   a) Explain how these two elements bond to form F_2O and name the type of bonding used. *[3]*
   b) **Explain** why the oxide of fluorine, F_2O, is a gas. *[2]* (WJEC)

26. The structure of three forms of the element carbon are shown.

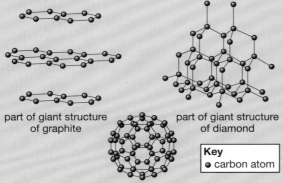

part of giant structure
of graphite

part of giant structure
of diamond

Key
● carbon atom

buckminster-fullerene molecule

a) Describe the bonding **and** giant structure of diamond. *[2]*

b) i) Graphite is soft.
Explain why. *[2]*

ii) Graphite can conduct electricity.
Explain why. *[1]*

c) Consider the diagrams of part of the giant structure of diamond and the buckminster-fullerene molecule.
How would you expect the melting point of buckminster-fullerene to compare with that of diamond?
Explain your answer. *[3]* (EDEXCEL)

27. The structure of silicon is similar to the structure of diamond.
Describe the structure of silicon and explain why it has a high melting point. You may draw a diagram if this helps. *[4]* (AQA)

28. In 1985 buckminster-fullerene, a new allotrope of carbon, was discovered. The diagrams in question 26 above show the arrangement of carbon atoms in diamond and buckminster-fullerene.

a) i) How are the structures of buckminster-fullerene and diamond similar? *[2]*

ii) How are the structures of buckminster-fullerene and diamond different? *[2]*

b) The melting point of buckminster-fullerene is much lower than that of diamond.
Use ideas about the structures of the two allotropes to explain this difference. *[4]*

29. Fullerenes can be joined together to make tube shapes called nanotubes.

a) Briefly describe two industrial uses of nanotubes. *[4]*

b) Nanotubes can be used as cages to trap other molecules inside.
In the future drugs used to treat cancer may be adminstered in this way.
What advantages may this use of nanotubes have in the treatment of cancer? *[3]*

30. Both diamond and graphite
A are electrical conductors
B are lubricants
C form the same products when burnt
D have the same hardness *[1]* (AQA)

31. You can use graphite to lubricate machines because its structure
A allows it to melt at low temperatures
B consists of layers that slide over one another
C consists of small, round atoms loosely bonded to each other
D contains atoms that are strongly bonded to four other carbon atoms *[1]* (AQA)

32. Carbon can exist in two forms, graphite and diamond. Explain why graphite can conduct electricity but diamond cannot. *[2]* (EDEXCEL)

33. Graphite is mixed with clay to make pencil leads.

a) i) Name the element of which graphite is one form. *[1]*

ii) Name one other crystalline form of this element. *[1]*

b) Sketch a diagram to show the arrangement of atoms in graphite. *[2]*

c) Suggest why this crystal structure of graphite enables it to leave a mark when a pencil is drawn across a sheet of paper. *[1]* (EDEXCEL)

▷ Metals and structures

34. Read the following account, taken from the *Express & Star*, and answer the questions which follow.

> # Warning on fake gold
>
> Trading standards chiefs in Wolverhampton today warned Christmas shoppers to beware of fake gold chains and jewellery.
>
> They advised people to avoid unlicenced street traders selling from suitcases who have been operating in the town centre.
>
> Police officers caught "traders" selling jewellery described as 18 carat gold.
>
> The fake gold was a mixture of nickel, copper and zinc with no trace of gold.

a) What is the name given to a mixture of metals? [1]

b) Use pages 390 and 391 to help you to complete the table. [3]

Name	Atomic number	Symbol	Density (g/cm³)	Melting point (°C)
nickel	28	Ni	8.9	1450
copper	29	Cu	i)	1084
zinc	30	Zn	7.1	ii)
gold	iii)	Au	19.3	1064

c) State why the fake gold chain would feel 'lighter' than a similar one made of real gold. [1]

d) i) Describe an experiment to show that fake gold conducts electricity. State what apparatus you would need, what you would do and what you would see. [3]

  ii) Why do metals conduct electricity? [2]

e) Explain why the experiment described in part d) would:

  i) prove that the chain was **not** made of plastic. [1]

  ii) **not** prove that the chain was made of gold. [1]

35. Copper metal is used for making electric wire, coins, pipes and saucepans.
For each of the above uses give a **different** reason why copper is chosen. The first is done for you.

Use of copper	Reason for choosing copper
Electric wires	Good conductor of electricity
Coins	a)
Pipes	b)
Saucepans	c)

[3] (NI)

36. a) Describe the structure and bonding in metals. [3]

b) Explain why metals such as nickel and platinum are good conductors of electricity. [2] (AQA)

37. The table gives some data about metals and metal alloys.
Use the data throughout this question.

Metal or alloy	Density/ g per cm³	Melting point/°C	Electrical resistivity/Ωm	Tensile strength/MPa
aluminium	2.7	934	2.6	80
iron	7.8	1810	10	300
steel	7.8	1700	15	460
aluminium alloy	2.8	800	5	600
titanium	4.5	1950	53	620

a) When an electric current passes through the metal filament of a light bulb, light is given out as the filament becomes 'white hot'.

  i) Suggest the best metal to use for a light bulb filament. [1]

  ii) Explain your choice. [1]

b) i) The low electrical resistivity of aluminium means that aluminium is a very good electrical conductor. Explain how the structure of aluminium enables it to conduct electricity. [2]

  ii) Discuss the properties that should be considered by an engineer when choosing a material for overhead power cables. [5] (OCR)

38. Nitinol is an alloy of nickel and titanium. It is a 'smart alloy'.

a) Explain what is meant by the terms:

  i) alloy [1]

  ii) smart alloy. [1]

b) Nitinol is used to make spectacle frames.

  i) What property of nitinol makes it particularly suitable for this use? [1]

  ii) What advantage do spectacle frames made from nitinol give to the wearer? [1]

39. a) The table gives some information about three materials.

Material	Reaction with oxygen	Reaction with acidic solution	Cost (£ per tonne)	Melting point (°C)	Density (g/cm³)	Tensile strength (MN/m)
Carbon steel	forms porous oxide layer on surface, flakes off	dissolves slowly	500	1539	7.86	250–400
Stainless steel	non-porous layer of chromium oxide on surface, re-forms if scratched	does not dissolve	2000	1440	7.8–8.0	500–1000
Aluminium	non-porous layer of aluminium oxide, re-forms if scratched	dissolves very slowly	1300	659	2.70	140–400

Discuss the relative merits of using carbon steel, stainless steel and aluminium for making
  i) car bodies and
  ii) exhaust pipes. *[6]*
b) An alloy can be made from two metals. In very many cases the atoms of the two metals differ in size. In these cases the alloy is harder than the pure metals it contains. Explain why this is so. *[3]* (EDEXCEL)

40. a) By reference to their structure, explain how the particles in a piece of metal are held together and how the shape of the metal can be changed without it breaking. (You may use a diagram in your answer.) *[5]*
b) Explain why metals are good conductors of electricity and suggest why this conductivity increases across the Periodic Table from sodium to magnesium to aluminium. *[4]* (AQA)

41. The following table shows the properties of five substances.

Substance	Melting point (°C)	Boiling point (°C)	Electrical conductivity when		Effect of heating in air
			solid	liquid	
A	800	1470	poor	good	no reaction
B	650	1110	good	good	burns to form a white solid
C	19	287	poor	poor	burns to form carbon dioxide and water
D	114	444	poor	poor	burns to form an acidic gas only
E	1700	2200	poor	poor	no reaction

Each substance can be used once, more than once or not at all to answer the following. Choose from **A** to **E** a substance which is :
a) a metal *[1]*
b) a non-metallic element *[1]*
c) a molecular covalent compound *[1]*
d) an ionic compound *[1]*
e) a giant covalent structure. *[1]* (EDEXCEL)

42. Questions a) to d) concern the arrangement of the particles in five different substances.

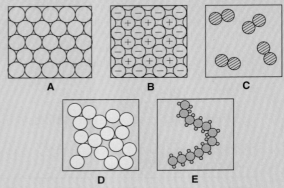

Choose, from **A** to **E**, the arrangement of particles at room temperature that corresponds to :
a) a substance that will conduct electricity when molten but not when solid *[1]*
b) the element mercury *[1]*
c) a substance whose boiling point is below room temperature *[1]*
d) a solid metallic element. *[1]* (EDEXCEL)

43. During the formation of a covalent bond the atoms …
A gain or lose protons
B share electrons
C gain or lose electrons
D share protons. *[1]* (AQA)

44. The formulae of the chlorides of some elements are shown in the table below.

I	II	III	IV	V	VI	VII	O
LiCl	BeCl₂	BCl₃	C …	NCl₃	OCl₂	FCi	No chloride formed
NaCl	Mg …	AlCl₃	SiCl₄	PCl₃	SCl₂	Cl₂	

a) What are the formulas of the two missing chlorides ? *[2]*
b) What type of structure will each of the missing chlorides have ? *[2]*

45. The diagrams below show different arrangements of particles in five substances at room temperature.
Use them to answer questions a) to d).

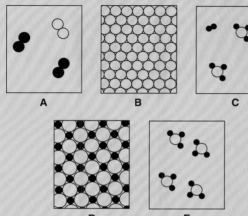

Choose from **A** to **E** a substance which is :
a) sodium fluoride [1]
b) ammonia [1]
c) copper [1]
d) a mixture of ammonia and hydrogen [1]
(EDEXCEL)

46. The following diagrams represent four different solid structures.

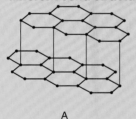

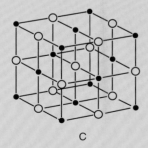
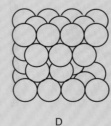

a) Which diagram represents :
  i) diamond ii) sodium chloride ? [2]

b) Draw a diagram to show the structure of sodium chloride after it has melted. [1]
c) Explain why sodium chloride has a relatively high melting point. [2]
d) Describe a test which you could do to show that silver is a metal. Include a diagram and state the result of the test in your answer. [4] (EDEXCEL)

47. Here is a list of compounds.
  calcium fluoride   carbon dioxide
  sulfur dioxide   potassium chloride
  sodium oxide   nitrogen dioxide
Choose from the list above :
a) 3 compounds which have a molecular structure, [1]
b) 3 compounds which have an ionic structure. [1] (OCR)

48. This question is about magnesium and its compounds.
a) The bonding in magnesium is metallic.
  i) Draw a diagram to illustrate metallic bonding. [3]
  ii) Use your understanding of metallic bonding to explain why metals can be pulled into wires [1]
b) Magnesium chloride $MgCl_2$ is a white crystalline salt similar to sodium chloride. Magnesium is manufactured by the electrolysis of molten magnesium chloride. Why does the magnesium chloride have to be molten and not solid? [2]
c) i) Draw a diagram to show the arrangement of the electrons in magnesium oxide, and show the charges on the ions. [3]
  ii) Suggest a reason why magnesium oxide is used to line the inside of furnaces. [1] (NI)

Questions 49–52
A simple molecular
B giant covalent
C giant ionic
D giant metallic
Choose from the list A to D the type of structure which is found in
49. zinc. [1]
50. substances that are gases at room temperature and pressure. [1]
51. sodium chloride. [1]
52. sulfur. [1] (AQA)

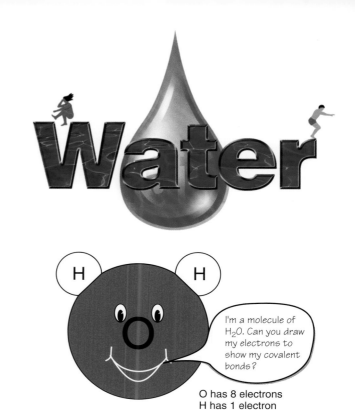

chapter 22

Life on Earth depends on water. Without it, we wouldn't be here!

The formula of water is **H₂O**.
The hydrogen and oxygen atoms are joined by *covalent* bonds.

It has a simple molecular structure. The forces *between* its molecules are not very strong. (Although the forces are stronger than you would expect for such a small molecule.)

Water freezes at 0 °C and boils at 100 °C. Fortunately for us, this means that water is a liquid at most temperatures on Earth.

I'm a molecule of H_2O. Can you draw my electrons to show my covalent bonds?

O has 8 electrons
H has 1 electron

The water cycle

The water cycle shows us how the world's water moves from place to place:

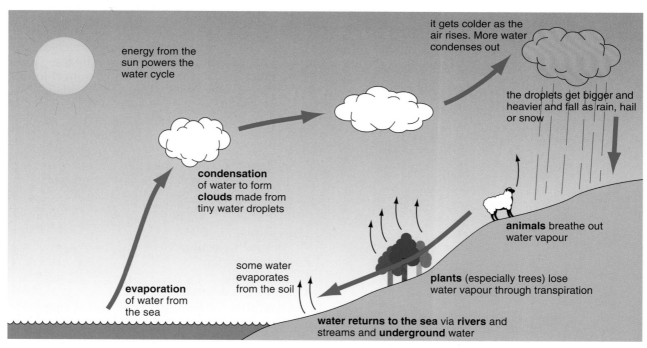

energy from the sun powers the water cycle

it gets colder as the air rises. More water condenses out

the droplets get bigger and heavier and fall as rain, hail or snow

condensation of water to form **clouds** made from tiny water droplets

animals breathe out water vapour

some water evaporates from the soil

plants (especially trees) lose water vapour through transpiration

evaporation of water from the sea

water returns to the sea via **rivers** and streams and **underground** water

- How does water vapour get into the atmosphere?
- Explain why it rains when clouds rise.
- Why is it called the water 'cycle'?

▶ Oceans

Oceans cover roughly 70 % of the Earth's surface.
They make up about 97 % of the world's water.
We believe that life on Earth was started in the oceans,
about 3 billion years ago. Since then the chemicals in
sea water have stayed about the same.
These are listed below.

salt	mass of salt in 100 g of sea-water (g)
sodium chloride	23
magnesium chloride	5
sodium sulfate	4
calcium chloride	1
potassium chloride	0.7

These salts have entered the oceans following
their weathering from rocks on land.
The dissolved salts are carried to the sea by rivers.

So why don't the proportions of salts build up over time?
This is explained by **chemical precipitation**.
As new salts enter the sea, other salts are being deposited
on the ocean floor as solids.
Equilibrium has been reached.

The oceans also contain dissolved gases. These come
from volcanic activity on the sea bed and from the gases
in the air.
The oceans play an important part in the carbon cycle
as reservoirs for carbon dioxide. (See pages 300 and 304.)

The carbon dioxide can form hydrogencarbonates
which are soluble in the sea water:

$$CO_2(g) + H_2O(l) \rightleftharpoons H^+(aq) + HCO_3^-(aq)$$
$$\text{hydrogencarbonate ions}$$

We also get carbonate ions formed:

$$HCO_3^-(aq) \rightleftharpoons H^+(aq) + CO_3^{2-}(aq)$$

These can form insoluble salts with calcium
or magnesium ions in the sea water:

$$Ca^{2+}(aq) + CO_3^{2-}(aq) \longrightarrow CaCO_3(s)$$

$$Mg^{2+}(aq) + CO_3^{2-}(aq) \longrightarrow MgCO_3(s)$$

These are called precipitation reactions. (See page 133.)

We are now producing so much CO_2 that the oceans can't get rid of it.
We are disturbing nature's balance in the carbon cycle. Many people
believe that global warming and changing weather patterns
are happening now because of the extra CO_2. (See page 149.)

*Water is the most abundant substance
on the Earth's surface*

*Some people blame the Greenhouse
effect for extreme weather conditions*

*The international conference on global
warming was held in Kyoto, Japan, in
1997. Politicians met again in 2000 to
decide on each country's reduction in
CO_2 emissions. However, the meeting
broke down with no agreement
reached. Then in Montreal (2005)
there were finally signs of progress
being made.*

▶ Chemistry at work : Uses of water

Where do we get our water from?

We can use water drawn from :

- lakes
- rivers
- aquifers (water held in underground rocks)
- reservoirs (large lakes created for storing water).

These are sources of fresh water, but they still have to be purified to make water fit for drinking. (See page 288.) Unfortunately, about 97% of the water on Earth is in the salty oceans. However, in hot countries they are changing sea water into usable water. (See page 294.)

This is a reservoir at Howden Dam, Peak District National Park. It serves Sheffield, Nottingham, Derby and Leicester.

Water in the workplace

A lot of water gets used by industry or in energy production as :

- **A raw material** – some manufacturing processes involve water, either in chemical reactions to make a useful product or for steam to generate high pressures. (See the Haber process on page 231.)
- **A coolant** – power stations use huge volumes of water. You've probably seen the massive cooling towers belching out steam. (See thermal pollution on page 296.)
- **A solvent** – water can dissolve many substances so it is used for cleaning, washing away waste, and for reactions that take place in solution e.g. brewing.

The fermentation of sugar in a brewery takes place in aqueous solutions

Water at home

The amount of water we use at home is more than you might think.
In the UK we use about 150 litres of water per person each day.
Look at how much water different activities take :

If you wash a car, it takes about 7 litres of water per bucket

Use	Approximate volume of water-use (litres)
shower	27
power shower	80
bath	80
flushing toilet	9
brushing teeth, washing hands	4
washing machine	80
dish-washer	30
washing dishes by hand	7.5
In the garden, using a hosepipe uses about 15 litres/minute and a sprinkler use about 9 litres/minute.	

► Chemistry at work : Uses of water

World-wide usage of water

Across the world, water is used as shown below :

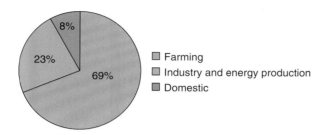

Farming — 69%
Industry and energy production — 23%
Domestic — 8%

The actual proportions differ from country to country.
For example, some African nations can use almost 90% of their water consumption in farming. That's because they need to irrigate their crops.
Whereas in European countries, the proportion used in industry and in producing energy is likely to be over 50%.
The percentage of water used in industry can be used as a measure of how developed a country is.

The individual usage of water (average per person) also varies widely.
A person in the USA is likely to use at least 20 times as much water as a person in Africa each day.

- Why do people in developed countries use a lot more domestic water than people in developing nations? Discuss the issues involved.

Water usage across the world reflects differing life styles

Conserving our water supplies

You might wonder: 'why bother trying to save water when it all gets recycled in the water cycle?'
However, once we have used water, there is always a price to pay before putting it back into the environment.
We can't just dump dirty water directly into rivers and seas.
Cleaning up our waste water costs us money.

As time passes there will be more and more people competing for the planet's water. So we need to conserve our water and do even more to restore its quality after use. By saving energy, we are also doing our bit to conserve water. That's because power stations use a high proportion of our water.

- Discuss how you could save water. Could governments help? What do water companies do when we have a dry summer? Does this affect your life at all?

What affects the reliability of our supply of water? This person is collecting water during a shortage.

▶ Drinking water

Did you know that about two-thirds of the mass
of your body is made up of water?
So it's not surprising that we should drink
between 6 and 8 glasses of water everyday!
You don't want to shrivel up like a dried prune!

We are lucky that we have drinking water 'on tap'.
Many people around the world have to collect
their own water, often from contaminated sources.
We looked at the important job that chlorine does
in making water safe to drink on page 55.
Now we can follow the journey of water,
from a reservoir to our taps at home.

Look at the diagram below:

Over 60% of your body mass is made up from water

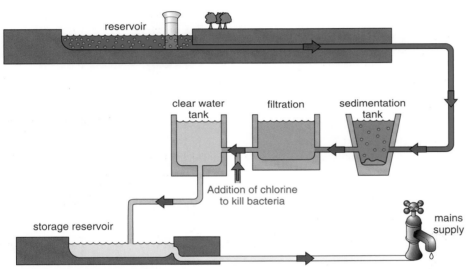

Let's look at the ways we treat water to make it safe to drink:

Aeration

The water is stirred up to let air (**oxygen**) mix with it.
This can also be done by letting it fall down steps.
This helps to remove iron(III) ions in the water. They precipitate out
as iron(III) hydroxide. If this happens in your water at home,
iron(III) hydroxide causes:

- vegetables to turn brown
- tea to have an inky colour and a bitter taste
- clothes to get rusty stains after washing.

Aeration prevents iron(III) hydroxide precipitating out of your water at home. However, some areas still have old iron pipes which can cause problems.

Sedimentation

Larger solid particles are allowed to settle at the bottom of the sedimentation tanks. But dirty water is a sol. There are colloidal particles of clay dispersed in the water.

The tiny particles are electrically charged (negative). They stay spread throughout the water because they repel each other. So water companies add **aluminium sulfate** to **coagulate** the clay. The particles are attracted to the highly charged Al^{3+} ions. They form clumps which get heavy enough to settle on the bottom of the tank.

The acidity of the water is controlled using **lime** slurry. This calcium hydroxide is alkaline so it raises the pH of the water. The aluminium sulfate coagulant is good at removing alkaline substances.

Aluminium sulfate coagulates clay particles. Some people in Camelford (Cornwall) were badly affected when too much aluminium sulfate was added to their water supply by accident.

Filtration

Then the water is filtered through layers of sand and gravel to make sure all solids are removed. These filter beds can also contain **carbon slurry** to get rid of substances that would give the water an odd smell or taste.

Chemical purification

Does your tap water ever smell of chlorine? **Chlorine** is added to kill bacteria in the water. This prevents diseases.

The water company will add enough chlorine to kill all the bacteria. But how would you like drinking water that tasted like it was from a swimming pool? They remove any excess chlorine by treating the water with **sulfur dioxide**. This reacts with the chlorine, getting rid of its smell and taste. The chlorine is reduced to chloride ions, $Cl^-(aq)$.

A little chlorine is left in the water to keep it free from germs in the journey to your tap.

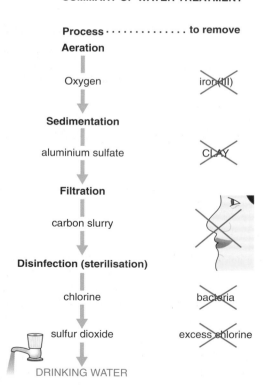

SUMMARY OF WATER TREATMENT

Process · · · · · · · · · · · · to remove
Aeration
Oxygen   iron(II)
Sedimentation
aluminium sulfate   CLAY
Filtration
carbon slurry
Disinfection (sterilisation)
chlorine   bacteria
sulfur dioxide   excess chlorine
DRINKING WATER

► Hard water

Do you know what **hard water** is?

If you live in an area which has hard water you certainly will!

You'll find it difficult to make a lather with soap.
Bits of white scum will float around in the water when you use soap.
However, it does have its advantages.
Look at this table:

Disadvantages of hard water	Advantages of hard water
Difficult to form lather with soap. Scum forms in a reaction which wastes soap. Scale (a hard crust) forms inside kettles. This wastes energy when you boil your kettle. Hot water pipes 'fur up' on the inside. The scale formed can even block up pipes completely.	Some people prefer the taste. Calcium in the water is good for children's teeth and bones. Helps to reduce heart illness. Some brewers like hard water for making beer. A coating of scale (limescale) inside copper or lead pipes stops poisonous salts dissolving into our water.

Let's see if we can find out why some water is hard:

Experiment 22.1   What causes hardness in water?

Use 10 cm^3 of each solution.
Add 1 cm^3 of soap solution to your boiling tube.
Put a bung in and shake.
See if you get a good lather (one that lasts 30 seconds).
A good lather means that the solution is **soft**.
A poor lather and white bits (scum) in your solution means that it is **hard**.

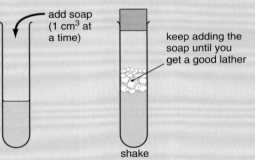

add soap
(1 cm^3 at a time)

keep adding the soap until you get a good lather

shake

Record your results in a table like this:

Solution	Volume of soap to get a good lather (cm^3)	Hard or soft?
sodium chloride calcium chloride potassium chloride magnesium chloride		

- Which substances make the water hard?
- Is it the metal or the chloride ions in these solutions which make the water hard? How can you tell?

What makes water hard?

If your water supply has flowed through chalk or limestone (**calcium carbonate**) it will be hard. Other rocks which contain calcium or magnesium, also cause hardness. Gypsum (**calcium sulfate**) is an example.

> **Calcium (or magnesium) compounds dissolved in water make it hard.**

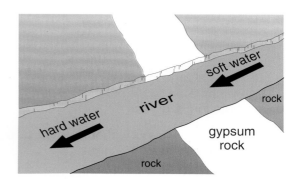
Water flowing over gypsum (calcium sulfate) becomes hard

Calcium sulfate is slightly soluble in water. When a river flows over gypsum, it dissolves some of the rock. Therefore, calcium ions get into the water as $Ca^{2+}(aq)$.

However, calcium carbonate is not soluble in water. Water does not dissolve chalk or limestone rock. But do you remember from page 109 how limestone caves are formed?

People often think that rainwater is the purest water you can get. This isn't quite true. On its way down, gases dissolve in the rain. One of these gases is carbon dioxide – a weakly acidic gas. It's the bubbles you see in fizzy drinks:

Limestone reacts with the weakly acidic rain and river water. Eventually underground caverns can be formed as the rock is weathered away.

$$\text{water} + \text{carbon dioxide} \longrightarrow \text{carbonic acid (a weak acid)}$$
$$H_2O(l) + CO_2(g) \longrightarrow H_2CO_3(aq)$$

This weakly acidic solution dissolves away the limestone or chalk:

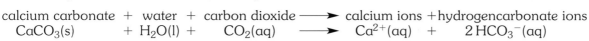
$$\text{calcium carbonate} + \text{water} + \text{carbon dioxide} \longrightarrow \text{calcium ions} + \text{hydrogencarbonate ions}$$
$$CaCO_3(s) + H_2O(l) + CO_2(aq) \longrightarrow Ca^{2+}(aq) + 2\,HCO_3^-(aq)$$

Look at the state symbols in the equation above. You can see that the **calcium ions** formed are soluble in water. Therefore the calcium gets into the water, making it hard.

You can test water from different places to see how hard it is.

Experiment 22.2   How hard?

Add 1 cm³ of soap solution to 10 cm³ of the water being tested. Stopper the test-tube and shake. Repeat this until you get a good lather (one that lasts at least 30 seconds).

Test different samples.
Record your results in a table:
- How could you display your results? Why? (See pages 12 and 13.)

Water tested	Volume of soap needed to get a good lather (cm³)
distilled water	
local tap water	
hard tap water	

▶ Removing hardness

You have now seen how hard water is formed. The calcium or magnesium in the water is present as charged particles called ions. The most common cause of hardness is the calcium ion, $Ca^{2+}(aq)$.

These calcium ions react with ions from soap (sodium stearate) to form **scum** :

$$\text{calcium ions (aq)} + \text{stearate ions (aq)} \longrightarrow \text{calcium stearate (s)}$$
$$\text{hard water} + \text{soap} \longrightarrow \text{scum}$$

If we can remove these Ca^{2+} ions from the water we will get rid of the hardness.

Temporary hardness

Do you remember how calcium ions from insoluble calcium carbonate get into hard water? Limestone reacts with acidic rainwater. It forms a solution of calcium ions and hydrogencarbonate ions.

This is called **temporary** hardness.

When this solution is boiled, the calcium and hydrogencarbonate ions react. They turn back into the insoluble calcium carbonate. Therefore *calcium ions are removed* from the water.

$$\text{calcium ions} + \text{hydrogencarbonate ions} \longrightarrow \text{calcium carbonate} + \text{carbon dioxide} + \text{water}$$
$$Ca^{2+}(aq) + 2\,HCO_3{}^-(aq) \longrightarrow CaCO_3(s) + CO_2(g) + H_2O(l)$$
$$\text{(limescale)}$$

The calcium carbonate formed is the **limescale** you get inside kettles and hot-water pipes.

Temporarily hard water can be softened by boiling.

Do you think this is a cheap way to get rid of hardness? Imagine your heating bills if you had to boil all your water before you could use your washing machine!

However, not all hard water can be softened by boiling. Other calcium compounds, such as calcium sulfate from gypsum, are not removed by boiling. These form **permanently** hard water.

There was a young boy called Arthur
Who tried to make a good lather
But try as he might
His bubbles weren't right.
"It's hard, soft lad!" said his father.

Limescale on the heating element in a kettle

*This hot water pipe has been almost blocked by scale (or limescale). We can remove the scale by adding a **weak acid**. This reacts with the calcium carbonate:*

$$CaCO_3(s) + 2\,H^+(aq) \longrightarrow Ca^{2+}(aq) + H_2O(l) + CO_2(g)$$

*The weak acid is called a **descaler**.*

Removing all types of hardness

We have seen how we can soften temporarily hard water by boiling. But permanent hardness can also be removed. Permanent means forever, but all types of hardness can be removed from water.

Washing soda removes hardness

1. Washing soda (sodium carbonate)
When you add washing soda to hard water, the calcium ions are removed. They react with the carbonate ions from the washing soda. This forms *insoluble* calcium carbonate :

calcium ions + carbonate ions $\longrightarrow$ calcium carbonate

$Ca^{2+}(aq)$ + $CO_3{}^{2-}(aq)$ $\longrightarrow$ $CaCO_3(s)$
(in hard water)   (from washing   precipitate
               soda)   (insoluble)

This is called an **ionic equation**. It only shows the ions which are affected in the reaction.

> **If a solid forms when two solutions are mixed, it is called a precipitation reaction.**

Most modern washing powders have their own water softeners added.

2. Ion-exchange column
This method is more suitable for large-scale treatment of hard water.
Look at the diagram opposite :
The column is filled with a resin which holds plenty of sodium ions (Na^+ ions).
Hard water goes in at the top. On the way down, the calcium ions are swapped for sodium ions. The calcium ions get stuck on the resin.
Sodium ions, which don't cause hardness, come out in the water at the bottom.

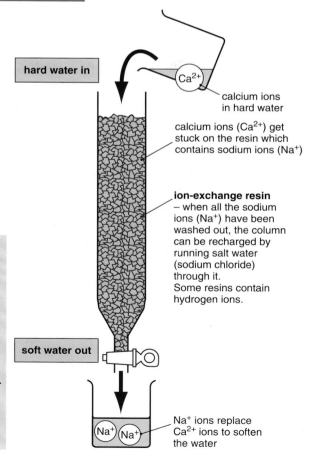

hard water in

Ca^{2+}

calcium ions in hard water

calcium ions (Ca^{2+}) get stuck on the resin which contains sodium ions (Na^+)

ion-exchange resin – when all the sodium ions (Na^+) have been washed out, the column can be recharged by running salt water (sodium chloride) through it. Some resins contain hydrogen ions.

soft water out

Na^+ Na^+

Na^+ ions replace Ca^{2+} ions to soften the water

Experiment 22.3   Removing hardness
Look back to Experiment 22.2 on page 291.
Use that method to test the hardness of your water samples before and after treatment :

1. Test some temporarily hard water.
   Now boil a fresh sample of the water and re-test it.

2. Test some permanently hard water.
   Now add a spatula of washing soda to a fresh sample. Make sure it dissolves, then re-test.

3. Test your samples of the hard water after they have passed through an ion-exchange column.

▶ Desalination

For many hot countries, getting a reliable supply of fresh water is difficult. With low rainfall, rivers and lakes run dry in hotter months. However, these countries sometimes have large coastlines. They have plenty of sea water, but this is unsuitable for most uses.

That is where **desalination** comes in. Special desalination plants can take most of the dissolved salts out of sea water (or from brackish water from marshes).

> **Desalination produces usable water from sea water by separating water from the dissolved salts.**

You have seen how distillation works on page 175. Water is evaporated, then cooled and condensed to separate it from dissolved solids.
The process used in some desalination plants is based on distillation.

You can drink the water from the sea water processed in this desalination plant

Experiment 22.4   Distilling salt water

Use the distillation apparatus on page 175 to collect pure water from a solution of sodium chloride (common salt).
You can use the glassware or the micro-scale equipment shown.
• How could you test that the water collected was pure water?

In a desalination plant, the pressure above the water is reduced. This lowers the boiling point of the sea water. The process is called **flash distillation**.
• Why is it important to lower the boiling point?
• What would be the main costs in this type of plant?

Another process called **reverse osmosis** is getting more popular. This uses a membrane to separate the water and the salts. There is no heating involved so it uses less energy than distillation. However, energy is still needed to pressurise the water. Look at the diagram:

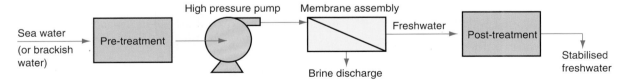

Today's membranes can remove 98% of dissolved salts from sea water. However, corrosion of pumps and pipes is a problem.

Desalination is used in Middle East countries that can use money from their oil to pay the costs of desalination. In the Virgin Islands, in the West Indies, 90% of water used is desalinated.

▶ Solubility curves

Saturated solutions

Only a certain amount of solute can dissolve in a particular volume of solvent at any given temperature. This amount varies from solute to solute.

> **A saturated solution is a solution in which no more solute will dissolve at a given temperature.**

We can describe how well a solute dissolves in water by its **solubility**.
The solubility of a substance tells us how many grams of it will dissolve in 100 g of water at a certain temperature.
For example, the solubility of sodium chloride is 36 g per 100 g of water at 25 °C.

Experiment 22.5   The solubility curve of potassium nitrate

Measure out 10 g of potassium nitrate.
Heat some water to 70 °C in a small beaker.
Transfer 10 cm³ of the hot water to a boiling tube.
Now add the 10 g of potassium nitrate to the boiling tube, and stir with a glass rod.
This should form a solution that is almost saturated.
Let the tube cool down. Note the temperature when the first crystals appear.
- At what temperature does the solution become saturated?
Now add another 5 cm³ of water to the boiling tube to make the volume up to 15 cm³. Stir the solution so that the potassium nitrate re-dissolves. Warm the tube in a water bath if necessary. Again cool the tube down and record the temperature at which the solid appears. Repeat for total volumes of 20, 25 and 30 cm³ of water.
- How will you convert your results into solubilities per 100 g of water? Remember 100 cm³ of water has a mass of 100 g. (You could use a spreadsheet.)
- Record your results and solubilities, in g/100 g of water, in a table. Think about how to make your results as reliable as possible.
- Once you have converted your results to solubilities per 100 g of water, draw a solubility curve for potassium nitrate.
- How does the solubility of potassium nitrate vary with temperature?

potassium nitrate solution

Solubility curves enable you to:
- read the solubility at temperatures you didn't investigate.
- scale up or down to work out how much solid dissolves in any volume of water.
- work out what mass of solid is crystallised from a saturated solution cooling between two given temperatures.

Example (using the graph opposite)

If 50 cm³ saturated solution of copper sulfate cooled from 70 °C to 30 °C, what mass of crystals would form?
To answer this:
1. Read the solubility at 70 °C and subtract the solubility at 30 °C i.e. (47.5 – 22.5) = 25 g/100 g of water.
2. So in 50 cm³ of water (which is the same as 50 g of water) we would get 25 ÷ 2 = **12.5 g** of copper sulfate crystallising out.

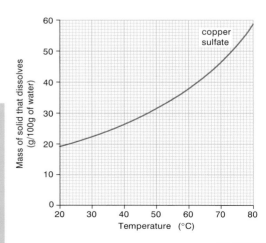

► Water pollution

Are you worried about pollution of our rivers, lakes and seas?
It is not possible to get absolutely pure water in nature
because it is so good at dissolving things. This is often
useful to us, but also makes it easy to pollute water.

Eutrophication

Do you remember the problem of **eutrophication**?
On page 240 we saw how *nitrate fertilisers* pollute
our rivers and can get into water supplies.
Detergents often contain phosphates which end up
in rivers and lakes. These also contribute to excess
growth of algae which covers the surface of the water.
Other life in the river cannot survive the competition.
As plants, including algae, and animals die
they are decomposed by micro-organisms.
These use up the dissolved oxygen in the river or lake.

Thermal pollution

Besides its use as a solvent, water is also used in industry
as a coolant. It transfers energy away from a reaction.
This raises the temperature of the water. An example
is its use to transfer energy in *power stations*. The water
is not polluted when it is passed out into a nearby river.
However, it is warmer than the river water.
This can affect the delicate balance of life
in the river. Remember that aquatic animals
rely on oxygen gas dissolved in the water.
We know that:

> **The higher the temperature, the less soluble
> a gas becomes in water.**

Power stations can cause thermal
pollution in rivers

So not as much oxygen dissolves in the warmer water
and animals die.
● How would you solve this problem?

Other waste

Chemicals discharged from factories can also pollute
our water, although rules are much stricter now.
Pesticides in our drinking water is another problem
which water companies are having to tackle.
Pesticide residues can get into waterways from crop spraying in
nearby fields. Our drinking water is now checked for acceptable
levels of pesticides.
● What are your views on this issue?

People in Minamata Bay, Japan
were poisoned by mercury which
got into the food chain

Lead in water

Lead is a toxic metal. Like other heavy metals, it tends to accumulate in organs once it gets inside your body.
It is particularly dangerous for children under the age of 6. This is because it hinders development of the brain.

So no wonder lead water pipes have been banned! Before we realised just how poisonous lead is, plumbers used lead because it is so malleable. It can be bent into shapes easily so was ideal for water pipes.

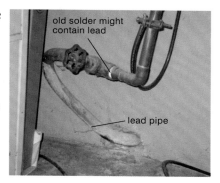

old solder might contain lead

lead pipe

Old houses had lead plumbing

Lead is not very soluble in water but a little does dissolve, and that is enough to cause problems. You get more lead ions in your water if the water has a low pH (acidic) or if the water is hot.
In the table on page 290 it says that hard water (which has a higher pH) can prevent lead or copper getting into the water.
Lead pipes in old houses should have been replaced by modern plastic pipes by now.

Even after it was banned, and copper pipes were introduced, lead still managed to get into drinking water. That's because the plumbers joined up copper pipes using solder containing lead. So houses built between 1930 and 1980 still have some risk of lead contamination.

However, there are some simple things to do to avoid the risk of lead:
- If you haven't used a tap for some time, run the water for about a minute before drinking from it. That's because water that has been in old pipes for a few hours has had a chance to dissolve more lead.
- Never drink, or brush your teeth with, hot water.
- Never drink water from the bathroom.

Water filters

More and more people are using filters for their drinking water. Even though our drinking water meets high standards, the filters are seen as an extra safety measure.

The filters contain **carbon**, impregnated with **silver**. The silver is an excellent bactericide. Although it is expensive, a filter will only contain about 0.07% silver. The filters can be joined to your cold water tap or you can pour your water through a filter jug.
Carbon absorbs organic compounds. Some filters also include an **ion-exchange resin** that gets rid of metal ions. However, some calcium ions in your water will bring health benefits. (See page 290.)

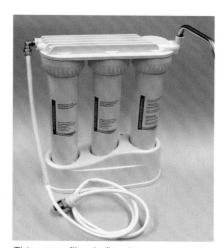

This water filter is fitted to your cold tap in the kitchen. You can buy taps with 'Cold', 'Hot' and 'Pure' handles to turn.

The best filters will absorb:
- heavy metals, such as lead, copper, cadmium, chromium, etc.
- bad smells and tastes
- chlorine
- most calcium and magnesium ions (removing about 60%)
- as well as killing 99% of bacteria.

Investigation 22.6   Purifying water

Your task is to purify a sample of muddy water.
You are given a yoghurt pot, sand and fine gravel.
If you want any other materials (except filter paper !)
you can ask your teacher.
See who can get the cleanest sample of water.

- Why wouldn't you try drinking your 'clean' water?
- How do water companies make water safe to drink?

Summary

- The 'water cycle' shows how water moves around the Earth.

- Water is a good **solvent**. It forms **solutions** with many **solutes**.
  This means that water from natural sources can have a range
  of pollutants dissolved in it.

- Drinking water is purified by physical and chemical means
  before it reaches our taps. Some people also use water filters
  at home.

- **Hard water** contains dissolved *calcium or magnesium* ions.
  We can soften the water by precipitating these ions out of solution
  or by passing it through an ion exchange column.

- We can plot the solubility (in g/100 g of water) of a substance
  at different temperatures on a '**solubility curve**'.

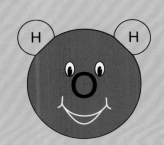

► Questions

1. Copy and complete:
   Water is good at . . . . many solutes. The water
   is the . . . . in the solution formed.
   Hardness in water is caused by . . . . or . . . .
   ions. These can be removed by a . . . .
   reaction or by using an . . . . column.

2. a) Explain why we sometimes describe
      oceans as 'reservoirs for carbon dioxide'.
      Use chemical equations in your answer.
   b) Why is it important that we reduce
      the amount of carbon dioxide
      in our atmosphere.
   c) Some countries argue that they should be
      allowed to produce more carbon dioxide
      because they have lots of forests. Explain
      their reasoning. What are your views ?

3. a) Why is shortage of water a problem for
      some hot countries even though they have
      large coastlines ?
   b) Explain the process of desalination.
   c) What is the main disadvantage of
      desalination using distillation ?
   d) Name another process that can be used
      instead of distillation. What is its
      advantage over distillation ?

4. Limestone is insoluble in water. However,
   water flowing through limestone becomes
   hard.
   a) Explain how calcium ions get into the water.
   b) List the advantages and disadvantages of
      hard water.
   c) Explain how adding washing soda removes
      hardness from water.
   d) How does an ion-exchange resin soften water?

5. Anna tested 3 samples of water with soap solution. She recorded how much soap was needed to get a permanent lather, before and after boiling:

Water sample	Soap solution added (cm³)	
	before boiling	after boiling
Sample A	8	7
Sample B	1	1
Sample C	7	2

a) What can you deduce about each sample of water?

b) Explain the chemical difference between permanent and temporary hard water.

c) Why is boiling not a good idea to remove temporary hardness from water on a large scale?

d) How can you remove limescale from hot water pipes or kettles?

6. This question is about purifying water from a river so that it is safe to drink.

a) What is aluminium sulfate used for in the treatment of water?

b) Explain why aluminium sulfate is better at this job than sodium sulfate.

c) Why is calcium hydroxide added to water?

d) How do we remove unpleasant tastes and odours from water?

e) Why is it desirable to remove iron(III) ions from water?

f) Typhoid and cholera are diseases that are carried in water. Why don't we have any cases of typhoid or cholera in this country?

g) What is added to water to get rid of excess chlorine before the water reaches our homes?

7. Imagine you are teaching a group of 12 year olds about how our drinking water is treated to make it safe. Draw a chart you could take with you to help explain.

8. Write a newspaper article on water pollution and how we can monitor water quality nowadays. Pages 364 and 365 can be the starting point of your research.

9. Look at the solubility curves below:

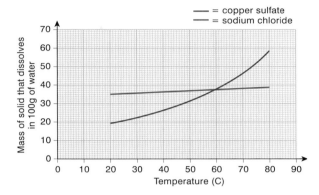

a) What do we mean by 'saturated' solution?

b) What mass of copper sulfate will dissolve in 100 g of water to form a saturated solution at 55 °C?

c) Compare the effect of temperature on the solubility of copper sulfate and sodium chloride.

d) At which temperature do copper sulfate and sodium chloride have the same solubility?

e) You add 100 g of sodium chloride to 100 g of water at 80 °C. What mass of sodium chloride remains undissolved?

f) What is the maximum mass of copper sulfate that will dissolve in 25 cm³ of water at 40 °C?

g) You cool 50 cm³ of a saturated solution of copper sulfate from 50 °C to 25 °C. What mass of copper sulfate crystallises out of the solution?

h) A student did an experiment to compare the solubility of 5 different solids at 20 °C. What type of graph would she use to display her results? Explain your answer.

Further questions on pages 322 and 323.

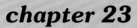

chapter 23

The Atmosphere

The Atmosphere

▶ **Gases in the air**

We live in a mixture of gases, called air.
The air makes up the Earth's atmosphere,
and it is essential for life on our planet.

Air is mainly a mixture of 2 gases – **nitrogen and oxygen**.
But do you know which other gases are all around us?
Look at the table opposite:

There are also small amounts of pollutant gases.
Pollution varies from place to place.

Gases in air	% of air
nitrogen	78
oxygen	21
carbon dioxide	0.04
water vapour	varies
argon	0.93
other noble gases	} 1

The composition of the air has been much the same for the last 200 million years

Solubility of gases

We saw in the last chapter (on page 296) how life in water
depends on dissolved oxygen gas.
The amount of oxygen dissolved depends on the temperature.
Look at the graphs below:

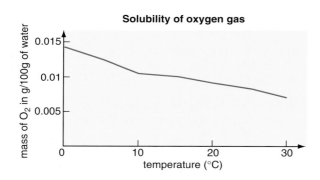

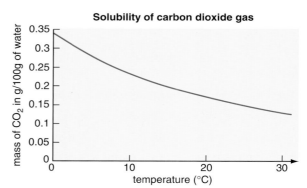

You can see that:

> **Gases get less soluble as the temperature of water rises.**

Look at the first graph:
● Can you remember about the thermal pollution of water?
  Why is this a problem to aquatic life?

Look at the second graph:
We also saw on page 285 how the oceans are important
in dissolving carbon dioxide gas (a 'greenhouse' gas).
● What will happen to the amount of carbon dioxide absorbed
  by the oceans as their temperature rises due to global warming?
  Why is this a problem?

► Chemistry at work : Uses of nitrogen

Nitrogen is separated from liquefied air by **fractional distillation**.
The air is turned into liquid by repeatedly compressing, cooling and letting
the gases expand. Carbon dioxide and water vapour are separated off as
solids first of all. Once the temperature is at about $-200\,°C$ most of the
gases have condensed into liquids. (Look at the table opposite.)
These gases are then allowed to warm up in a fractionating column.
The nitrogen boils off first at $-196\,°C$. It is collected from the top of the
column and liquid oxygen comes out at the bottom.

Gas	Boiling point/°C
nitrogen	-196
argon	-186
oxygen	-183

Nitrogen for freezing

Liquid nitrogen is very cold. It's almost $-200\,°C$!
Therefore, it can be used to freeze things quickly.
Food can be frozen on a conveyor belt
as it is produced.
Hospitals can use it to store tissues
for many years.

It can also be used to mend leaking pipes.
You can pour liquid nitrogen on the pipe.
It freezes the liquid inside the pipe while
you repair the leak.
This saves money because you don't
have to drain the whole pipe.

Liquid nitrogen can also freeze
marshy ground that is
too wet for mechanical diggers.

*Sperm can be frozen and
stored in liquid nitrogen*

Nitrogen is not reactive

As well as freezing food, nitrogen helps to stop
food 'going off' in another way.
Nitrogen is an un-reactive gas.
So when we pack foods, nitrogen is used
inside the sealed packaging.
This keeps the reactive oxygen gas in the air
away from the food. Bacteria cannot multiply,
and the food stays fresh longer.

There is a risk of fire on the giant ships
that carry crude oil. The vapour from the oil is dense,
and can form an explosive mixture with the air.

It is especially dangerous when oil is pumped ashore,
or when the ship's tanks are cleaned out.
Nitrogen gas is pumped into the tanks to remove any oxygen.
Then an accidental spark will not result in disaster.

Packing peanuts on a production line

▶ Oxygen gas

Do you know which is the reactive gas in the air?
Think of some reactions we have seen, such as
combustion or oxidation. Which gas is needed
for these reactions to take place?

Oxygen is used up when things burn or
during respiration in our cells.

Look at the reactions below:

**Remember the test for oxygen:
a glowing splint re-lights.**

Experiment 23.1   Oxygen and burning

Try the experiment shown with the night-light.

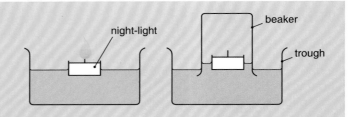

- What happens to the flame?
- What happens to the level of the water in
  the beaker?
- Explain why this happens.
- Wax is a hydrocarbon. What is formed
  as it burns? (See page 147.)
- Why *can't* we use this experiment to find out the proportion of oxygen in air?

Experiment 23.2   Oxygen and rusting

Set up the apparatus as shown:

Leave it for a week.

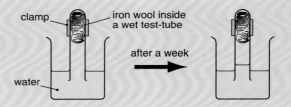

- What happens to the level of water in the test-tube?
- Explain why this happens. (See page 76.)

Experiment 23.3   How much oxygen?

Set up the apparatus as shown:

Heat the copper. At the same time,
pass air over it, from syringe to syringe.
Carry on heating until the volume does not
go down any further.
Let the apparatus cool, then read the final volume
of air left.

- How much air is left at the end?
- Explain what has happened (include the word equation:
  copper + oxygen ⟶ copper oxide).
- Why should you let your apparatus cool down
  before taking the final reading on the syringe?
- What are the sources of error in this experiment?

► Chemistry at work : Uses of oxygen

About 20 % of the air is made up of oxygen.
Do you know the main difference between
oxygen and nitrogen ?
Let's look at some uses of the reactive gas
oxygen :

Oxygen for breathing

We all need oxygen to breathe.
Can you think of any places where you need to take
your own supply of oxygen to survive ?
Look at these photos :

Premature babies sometimes need oxygen to help them breathe

This welder is burning a mixture of ethyne gas (a hydrocarbon) and pure oxygen

The air gets thinner as you go higher

Oxygen from liquid air

We can separate the gases in air by the **fractional distillation**
of liquid air. After removing water vapour and carbon dioxide,
the air is cooled and compressed. The cold, compressed air is
allowed to expand into a larger space. This cools the gases
down even more. By repeating the process several times the
temperature drops to $-200\,°C$. The oxygen and nitrogen in the
air liquefy at this temperature.
When this mixture is warmed, nitrogen boils off first as it has a
lower boiling point than oxygen. This takes place in a
fractionating column.

Gas	Boiling point (°C)
nitrogen	−196
argon	−186
oxygen	−183

► Carbon dioxide

You can see from the table on page 300
that air contains about 0.04 % carbon dioxide.
This doesn't seem a lot!
However, 250 years ago there was only 0.028 %.
As you know from page 149, carbon dioxide
is a 'greenhouse' gas. It helps to keep the Earth warm.
But can you remember why people are worried
about the increasing amount of carbon dioxide?

CO$_2$

- is colourless
- is slightly soluble in water (forming a weakly acidic solution)
- is denser than air
- puts out fires
- turns limewater milky

Carbon cycle

Carbon dioxide is a vital part of the **carbon cycle**.
The carbon cycle shows how carbon is moved
from place to place around the Earth.
Look at the cycle below:

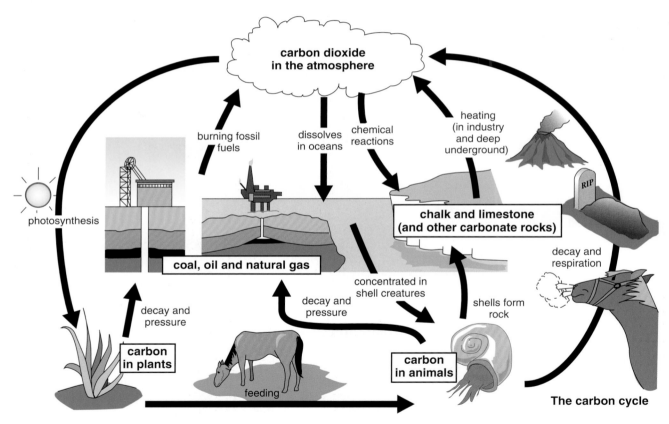

The carbon cycle

Most of the carbon on Earth is in rocks.
The carbon trapped in fossil fuels
is released when we burn the fuels.
The oceans absorb a lot of the CO_2 produced.

Insoluble carbonates and soluble
hydrogencarbonates are formed. (See page 285.)
However, the oceans cannot remove all
the extra CO_2 we are now producing.

► Chemistry at work : Uses of carbon dioxide

Fizzy drinks

Carbon dioxide gas is slightly soluble in water.
It forms a weakly acidic solution.
Look at the photos :

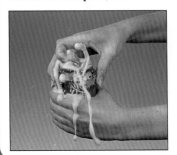

Carbon dioxide puts the 'fizz' in drinks. It is dissolved under pressure into the drink to form 'carbonated' water.

The higher the pressure the more soluble a gas becomes.

What happens when you release the pressure by opening a can or bottle ?

Champagne and sparkling wine is bottled before fermentation has finished. (See page 166.) The carbon dioxide builds up pressure in the bottle.

'Dry ice'

If you cool down carbon dioxide gas, it turns into a solid. It doesn't form a liquid.
The solid carbon dioxide is known as 'dry ice'.
It is much colder than ice. At room temperature, it sublimes (turns from the solid into a gas).
Look at these uses of 'dry ice' :

Ice cream samples can be kept cold with 'dry ice'

Lumps of 'dry ice' sublime. This cools down the water in the air to give smoky effects on stage.

Fire extinguishers

Most things won't burn in carbon dioxide gas.
For example, it puts out a lighted splint.
However, this is *not* the test for carbon dioxide.
Other gases, such as nitrogen, do the same thing.
Do you know the test for carbon dioxide ?
Can you remember why the limewater goes milky ?
(See page 111.)

Carbon dioxide is denser than air. It forms a 'blanket' over a fire. This cuts off the oxygen supply for the combustion reaction.

Carbon dioxide puts out most fires

▶ Chemistry at work : Ozone

Ozone molecules are made up from 3 oxygen atoms, O_3.
They are found mainly in our upper atmosphere.
Ozone molecules absorb harmful ultraviolet light
from the Sun. They protect us and without it life on Earth
would not have developed as it has. (See the next page.)

*Ozone shields us
against harmful
u.v. light*

Have you heard about the hole in the ozone layer?
On page 55 we saw how CFCs (chloro-fluoro-carbons)
are damaging the ozone layer around our planet.
CFCs are unreactive organic compounds containing
chlorine, fluorine and carbon atoms e.g. CCl_3F.
They were used in aerosol sprays and in fridges until scientists
discovered their effect on the ozone layer in the 1980s.

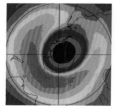

*The ozone layer is
under attack.
This shows the hole
over Antarctica.*

In the upper atmosphere the CFCs are broken down
by energy from sunlight. A chlorine atom forms
when its covalent bond, to carbon, breaks :

$$CCl_3F \xrightarrow{\text{energy from sunlight}} CCl_2F^{\bullet} + Cl^{\bullet}$$

Remember that a covalent bond is a pair of electrons.
The covalent bond between carbon and chlorine
breaks 'evenly'. Carbon takes one electron,
and chlorine takes the other:

$$-\overset{|}{\underset{|}{C}} \!:\! Cl \longrightarrow -\overset{|}{\underset{|}{C}}{}^{\bullet} + Cl^{\bullet}$$

(a chlorine atom or free radical)

FREE
RADICAL

*Free radicals are
very reactive!*

Atoms or molecules with a free (spare) electron are called free radicals.

Free radicals are *very reactive*.
The chlorine free radicals attack ozone molecules
in rapid chain reactions, destroying our protection.

As more ultraviolet light reaches us on the surface,
we will see more health problems, such as:

- increased risk of sunburn
- faster ageing of our skin
- skin cancer
- damage to our eyes, such as cataracts.

*Do you protect your skin from
dangerous u.v. light?*

Fortunately many countries have now banned the use of CFCs.
Chemists have found other compounds to use instead.
HFCs (hydro-fluoro-carbons) and alkanes do not damage the
ozone layer.
However, other developing countries are still producing CFCs.
- Why do you think they still use them?
- How could nations help each other to solve this problem for us all?

▶ Where on Earth did our atmosphere come from?

It is hard for us to imagine how old the Earth really is. It might help you to think of the Earth's history as a 24-hour clock. On this scale, humans arrived on Earth at one second to midnight!

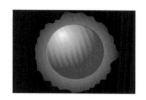

When the Earth was formed (about 4600 million years ago) its atmosphere was probably made from hydrogen and helium for a short time, just like 99% of the universe. The Earth itself was a mixture of molten material including iron.

As the Earth cooled down, a solid crust formed on the outside. The molten rock underneath often burst through the thin crust. Volcanoes were erupting all over the Earth's new surface. These volcanoes gave out gases, just like volcanoes do today. Some theories suggest that the main gas was carbon dioxide, just like the atmospheres of Mars and Venus today. There was also some water vapour and nitrogen plus small amounts of methane and ammonia in this early atmosphere.

volcanic gases

As the Earth cooled down even more, the steam condensed and fell as rain. It filled up hollows in the solid crust, and the **oceans** were formed.

first seas and oceans formed

The first **living things** are thought to have developed in the oceans. One theory is that life started near volcanoes on the ocean floor. All the elements needed for simple cells to evolve were there. The first organisms evolved into simple plants, like algae. These used up carbon dioxide during photosynthesis and made the first molecules of **oxygen** gas (O_2). The amount of oxygen in the atmosphere built up slowly over millions of years.

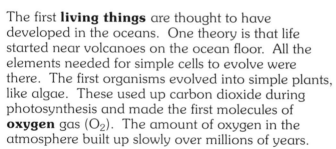

The oxygen was important because some of it turned into **ozone** (O_3). Ozone stops harmful rays from the Sun reaching the Earth's surface. (See page 306.) So it became possible to live out of the water.

Some of the oxygen reacted with ammonia, giving off **nitrogen**. More nitrogen was formed by bacteria living in the soil. The levels of methane began to fall as it too reacted with oxygen.

Eventually, about 200 million years ago, the atmosphere reached the mixture we have today of roughly 20% oxygen and 80% nitrogen. Much of the carbon from CO_2 in the original atmosphere is now trapped in fossil fuels and carbonate rocks. (See the next page.)

Summary

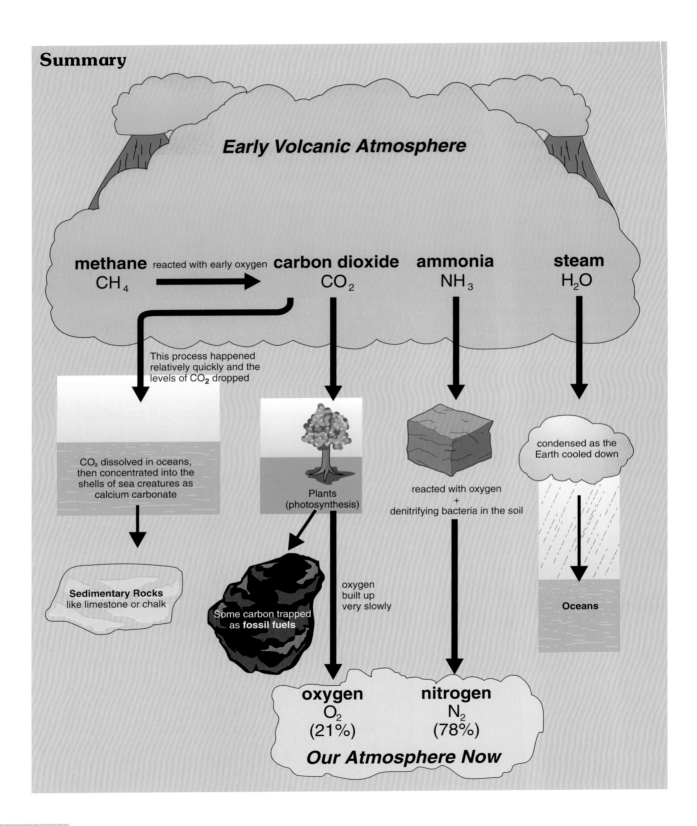

Early Volcanic Atmosphere

methane
CH_4 → reacted with early oxygen

carbon dioxide
CO_2

ammonia
NH_3

steam
H_2O

This process happened relatively quickly and the levels of CO_2 dropped

CO_2 dissolved in oceans, then concentrated into the shells of sea creatures as calcium carbonate

Plants (photosynthesis)

reacted with oxygen
+
denitrifying bacteria in the soil

condensed as the Earth cooled down

Sedimentary Rocks
like limestone or chalk

Some carbon trapped as **fossil fuels**

oxygen built up very slowly

Oceans

oxygen
O_2
(21%)

nitrogen
N_2
(78%)

Our Atmosphere Now

308

► Questions

1. Copy and complete:
The Earth's atmosphere contains about four-fifths . . . . and one-fifth . . . . .
However, its early atmosphere was made up of gases from . . . . These included carbon dioxide, ... , ... and steam. The first oceans formed when the steam . . . . as the Earth cooled down. The first . . . . gas was formed by photosynthesis in plants.

2. Sami did an experiment to find out how much oxygen is in the air.

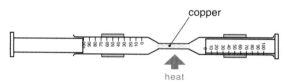

She started with $100\,cm^3$ of air. She passed the air back and forth over the hot copper by pushing each syringe in turn. She saw the copper turning to black copper oxide.

a) Use a word equation to show how Sami has removed the oxygen from the air.

b) Sami carried on with her experiment until there was no further change in the volume of gas left. Not all the copper had turned black. What volume of air would you expect to be left?

c) Which gas would make up most of the sample left at the end of her experiment?

d) Why should Sami let the gas cool down before taking her final volume reading?

e) What would happen if Sami started with too little copper in the heated tube?

f) One of the syringes had a plunger that did not move smoothly in and out. Explain how this might affect the results of the experiment.

g) Wayne said that it would be better to use a candle standing in water, covered by a beaker to find out how much oxygen is in the air. List as many points as you can to persuade Wayne that he is wrong!

3. a) Why does the amount of water vapour in the air vary?

b) Draw a flow diagram of the water cycle. Include these words:

> clouds, sea, rivers, rain
> evaporates, condenses

c) How do the i) soil, ii) plants and iii) animals, add to the water vapour in the air?

d) Where does the energy for the water cycle come from?

4. Look at the carbon cycle on page 304.

a) Why does the amount of carbon dioxide in the air vary from place to place?

b) The word equation for photosynthesis is:

carbon + water ⟶ glucose + oxygen
dioxide

Explain how the carbon in carbon dioxide gets into plants.

c) Explain 4 ways in which the carbon in plants can get back into the atmosphere.

d) What happens to the carbon dioxide absorbed in the oceans? (Also see page 285.)

5. a) Which volcanic gases made up the Earth's early atmosphere?

b) Which planets today have atmospheres similar to Earth's early atmosphere?

c) Describe the processes that removed the original gases from the atmosphere. How did this increase the proportions of oxygen and nitrogen in the air?

d) Why was ozone important to the development of life on Earth?

e) i) How has pollution damaged the ozone layer?
   ii) What are we doing to tackle the problem?

6. Draw spider diagrams showing some uses of nitrogen, oxygen and carbon dioxide.

Further questions on pages 324–325.

chapter 24

The Earth

► Journey to the centre of the Earth

As you know, the Earth is surrounded by a thin layer of gases. But have you ever wondered what is inside our planet?
If you started digging a hole straight down, about 13 000 kilometres later you would come out in Australia!
But what would you find on the way?

First of all you would go through the Earth's thin **crust**. All of our resources for raw materials come from the crust, the oceans and the atmosphere.

It can be as thin as 5 km under the oceans, going up to about 70 km under the continents. The thin crust is the least dense of the Earth's layers.

Under the crust you find the **mantle**. This layer goes down almost half way to the centre of the Earth.

The mantle is almost entirely solid. However, there is a small amount of molten material under the **lithosphere** (crust and uppermost part of the mantle). This zone, under the lithosphere, is called the **asthenosphere**.

Next, you reach the **outer core**. This is a dense liquid, made of molten iron and nickel. Both of these metals are magnetic. They make the Earth behave like a giant magnet itself.

Finally there is the **inner core**. This is the densest part of the Earth. Unlike the outer core it is solid because of the very high pressure. It is also made of iron and nickel.

The outer and inner core make up just over one-half of the Earth's radius.

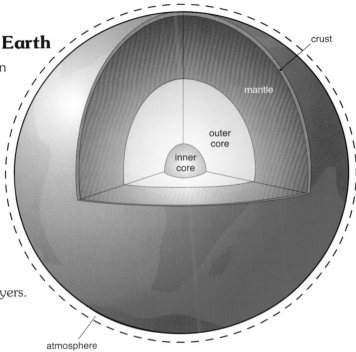

The crust and upper part of the mantle are called the **lithosphere**. *The overall density of the Earth is greater than the average density of its crust. This shows that the materials inside the Earth must be different to, and more dense than, the rocks in the crust.*

There was a young man from Perth,
Who tunnelled for all he was worth.
He became a real bore
As he shot through the core
And came out t'other side of the Earth!

▶ Igneous rocks

When molten rock cools down, it usually forms crystals. These make up **igneous** rock. The word igneous comes from the Latin word for 'fire' (*ignis*).

The molten rock comes from deep underground. It tends to rise to the surface through cracks in the Earth's crust. If the molten rock, or **magma**, reaches the surface, it forms a volcano.

Some igneous rocks are made from small crystals, like the rock basalt. Others have larger crystals, like granite. So what affects the size of the crystals in an igneous rock?

*"This 'hard case' has grains I see well,
A rock that has risen from the fires of hell.
The rock set in crystals as the magma rose higher,
Forming **igneous** rock – from the Latin for 'fire'."*

Sherlock inspects granite rock

Experiment 24.1   Crystal sizes

You cannot melt rocks with a Bunsen burner. In this experiment you will use a solid called 'salol' to model the behaviour of rock. If you look carefully you can see how the rate of cooling affects crystal size.

Watch the crystals form on the two slides. You may need to start off the crystals on the warm slide with a small 'seed' crystal of salol. You can use a hand lens to observe the size of the crystals.

- What do you conclude from this experiment?

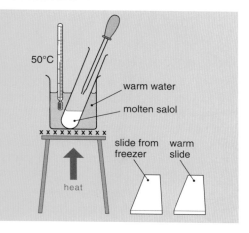

50°C — warm water — molten salol — heat — slide from freezer — warm slide

---

Molten rock that erupts from a volcano is called lava. It cools down quickly. This forms rocks with small crystals, called 'fine grained' rock.
Basalt is the main example. It is a dense rock, rich in iron minerals. Most ocean floors are made from basalt.
It is an **extrusive** igneous rock. **Rhyolite** is another extrusive igneous rock. This is rich in silica and is less dense than basalt.

Other molten rock rises towards the surface but never reaches it. It cools down slowly, surrounded by rocks under the ground.
The solid rock formed has large crystals, like **granite**.
It is said to be 'course grained'.
It is an **intrusive** igneous rock. **Gabbro** is another course grained rock. But whereas granite is rich in silica, gabbro is rich in iron.

The granite formed beneath the surface can later be exposed by Earth movements or by weathering and erosion of the softer surrounding rock.

You can find out more about igneous rocks on page 318.

Basalt : quick cooling ⟶ small crystals

Granite : slow cooling ⟶ large crystals

▶ Our changing planet

We can use the natural radioactivity in rocks
to find out how old the Earth is.
It is thought to be about 4 600 000 000 years old !
This vast time-scale is hard for us to imagine.

There has been plenty of time for changes
caused by :
- weathering
- erosion
- volcanoes,
- movements of the Earth's lithosphere.

These have re-modelled the surface of the Earth.

New mountain ranges have formed to replace
older mountains worn away by weathering.

Drifting continents

In 1915, a German scientist called Alfred Wegener
put forward his ideas about the history of the Earth.
It had already been noticed that Africa and
South America looked like two pieces in a jig-saw.
Alfred thought that the two continents were once
joined together. Over millions of years they
became separated and then drifted apart.

He found that the types of rock match across
Africa and South America. So did the fossils.

But he could not explain *how* the continents had
moved.

	ancient rocks (over 2000 million years)

area where fossils of Mesosaurus (a reptile) are found

*The shapes of South America and Africa slot together as in
a jig-saw. This gives us evidence that the continents were
once joined and must have drifted apart.*

- *What other evidence is there ?*

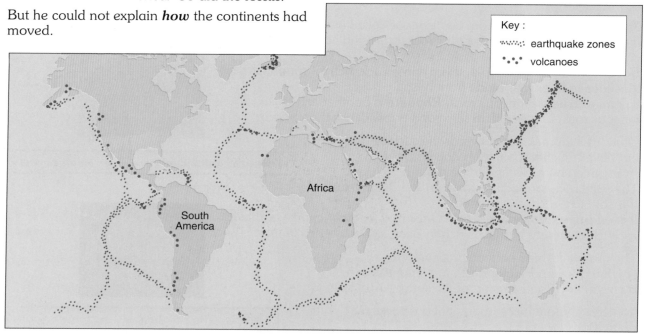

Key :
- ⌇⌇⌇ earthquake zones
- •••• volcanoes

▶ Plate tectonics

Perhaps it was not surprising that other scientists didn't think much of the 'drifting continents' theory.

People believed that features like mountains, were formed when the Earth's early crust shrank and wrinkled as it cooled down. (See page 307.) They also thought that the continents were once joined by 'bridges' of land that had since sunk below the oceans.
However, after Alfred Wegener's death, scientists started to notice things that supported his theory.

The same rock and fossil features were found between **other pairs** of continents. Then explorers of the oceans discovered evidence that the sea-bed was spreading. In other words, continents **were** actually moving apart.
At last, people had to take the ideas of Alfred Wegener seriously.
The theory of **plate tectonics** was formed.
This says that the Earth's crust and upper mantle is made up from huge slabs of rock, called tectonic plates. It also explains how these massive chunks of lithosphere move.

The plates are still moving today, although only a few centimetres a year. This small movement is enough to cause earthquakes and volcanoes.

Large earthquakes happen where plates slip a few metres past each other. Volcanoes appear where molten rock from beneath the crust rises to the surface through joints between the plates.

Look at the map on the last page :
It shows areas affected by earthquakes and volcanoes.
You can use these to find the shape of the plates below :

All the continents were once joined together. The huge land mass was called Pangaea.

Key :
- ▲▲▲ mountain ranges
- —— plate boundaries
- ➤ direction plates are moving

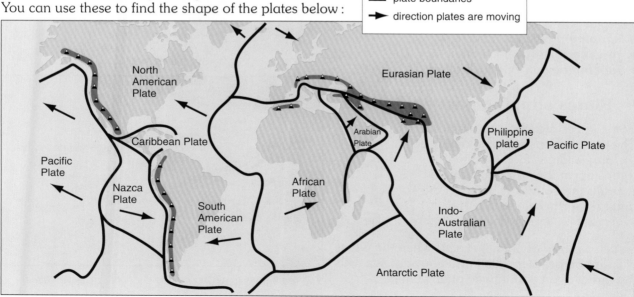

North American Plate

Eurasian Plate

Caribbean Plate

Arabian Plate

Philippine plate

Pacific Plate

Pacific Plate

Nazca Plate

South American Plate

African Plate

Indo-Australian Plate

Antarctic Plate

▶ Moving plates

Nowadays, we have developed theories to explain how the plates move.
Do you remember what lies underneath the lithosphere?
The asthenosphere contains up to 10% molten material which forms a
film around the edges of crystals. This allows solid material to flow
very slowly under the lithosphere. But even the solid mantle beneath
the asthenosphere can flow too. The plates move because of the
huge forces caused by **convection currents**.
These are set up in the slowly flowing mantle under the tectonic plates.

Experiment 24.2   Convection currents

Set up the apparatus as shown :
Use tweezers to place a crystal of
potassium manganate(VII)
at the bottom of the beaker.

⚠ potassium manganate(VII)

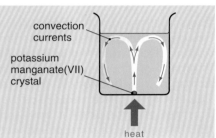

* Draw a diagram to show what happens.
* Explain how two of the Earth's plates might move apart.

We know it gets hotter the deeper into the Earth we go.
But where does the heat come from ?
Natural radioactive atoms are decaying inside
the Earth all the time.
These nuclear reactions give out lots of energy. One theory
suggests that they give out enough energy to set up
the convection currents that move the Earth's plates.

Look at the map on page 313 :
It shows the plates and the direction they are moving.

The plates in different parts of the world can be :
* *slipping past* each other,
* *moving towards* each other, or
* *moving away from* each other.

*This is the San Andreas
fault. You can see where
the two plates meet.*

▶ Plates slipping past each other

Some plates are trying to slip past each other.
The most famous example of this is
the San Andreas fault.
It runs down the west coast of the USA.
It marks the boundary between the Pacific plate
and the North American plate.
When the plates slip, there is an earthquake.

As yet, we cannot predict exactly when the plates
will slip. So people in California have to live with
the threat of an earthquake at any time.
(See page 320.)

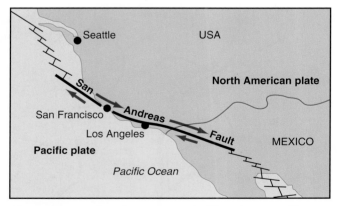

*The red arrows show the movement of the plates on either
side of the fault*

Earthquakes

Have you seen the TV pictures on the news after an earthquake?
Look at the photos of the damage they can cause:
As you can see, the shock waves from
an earthquake can be very strong.

Friction stops two plates from sliding past
each other smoothly.
During an earthquake, the two plates
overcome the friction. They slip suddenly.

This happens at the earthquake's **focus**.
The focus may be deep underground.
Then the shock waves spread out in all directions.

The place directly above the focus on the surface
is called the **epicentre**.
The waves get weaker as they travel further
from the epicentre, but they can still be
detected on the other side of the world!

Scientists use a finely balanced instrument
called a seismograph to record these vibrations.

Japan lies on the boundary of two plates.
Tall buildings in Japanese cities are now built on giant
shock absorbers. Special computers can adjust the
shock absorbers to reduce the movement of the building.

On Boxing Day, 2004, an earthquake under the Indian Ocean caused a huge tidal wave. The 'tsunami' killed over a quarter of a million people. An early warning system is now being put in place. However, scientists cannot predict when an earthquake will take place accurately yet. (See page 319.)

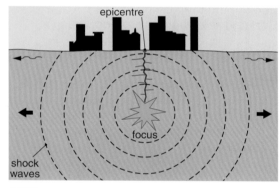

Some buildings in Japan are designed to withstand an earthquake

Around 100 000 people died in this earthquake in Pakistan

In 2005, an earthquake in Pakistan killed around 100 000 people.
Most people were crushed as buildings collapsed.

- Why do you think so many buildings failed
  to stand up to the shock waves?

You can read about predicting earthquakes on page 320.

▶ Plates moving towards each other

There are 2 types of material forming the plates.
In some places the plates are thin and dense.
These are found under the oceans (basaltic crust).
Other plates are thicker and less dense.
These make up the continents (granitic crust).

In some places 2 plates are moving towards each other. Where they meet, the denser oceanic crust slips under the continental crust. This is called **subduction**.

The friction between the plates can cause earthquakes.
The rock can even get so hot that it melts and rises to the surface to form volcanoes.

Look at the map on page 313 :
Can you see where this is happening ?

Eventually, the whole oceanic part of the plate slips under the continental crust.
Then two areas of continental crust collide.
The crust folds upwards, forming mountains.
Metamorphic rocks can form under this great pressure.
The Himalayas and the Alps were formed like this.

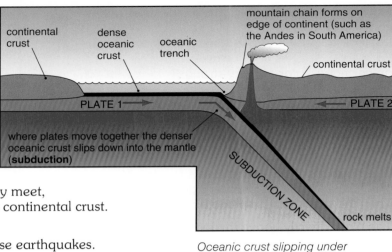

Oceanic crust slipping under continental crust (subduction)

eventually

Continental crust colliding

Folds and faults

The great forces involved when plates move can cause layers of rock to snap. This forms a **fault**.
Sometimes the layers bend, forming a **fold**.
Layers can even be turned upside down!

> *Experiment 24.3   Smashing plates !*
>
> Use coloured strips of modelling clay to show the layers of rock.
> Move 2 blocks of wood together to show how the layers can fold.
> Use a knife to cut the layers, then move them slightly to form a fault.
>
> - Can you think of a reason why rocks sometimes form a fold and sometimes a fault ?

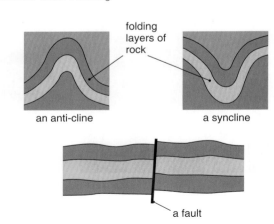

▶ Plates moving away from each other

Where plates move apart, magma rises to the surface.
This usually happens under the oceans.
As the molten rock sets, a ridge forms.
An example is the mid-Atlantic Ridge.

Look at the map on page 313.
Can you see where the mid-Atlantic Ridge is?

Look at the photo of Iceland:
Iceland is part of the mid-Atlantic Ridge.
Islands are made when the new rock builds up
above the level of the sea.

Sea floor spreading

You have already seen how the continents
were once joined together. Studies of the
ocean floor have given us more evidence
that some plates are moving apart.

The rocks at the ridge contain a lot of
iron compounds. As you know, iron is magnetic.
Its atoms line up when the molten rock sets
on the sea bed. They point towards the North Pole.

However, every few hundred thousand years,
the Earth's magnetic field is reversed.
The South Pole becomes magnetic north.
The rocks on the sea floor show
when these changes happened.

Look at the bands of rock in the diagrams:

They give us evidence that the sea floor is spreading.
In fact, each year you have to travel
about an extra 2 cm to get to America!

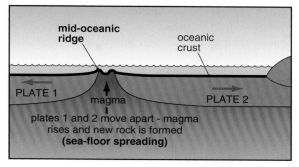

*Iceland is part of the
mid-Atlantic ridge*

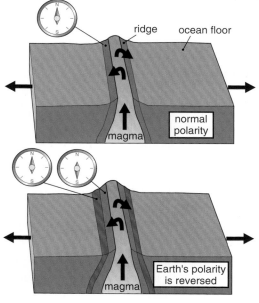

*Stripes of rock with reversed magnetism have
built up on the sea floor. (See page 321, Q7.)*

► Volcanoes

You have seen how molten rock, called magma, rises through faults in the Earth's crust.
If the magma breaks through to the surface, it is called **lava**.

Volcanoes form where the lava sets into rock.
Do you think that the rock formed will be **igneous**? (See page 311.)

Lava varies in different places. Some lava is runny, some is thick. This is because magma can have different mixtures of minerals in it.

It can also arrive at the surface at different temperatures, or with different amounts of gases dissolved in it.

All of these affect the thickness (viscosity) of the lava. You can try this investigation:

Mount St Helens in the USA. It erupted in 1980, killing more than 60 people.

This lava has cooled down and has almost stopped flowing.

Investigation 24.4   *How does temperature affect the thickness of lava?*

You can use treacle instead of lava!

- Plan your investigation.
- Make sure it is a fair and safe test.
- How will you measure the rate of flow?
- Check your plan with your teacher before you start!

The shape of a volcano depends on the thickness of its lava. **Iron-rich basalt** lava forms this type of 'safe' volcano shown opposite:
Do you think that runny lava will set nearer to a volcano's vent (outlet)?
Thin, runny lava forms a volcano which is very wide, with shallow sides. We don't usually get violent eruptions in this type of volcano.

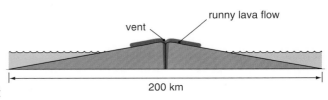

Mauna Loa, Hawaii (a shallow-sided volcano)

Look at the bottom diagram:

Thicker lava sticks nearer the vent. **Silica-rich rhyolite** forms like this.
It forms a steep sided volcano.
This type of volcano is more dangerous.
The vent can get blocked up by the thick lava.
The pressure builds up inside. Suddenly, we get a violent eruption as the lava bursts through.

The eruption throws out 'bombs' of molten lava and ash rises high in the air. We also get pumice (a light rock full of gas bubbles) formed as the lava cools down. When the ash settles, we can get sedimentary rock forming over time as layers build up.

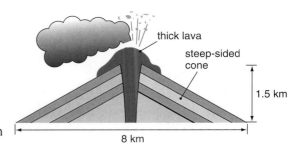

Mount Pelée, West Indies (a steep-sided volcano)

► Chemistry at work : Predicting volcanoes

Volcanoes

Volcanologists have a range of data they use to try and predict eruptions. They can give us the *probability* of an eruption, but they are still vague. It's a bit like weather forecasters saying you have a 30% chance of rain. We find that people live near volcanoes because the land is fertile. So giving a warning of an eruption can save thousands of lives.

The scientists carefully monitor a volcano, looking for signs of abnormal activity.
Here are some techniques volcanologists use to predict as best they can :

- They use **tiltmeters** – these are a bit like spirit levels. They can detect slight changes on the sides of the volcano. As magma builds up before an eruption, the sides start bulging. Look at the diagrams opposite:

- They can measure the distance across the top of the volcano accurately using **laser beams** or surveying instruments. If the cone gets wider, that is a sign of magma building up. The latest **GPS** (global positioning satellite systems) are also now used to monitor movement.

- Volcanologists have found that as magma rises nearer the surface, the mixture of gases normally given off changes. For example, there may be a sudden increase in the percentage of sulfur dioxide released.

- They can monitor the vibrations at the surface, looking for signs of increased activity. Before an eruption, **seismometers** pick up hundreds of tremors per hour. Every volcano is different. But by studying a volcano over a long period of time, you can get to know its own signs of an eruption. Even then we can't be sure of the exact size.

Volcanologists still collect data first-hand, despite the use of satellites nowadays

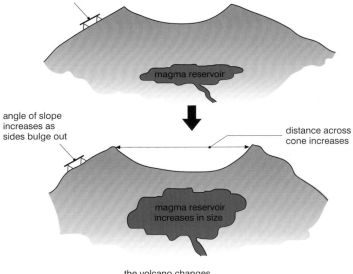

tiltmeter measures angle of the slope (this sensitive instrument could detect a 1 mm rise in a plank 1 km long!)

magma reservoir

angle of slope increases as sides bulge out

distance across cone increases

magma reservoir increases in size

the volcano changes before an eruption

We can monitor tiny earth tremors using seismometers

► Chemistry at work : Predicting earthquakes

Earthquakes

As we know from page 315, earthquakes can be devastating.
But again, making accurate predictions of when they will take place is difficult.

The same technology is used as volcano forecasts. Markers placed across
a plate boundary can be monitored for movement. The latest satellite images
can detect movement of plates down to 1 mm per year.
It has also been found that rocks heat up before earthquakes as a result of
extreme compression. So satellites with infra-red cameras can monitor the
Earth's surface for unexpected rises in temperature.

The use of new and improved techniques is gradually improving our ability to
predict the unpredictable !

An earthquake in Haiti

Summary

- The Earth is made from
  - a thin outer **crust**,
  - the **mantle**,
  - the liquid **outer core**, and
  - the solid **inner core**.
- The Earth gets hotter the deeper you go. One theory suggests that the energy comes from natural radioactive processes inside the Earth.
- The Earth's **lithosphere** (crust and upper part of the mantle) is made from huge slabs of rock called **plates**.
  The plates probably move because of convection currents in the mantle.
- Movement at the boundaries between plates causes earthquakes and volcanoes.

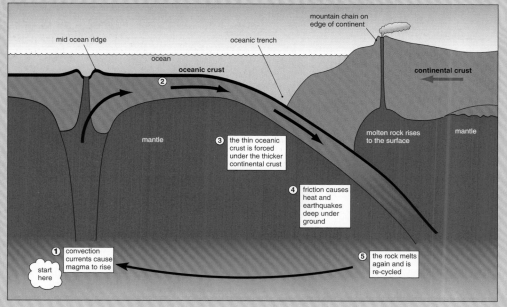

► Questions

1. Copy and complete:
   The Earth is made from a thin outer . . . . ,
   the . . . . , an . . . . liquid core and an inner
   core which is . . . . . The Earth's crust is made
   from massive slabs of slowly moving rock
   called tectonic . . . . . They move because of
   . . . . currents in the . . . . . Where they meet
   you get . . . . and volcanoes.

2. Imagine that the Earth is like an egg!
   a) Which part of the egg is like the Earth's
      crust?
   b) Which part of the Earth is represented by
      the white of the egg?
   c) Give one way in which the egg's yolk is like
      the Earth's core.
   d) Give one way in which the egg's yolk is
      different from the Earth's core.
   e) How could you represent the Earth's plates
      on the egg?

3. a) What evidence did Alfred Wegener put
      forward for his theory of 'drifting
      continents'?
   b) At one time all the continents were
      thought to be joined together in a huge
      land mass called Pangaea. What evidence
      is there in this diagram to support this
      theory?

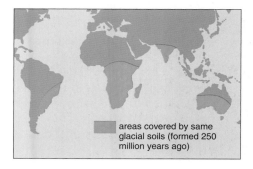

   areas covered by same
   glacial soils (formed 250
   million years ago)

4. a) Trace or get your own printed map of the
      world and mark on the plate boundaries.
   b) What evidence is there that these
      boundaries are the places where plates
      meet?

5. a) Why are San Francisco and Los Angeles
      prone to earthquakes?
   b) What is the focus of an earthquake?
   c) What is the epicentre of an earthquake?
   d) What precautions could you take against
      earthquakes if you lived in Los Angeles?

6. a) What is 'subduction'?
   b) How were chains of mountains, like
      the Alps and Himalayas formed?
   c) What 'powers' the movements of
      the Earth's plates?

7. a) Explain what we mean by 'sea-floor
      spreading'.
   b) Which type of rock will form on the sea-
      floor at a mid-ocean ridge?
   c) Look at the diagram below:

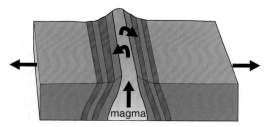

   magma

   Explain the pattern of stripes in the rocks
   on either side of the ridge.

8. Lava from different volcanoes has different
   thicknesses. The temperature, dissolved
   gases, and chemical composition of the lava
   affect the flow. Look at these two volcanoes:

   volcano A   volcano B

   a) Explain the different shapes of the two
      volcanoes.
   b) How do volcanologists try to predict when
      volcanoes will erupt? List the techniques
      available.
   c) Why is predicting eruptions an 'inexact
      science'?

Further questions on pages 326 and 327.

► Water

1. a) The solubility of a solid changes as the temperature changes. What do you understand by the term 'solubility' ? [4]
    b) A student investigated the solubility of two different solids at various temperatures. The table shows the results obtained.

Temperature (°C)	Solubility of solid (g/100 g water)	
	Sodium chloride	Potassium chlorate
0	33.0	3.5
20	33.5	7.5
40	34.0	14.0
60	34.5	24.0
80	35.0	37.5

    i)   Plot the results for the two different solids on a piece of graph paper. [8]
    ii) Compare the effects of increasing the temperature on the solubility of **each** solid. [2]
    iii) At which temperature are the solubilities of the two solids the same? [1]
    iv) What is the solubility of the two solids at the temperature in part (iii) ? [1]
    v)   If a saturated solution of potassium chlorate containing 50 g of water is cooled from 70 °C to 30 °C, what mass of the solid would crystallise? [4]
    vi) If a solution is made by dissolving 15 g of each solid in the same 100 g of water at 70 °C and the solution is cooled to 20 °C, crystals are observed.
         Which solid crystallises ? Explain your answer with reference to both solids. [4]   (NI)

2. The following extract is taken from an information sign near a reservoir.

> Until 1909, the water was simply stored and piped to customers but complaints about taste and odour led to the installation of filter beds. A quote of the time read 'the appearance of the public water supply was such that the poor used it as soup, the middle class for washing their clothes and the wealthy for watering their gardens'.

In addition to filtration, two other processes are now carried out to make water suitable for domestic use. These processes are sedimentation and chlorination.
Explain what is meant by sedimentation. [2]   (EDEXCEL)

3. A bottle of mineral water has this information on the label.

Ion	Concentration in mg/l
calcium	73
magnesium	8
potassium	1
sodium	18
chloride	28
hydrogencarbonate	5
sulfate	388

    a) A sample of this water was evaporated and a solid left behind.
       i) What is evaporation ? [1]
       ii) Name the chemical compound which would make up most of the solid left behind. [2]
    b) This mineral water is **hard water**. What is meant by **hard water** ? [1]
    c) Explain why hard water should not be used in hot water systems. [2]
    d) Give one advantage of hard water. [1]   (AQA)

4. The table at the top of the next page gives some information about the waters of the Old Sulphur Well in the North Yorkshire spa town of Harrogate. These mineral waters are from the strongest sulphur spring in England.
    a) i) Suggest how this water came to contain minerals. [2]
       ii) Use page 391 to help you answer this question.
           Write the formula of the barium compound present in this water. [1]
    b) i) Carbonated drinks are fizzy.
           How could this mineral water be made fizzy ? [2]
       ii) Warm fizzy drinks 'go flat' faster than cold fizzy drinks. Explain why. [2]

Old Sulphur Well

(Chemical analysis by Thorpe in 1875)

minerals	milligrams per litre
sodium chloride	12,738.63
barium chloride	93.59
strontium chloride	a trace
calcium chloride	621.98
magnesium chloride	602.68
potassium chloride	136.72
lithium chloride	10.73
ammonium chloride	14.69
magnesium bromide	32.54
magnesium iodide	1.61
calcium carbonate	424.32
magnesium carbonate	84.85
silicon	9.99
sodium sulphydrate	74.34

(Data: Harrogate Museums and Arts)

c) i) Describe a simple test that you could do to show that this mineral water is hard. Give the result of the test. *[2]*

ii) Give **one** example which shows that hard water is good for health. *[1]*

iii) Describe **one** problem caused by hard water. *[1]*

iv) Use the table to name **one** mineral which makes this water hard. *[1]*

v) Ion exchange columns remove hardness from water. Explain how. *[2]* (AQA)

5. The following label is found on a bottle of water.

ion	mg dm^{-3}	Bottled at source
hydrogencarbonate	450	**SINCLAIR WATER**
calcium	160	Natural Mineral Water
chloride	75	
sodium	62	SINCLAIR Natural Mineral Water fell as rain or snow before the age of pollution. Every drop of SINCLAIR Natural Mineral Water has naturally filtered through white sandstone for at least 3000 years.
magnesium	28	
sulphate	25	
potassium	5	
nitrate	0.3	
fluoride	0.2	

a) i) How could you test, in the laboratory, whether a sample of water is hard or soft? *[4]*

ii) What would you **see** if you carried out the test on Sinclair water? Use evidence from the label to explain your answer. *[2]*

b) The following information is from the website of Wessex Water, a supplier of tap water.

Water naturally contains a range of trace substances at levels suitable for drinking: many of these substances are essential to our health. The content of water is complex and varies from area to area, usually because of the different rocks and soils through which the water filters. The quality of drinking water in the Wessex Water region is extremely high and is among the best in Europe. We carry out 300 000 tests on drinking water every year. Our drinking water completely meets the European Union and UK standards.

www.wessexwater.co.uk

Use evidence from the passage to give **two** reasons why it is as safe to drink water from Wessex Water as to drink bottled mineral water. *[2]*

c) Another extract from Wessex Water's website also gives the following information.

A number of different processes can be used to purify water to the high quality required by law. The water arriving at a treatment works flows through fine filters to remove algae, insects and any other objects. The steps are:

1. **Sedimentation**
   A small amount of chemical coagulant is added to help bind together any impurities so that they will be trapped in the sand filters.
2. **Sand filtration**
   The water flows into fine sand filters which trap particles of impurity.
3. **Chlorination**
   Filtered water is treated with lime and chlorine.

i) What would happen if **chemical coagulant** was not added to the water? *[1]*

ii) Suggest why the sand used in the filters is **fine** (small grained). *[1]*

iii) Why is chlorine added to the water? *[1]* (EDEXCEL)

323

▶ The atmosphere

6. The chemical industry uses raw materials. These may come from the air or the ground. Air is a mixture of gases which can be separated by fractional distillation.

a) The pie chart shows the composition of dry air.
Copy and complete the labelling of the pie chart.

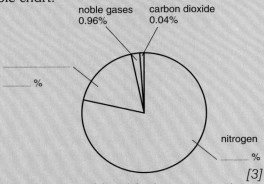

noble gases 0.96%   carbon dioxide 0.04%

_____ %

nitrogen _____ %

[3]

b) Combustion of fossil fuels increases the amount of carbon dioxide in the atmosphere.
Describe two natural processes that remove carbon dioxide from the atmosphere. [3]

c) Nitrogen is extracted from the air and, in the Haber process, is reacted with another element to produce ammonia, NH_3.
Name the other element. [1]

(EDEXCEL)

7. The diagram below is of the carbon cycle.

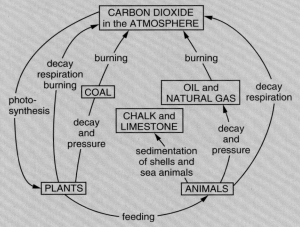

a) Which gas, present in air, provides the plants with the carbon they need? [1]

b) Name the process that plants use to change this gas into carbon compounds. [1]

c) How do animals get the carbon compounds they need? [1]   (AQA)

8. The table below gives information about the Earth's atmosphere at various times in its history.

Time	Atmosphere
4500 million years ago	Hydrogen and helium.
3800 million years ago	Mainly carbon dioxide, steam and hydrogen, with small amounts of methane, ammonia and hydrogen sulfide.
2600 million years ago	Carbon dioxide begins to decrease.
1800 million years ago	Oxygen begins to build up.

a) State the origin of the gases present in the atmosphere 3800 million years ago. [1]

b) Suggest what caused:
 i) the decrease in carbon dioxide concentration in the atmosphere 2600 million years ago
 ii) the build up of oxygen in the atmosphere 1800 million years ago. [2]

c) Use the table above to explain how the oceans formed, and suggest an approximate date for their formation. [3]

d) The table below shows the concentration of carbon dioxide in the air at the Mauna Loa observatory in Hawaii over a period of 30 years from 1958.

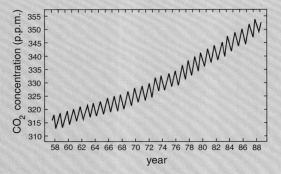

Give, and explain, **two** reasons why the carbon dioxide concentration has been increasing since 1958. [4]   (EDEXCEL)

9. For many years scientists have measured the amount of carbon dioxide in the atmosphere. They have shown that the amount of carbon dioxide is increasing.

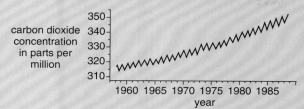

carbon dioxide concentration in parts per million

a) An increase in the Earth's temperature causes carbon dioxide to be given off from oceans.
Suggest why carbon dioxide is released from oceans when they become warmer. *[1]*

b) Suggest why some scientists do not accept that this explains the reason for the increase of carbon dioxide in the atmosphere. *[1]*

c) Carbon dioxide can be formed by burning carbon.
Write the balanced equation for this reaction. *[2]* (EDEXCEL)

10. a) During the first billion years of the Earth's existence, there were many active volcanoes. The volcanoes released the gases that formed the early atmosphere.

i) Describe how volcanoes caused the oceans to be formed. *[2]*

ii) Most of the early atmosphere was carbon dioxide.
Give **one** way in which carbon dioxide is removed from the atmosphere. *[1]*

b) The atmosphere on Earth today is very different from the early atmosphere. The pie chart shows the amounts of different gases in the air today. Choose gases from the box to label the pie chart.

argon	carbon dioxide
hydrogen	nitrogen oxygen

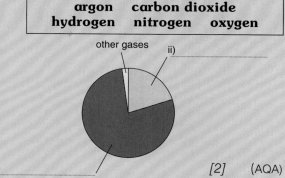

other gases   ii) _____

i) _____   *[2]* (AQA)

11. About 200 years ago Joseph Priestley carried out experiments on different gases. He called one gas that he investigated 'fixed air'. We now call this gas carbon dioxide.

a) How do we know that Priestley used the words 'fixed air'? *[1]*

b) When Priestley put a mouse in 'fixed air' (carbon dioxide) it had difficulty breathing. The mouse would have died if he had not removed it.
Priestley then put a green plant into 'fixed air' expecting it to die.

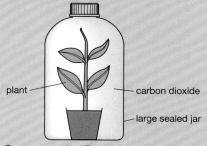

plant — carbon dioxide — large sealed jar

i) Suggest why Priestley expected the plant to die. *[1]*

ii) Two weeks later the plant still looked healthy. Explain why the plant was still healthy. *[2]*

c) As a result of his experiments, Priestley suggested that everyone should plant trees by their houses to 'keep the air good to breathe'. What gas do the trees produce 'to keep the air good to breathe'? *[1]* (AQA)

▶ The Earth

12. The theory of plate tectonics is believed to explain why mountains are formed.
a) What are tectonic plates? [2]
b) Explain, as fully as you can, **one** way in which tectonic plates cause mountains to be formed. [3] (AQA)

13. a) Give **three** features that may be observed at a plate boundary. [3]
b) What causes the Earth's plates to move? Explain your answer. [4] (AQA)

14. The diagram shows how a line of underwater mountains is forming in the middle of the Atlantic Ocean. This is called a mid-ocean ridge.

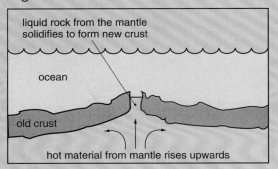

a) Solid rock forms when the liquid from the mantle layer reaches the water.
  i) Which type of rock will form: **igneous**, **metamorphic** or **sedimentary**? [1]
  ii) Explain why the crystals formed in this rock are very small. [2]
b) Europe and North America are at opposite sides of the Atlantic Ocean. Explain why Europe and North America move a little further apart each year. [2] (OCR)

15. At one time it was believed that mountain ranges on the Earth's surface were the result of shrinking of the crust as the Earth cooled down.
Mountain ranges are thought to form as tectonic plates move.
Give **two** other pieces of evidence which suggest that tectonic plates move. [2] (AQA)

16. Tony was studying rocks in a quarry. He made this sketch of the side of the quarry.

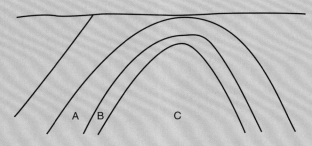

a) The rock at C is very hard rock containing large crystals. Describe how this rock was formed. [2]
b) Rock A and rock B have the same chemical composition but are different types of rock. Describe how rock B was formed. [3] (EDEXCEL)

17. Complete the following sentences by filling in the missing words.
The Earth's crust and upper mantle are split into a number of large pieces called . . . . plates.
Heat produced by natural . . . . processes causes . . . . currents in the Earth's mantle which in turn makes the plates move.
This movement can cause sudden and disastrous events at the plate boundaries such as . . . . or . . . . . [5]

18. **Earthquakes** often strike without any warning and can cause the destruction of many buildings.
Earthquakes are caused by the movement of **tectonic plates**.
a) What are **tectonic plates**? [2]
b) Explain, as fully as you can, what causes the movement of tectonic plates. [3]
c) How does the movement of tectonic plates cause an *earthquake*? [1]
d) Explain why scientists cannot accurately predict when an earthquake will take place. [2]

19. Scientists once though that mountains and valleys were formed because the Earth had shrunk in size.
Scientists today think that mountains are formed because of the movement of tectonic plates.
a) Why did scientists think that the Earth had shrunk in size? [1]
b) What are tectonic plates? [1]
c) Suggest why many scientists did not at first accept the idea of 'continental drift'. [2]

20. A tsunami is a large wave which can cause the flooding of coastal regions.
Tsunamis can be caused by an earthquake taking place under the sea.
The diagram shows how an earthquake can cause a tsunami.

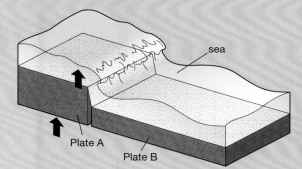

a) What is an earthquake? [2]
b) Explain, as fully as you can, why earthquakes take place. [3]
c) Explain, with the help of the diagram, why this earthquake caused a large tsunami wave to form. [2]

21. In some places on the Earth's surface magma from the mantle rises up through the crust and erupts as lava from volcanoes.
a) Use ideas about the density of magma to describe how it rises up through the crust. [2]
b) Magma can have different compositions that result in differences in the lava from volcanoes and the rocks that form when the lava cools.
These differences are summarised in the table below.
Complete the table using words from the list below.

granite   **iron-rich**
runny lava   **silica-rich**

Magma composition	Type of eruption	Material ejected	Rock formed from lava
i)	flowing	iii)	gabbro
ii)	explosive	ash and pumice	iv)

[4]

c) Some geologists study volcanoes. Suggest why this study is important. [2]
d) Many people live near to volcanoes, despite the danger of an eruption. Suggest advantages of living near to a volcano. [2]

chapter 25

▷ Chemical industry

From your trainers to your toothpaste, you can thank the chemical industry for providing the materials we all take for granted.

We can think of the chemicals they make in two parts:
- traditional bulk chemicals
- specialist (fine) chemicals.

Traditional **bulk chemicals** would include the manufacture of:
- ammonia
- sulfuric acid
- iron and steel
- nitric acid
- fertilisers
- products from oil
- products from salt – sodium chloride
- products from limestone.

These operate on a large scale in huge plants. The industries tend to run their processes continuously. They can make their product 24 hours a day, 7 days a week.

These **continuous processes** are economical. That's because they can keep conditions ideal for the reactions, saving energy, and make a constant supply of product.

Road tankers deliver many products for the chemical industry

These products have been in demand for a long time so companies have long-established customers. However, developing countries are now competing. Their prices can be lower because of reduced labour costs. That's why steel production has decreased sharply in the UK.

We also have companies who make **specialist (fine) chemicals**. These include some pharmaceutical companies. They make drugs on demand for their customers in **batch processes**. They use just the right amounts of reactants to make the order.

Others companies, like brewers, will also operate batch processes because they need to leave the reactants a while to react. In a brewery the fermentation is carried out in large vats. When the reaction to make the alcohol (ethanol) is finished the vat is emptied. Its contents go on for further processing and the vats are re-filled with fresh reactants.

A batch process is not as efficient as a continuous process in general. Batch processes take more workers to operate them, which adds to costs.

The traditional chemicals do not need a lot of money spending on research and development. Their processes have been refined over many years. However, specialist drugs companies spend much more on developing new products.

Fermenting sugar is a batch process

Developing a new drug

In the case of a developing a new drug, the research starts with experiments to make a compound that works. In the past a research chemist might make one or two hundred new compounds in a year. However, now they can make thousands. That's because, since the 1990s, more and more chemists are using machines to help.

The machines can make thousands, or even millions, of compounds a year. They can automatically mix all possible combinations of promising reactants. This is called **'combinatorial chemistry'**.
These can then be screened to see if they are effective drugs.

Developing and testing new drugs takes years and costs millions of pounds

So this can save time and labour costs, but then the long process of testing the new drug has to start. It usually takes between 7 to 10 years from discovery to general use. It is estimated that this costs about £400 million pounds for a new drug. This can include animal tests and several rounds of clinical trials on human volunteers.

Those people against testing on animals argue that effects on animals do not predict effects on humans anyway. Many drugs that pass animal tests then go on to fail their clinical trials.

The clinical trials will involve the use of a **placebo** test group. These people believe they are taking the new drug. In fact their dose contains no drug at all. These act as a control group and their results are compared to the group taking the drug.
Researchers will see if any differences are significant.

Here is a simplified flow chart to show what happens :

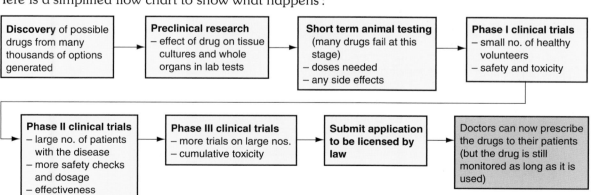

Developing a new drug

So a drugs company needs to get back all its development costs before it starts making a profit. However, a successful new drug can make billions of pounds !

▷ Drugs

What do you think of when you hear the word 'drugs'?
Drugs are substances which affect the body in some way.
That effect might be good or bad.
Drugs can help us or might harm us if abused.
● Can you name any drugs that are addictive?

Let's look at the development of a class of drugs
called **analgesics**. These are drugs that relieve pain.

We can start the story at the time of the ancient Egyptians
and the ancient Greeks. The Epyptians used a bush
called myrtle to make a treatment for muscle pains.
Then Hippocrates, the father of modern medicine,
used the leaves and bark of willow trees to ease pain.
He made it into a tea and gave it to mothers to help with childbirth.

Hippocrates – the ancient Greeks used willow to relieve pain

The next time that willow was recorded as a drug
was in 1763 – a gap of about 2000 years!
The Reverend Edward Stone wrote to the Royal Society
telling them of the effects of chewing a willow twig.
This led to the search for the active ingredient.

In 1835 another plant, a flower from the riverbank,
was also discovered to have these analgesic effects.
Both plants were used to make **salicylic acid**.
It was named after the Latin word for willow, *salix*.
Look at its structure below:

where ⬡ is a benzene ring

However, there was a problem with salicylic acid.
It caused painful ulcers in the mouth and attacked
the lining of the stomach.
Chemists tried the sodium salt of salicylic acid too.
This was soluble in water because of its ionic structure.
It was not so bad for your mouth and stomach,
but tasted just awful!

A French chemist called Charles Frederic Gerhardt
made a safer drug in 1853.
Look at its structure opposite:

aspirin

330

But the drug was tricky to make and it was forgotten until 1899. It was then that Felix Hofmann rediscovered Gerhardt's method. He was searching journals for painkillers for his father who had arthritis. This one worked a treat, without the side effects.

Felix worked for a large German chemical company called Bayer. He told them of the drug's success and convinced them to manufacture it. It was given the trade name **Aspirin**. At first it was produced as a powder, but in 1915 the first tablets were sold.

Aspirin is an analgesic

Germany had to give up the rights to the name Aspirin at the end of the First World War. This was part of the Treaty of Versailles. In fact, Bayer bought back the rights in 1996 for many millions of pounds.

Aspirin is used for headaches, fevers, and in larger doses for arthritis. It is now also used in small doses to reduce the risk of heart disease.

However, the long term use of the tablets every day has brought with it the old problems. People are worried about the side effects, such as internal bleeding. About 6 % of aspirin users suffer from bleeding intestines.

The search for new analgesics did not stop at aspirin. In the 1950s a new drug called **paracetamol** was introduced. This did not cause internal bleeding. However, it did not reduce the swelling from arthritis like aspirin.

Ibuprofen was developed by chemists at Boots

Then in the late 1980s pharmacists could sell **ibuprofen** 'over the counter'. This was much better at reducing the swelling, as well as the pain, from arthritis and rheumatism.

Other related drugs were developed in the 1980s. An example was **benoxaprofen**. This was sold to about half a million patients in the USA in its first six weeks. But then the reports of bleeding, and liver and kidney damage started. It was banned in 1982 and was implicated in 80 deaths in the UK and 30 deaths in the USA. This raises questions about how we test new drugs.

Laboratory animals are tested with larger doses than humans would take to look for side effects. Then field trials are carried out with volunteer patients before a drug is passed as safe for the public. (See page 329.)

● *What are your views on testing drugs on animals ?*

▷ Paints and pigments

Have you ever painted a room? Why do people paint their homes?
Paint protects materials, but what else is important when you
choose which paint to use?
That's where pigments come in. They are the finely ground solids
that give paints their colour.

What's in a paint?

In general paints contain:
- **A pigment** (to provide colour).
- **A binder** (to help the paint attach itself to an object and to form
  a protective film when dry).
- **A solvent** (to help the pigment and binder spread well during
  painting by thinning them out).

Do you remember looking at emulsions on page 176?
They were made of two liquids – oil and water – mixed thoroughly
together.
An **emulsion** is an example of a colloid.

> **A *colloid* is a finely dispersed mixture of two or more
> substances that do not dissolve in each other.**

You've probably heard of emulsion paints. These are water-based,
but the binder is oily so when stirred up with the water, we get an
emulsion.

The pigment is sometimes soluble in the solvent but is often
insoluble.
It is made of tiny specks of solid. They are so small you can't see
them individually. They are suspended in the paint. They are so
tiny that they don't settle out easily. We call this type of colloid a
sol or a **suspension**. You can tell a sol from a solution because
light will pass easily through a solution. It looks clear.
However, light is scattered by the bits of solid in a sol. It looks
translucent.
So a paint can be an emulsion and a sol.

*Emulsion paint is an oil-in-water
emulsion*

Experiment 25.1   Looking at sols and solutions

We will try dissolving different powders in water.
Shake a little of the solid in half a test tube of water.
Line the tubes up in a test tube holder.
Place a lamp on the other side or put the holder
on a window sill.

- Which mixtures are sols and which are solutions?

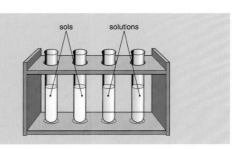

Gloss paints are the oil-based paints that give a hard, shiny finish. The pigment is dispersed through an oil (such as linseed oil), which can be thinned with a solvent (such as turpentine).

Paint drying

Have you ever touched wet paint by mistake? Paint takes a while to dry.

Some paints dry as the solvent evaporates. The binder and pigment are left behind to coat the surface. There is no chemical reaction involved here. (Remember evaporation is a physical change – see page 25.) The binder is likely to be made of long molecules that get tangled up together once the solvent has evaporated.

Other paints dry by means of a chemical reaction. Some paints contain plant oils. We have mentioned the linseed oil in gloss paint. It is extracted from the flax plant. Another common oil is tung oil. This is extracted from the nuts of a tung tree.

We have looked at plant oils in Chapter 14. Remember that they contain unsaturated hydrocarbon chains within their molecules. It is the carbon–carbon double bonds that react with oxygen when the paint is exposed to air.

Have you ever opened a tin of paint and found it had a skin across the top? This is the hardening process, started because some air has got into the tin and reacted with the paint.

The oxidation of the oils forms cross-links between oil molecules. Gradually we get a polymer built up. (See page 153.)
It is this polymer that forms the tough shiny coating on the surface. This is how **varnishes** harden and protect wood.

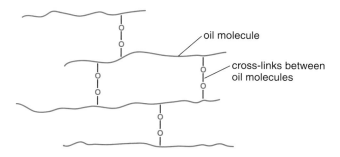

Tung oil hardens faster than linseed oil because it has an extra double bond in each chain.

Varnish forms a hard shell to protect wood

▷ Smart pigments

Have you ever used a 'strip thermometer'? They give a visual display of the temperature as numbers 'magically' light up.

Some substances respond to temperature changes by changing colour. They absorb heat (infra-red radiation) and give out light. We can use these **thermochromic pigments** to colour:
- paints
- plastics
- paper
- fabrics.

Thermochromic pigments can be used on a thin plastic film to make a thermometer

They are useful wherever you need a visual warning of a change in temperature.
For example:
- cups for hot drinks
- electric kettles
- any surfaces that might get hot
- storing food (e.g. frozen food that gets too warm – see active and intelligent packaging on page 159).

You can also use the thermochromic pigments to test fuses or batteries. This will work by a current heating a wire in the test circuit, which produces a change in colour.

The pigments are soluble in acrylic, so we can use them in paints to produce interesting effects.
Look at the diagram opposite:

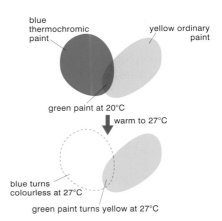

So a green paint will turn yellow as it gets warm. Can you think of a use for this?

Experiment 25.2   Thermochromic paints

Mix a very small amount of thermochromic pigment into acrylic and water down to make your paint. Apply this to a thin plastic cup. Allow to dry. Then pour in some hot water.
- What happens?
Pour the hot water out.
- What happens?

We also have pigments that will glow in the dark. These are called **phosphorescent pigments** (used in photochromic paints).
They absorb energy from light and then release it as glowing light.
They have taken the place of materials like zinc sulfide.
The new pigments will glow ten times as long. So having been exposed to light all day, the pigments will release light and glow all night.

They have replaced hazardous radioactive materials used on old watch and clock hands.
In the 1920s, American women were employed to paint luminous numbers on watches. This was delicate work and the women would use their lips to make a fine point on their brushes.
They didn't realise the dangers. The paint contained radioactive radium! Unfortunately many developed cancer of the jaw, and other cancers later in life.

We now see phosphorescent pigments on road and emergency signs

▷ Dyes and dyeing

The pigments we have looked at can also be used on fabrics. The smart pigments can make dyes that will change colour when you wear them. But there are lots of natural dyes that we have been using for thousands of years. We can extract these from plants, earth or even insects.

The problem with natural dyes is that they fade quickly. They are easily washed out of fabrics. That's where **mordants** can help. A mordant helps a dye attach itself more firmly onto the molecules in a fabric. It improves a dye's colour-fastness.

Roman emperors wore special purple togas. It took 10 000 small sea molluscs to make enough dye for one toga.

Experiment 25.3   Make your own dye

Boil up some strongly coloured plant material for about 15 minutes in a beaker.
Add more water if necessary to stop it boiling dry.
You want a deeply coloured liquid to test on your white fabrics.
You can test:

a. The effect of different mordants
Soak separate samples of your fabric into a beaker containing different mordants.
Try:
  i   125 cm^3 of vinegar
  ii 125 cm^3 of water with a teaspoon of cream of tartar added
  iii 125 cm^3 of saturated sodium chloride solution.
Then wring out your fabric samples and soak them in your dye.
Rinse with cold water and leave to dry.
Also dye a sample with no mordant added to act as a control.
Test them next lesson – are they colour-fast?
• Which mordant is most effective?
b. The colour-fastness on different fabrics
c. Different plant material
Make your tests as fair and as safe as possible. Let your teacher check your plan before starting.
• Which factor will you vary and which will you control in each investigation?
• Try to think of a way to generate quantitative data.
• Evaluate your investigation.

There is a wide variety of synthetic dyes now available to manufacturers

Synthetic dyes

In 1856, a young chemist called William Perkins made a surprise discovery.
He was trying to make quinine – a drug to fight malaria.
However, he made a purple dye by mistake. It was the first synthetic dye.
His raw material came from coal tar, and this signalled the start of the chemical industry.

He tried his purple dye on silk and found it far better than existing dyes.
At this time, a lot of trial and error was involved in looking for new materials.
Perkins' discovery led to a race for more dyes. German chemists were successful in developing many new dyes.

William Henry Perkins (1838–1907) discovered the first really synthetic dye

▷ Chemistry at work : What is nanotechnology?

Nanotechnology is the design and creation of machines that are so small we measure them in nanometres.

It is difficult for us to imagine a nanometre (nm).
The tiny sizes involved in nanotechnology are mind-boggling.
The nanometre is the unit of measurement used on an atomic level :

$$1 \text{ nm} = 10^{-9} \text{ m}$$

Nanoscientists study particles that are between 1 and 100 nm in size.

In other words, a nanometre is a billionth of a metre.
You could get about a million nanometres across a pinhead.

Nanotechnologists are now making machines that are less than 100 nm in size. They have the techniques to move individual atoms or molecules. Then they can place them where they want them.

At the moment, research is in its early stages.
There are two approaches to making the molecular machines.

You can :

- **Sculpt** or 'chisel away' at materials until you are left with the molecules or atoms that you want on the surface.
  This method is used in **microelectronics**.

- **Build** your machine up from individual atoms or molecules.
  You can do this by physically moving the molecules using **atomic force microscopes**, as in the photos, or by chemical reactions in solution.

The chemical technique seems to offer the best way forward.
Physically moving molecules around takes too long, so any products made would cost too much money.

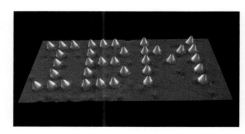

Scientists at IBM placed 110 molecules of carbon monoxide on the surface of copper. They used the latest atomic force microscopes to position the molecules.

This atomic force microscope can move and re-position atoms or molecules

▷ Chemistry at work : Applications of nanotechnology

Sun-screens

You can now find nanotechnology improving sun-screens.
Scientists can coat individual specks of titanium oxide powder with
a complete coating of silica. The thickness of the silica coating can be
adjusted at an atomic level. These 'nanoparticles' seem to be more
effective at blocking the Sun's rays than conventional u.v. absorbers.

So nanotechnology will save many people from damaged skin and
cancers caused by too much ultraviolet light.

Future applications?

Some say that nanotechnology will give us health-care that can
treat individual cells. Perhaps we'll be able to monitor our health
on a cellular level with highly selective sensors.
This should increase our life spans.
Nano-bucky-tubes (see page 263) are now being developed that
will one day search for cancerous cells.

Or imagine tiny machines patrolling your blood vessels, cleaning off
any fat deposits from their walls. This would help those at risk from strokes.
Imagine computers with memory capacities and speeds that we can only dream about now.
The tiny machines built so far include 'molecular model kit' rotors, 'robots', motors and circuits.
And just imagine the surface area possible using nanoparticles as catalysts !

Will nanotechnology harm us?

Cosmetic companies are developing anti-wrinkle creams with
nanoparticles that can pass through skin to its lower layers.
There, the active ingredients can stimulate the production
of new cells.
However, people say that there has been little research into the
effects of nanoparticles that could penetrate into the bloodstream.
They say that these cosmetics should have the same testing and
licensing process as pharmaceutical drugs. (See page 329.)

*K. Eric Drexler wrote a book called 'Engines of Creation'. In the book
he speculates that one day we might invent a nanomachine that can
reproduce itself. Then the world could be overrun by 'grey goo', as
he calls it. Some people are so worried that they have called for a
halt in nanotechnology research.*

● What are your views on these issues ?

337

▷ Detergents

Detergents are cleaning agents.
See page 178 for the emulsifying effects of detergents.
They help to remove grease and oily stains.
They also help water soak into fabrics.
The experiment below shows how a detergent affects
the forces between molecules at the surface of water.

*Birds like this can be rescued by cleaning
the oil from their feathers with detergents*

Experiment 25.4   Detergents and surface tension

Carefully float a needle on the surface of a beaker of water.
Start with the needle on a small piece of paper towel, which will sink as it soaks up water.
Now gently add a drop of detergent to the water in the beaker.

- What happens to the needle? Try to explain why.

The needle floats because of water's surface tension.
There are relatively strong forces between the molecules of water
at its surface. If we can weaken these forces, water will soak into
fabrics more effectively.

Experiment 25.5   Hard water

Add a few drops of soap solution to half a test tube of
hard water. Put a bung at the top and shake the tube.

- What happens?

Now repeat the experiment but use a *soapless* detergent.
- What difference do you notice?

scum

lather

hard water +
soap solution   hard water +
soapless detergent

Washing-up liquids generally contain:

- A **surfactant** (the actual detergent that removes the grease).
- **Water** (to thin out the mixture so it can squirt more easily from the bottle).
- Colouring and fragrance **additives** (to improve the appeal of the product to potential customers).
- **Rinse agent** (to help water drain off crockery).

DISHWASH

Investigation 25.6   Comparing detergents

Design an investigation to compare the effectiveness of either:

- Washing-up liquids – eco-friendly against conventional brand.
- Washing powders – biological against non-biological brand.

Let your teacher check your plans before you start your practical work.

Summary

- The chemical industry provides us with the substances we need to manufacture new materials.
- Bulk chemicals are usually made in **continuous processes** designed to be as economical as possible.
- Drugs and medicines are often made in **batch processes**.
- Developing new drugs involves a lot of research and testing before they can be used by the general public. This is very expensive.
- Pigments are coloured substances used in paints and dyes.
- **Smart pigments** respond to changes around them. For example, **thermochromic** pigments change colour at different temperatures.
- **Nanotechnology** is an exciting new development. Individual atoms or molecules can be manipulated to make tiny machines or 'nanoparticles'.
- **Detergents** are cleaning agents. They are the essential part of mixtures such as washing-up liquid and washing powder.

▶ Questions

1. Copy and complete:
   In the chemical industry, processes that run 24 hours a day are called . . . . processes. On the other hand, processes that produce quantities of product on demand are called . . . . processes.
   Coloured substances in paints and dyes are called . . . . .
   Some can change colour depending on the temperature and are called . . . . .
   Scientists can now move and re-position atoms and molecules in a new development called . . . . .

2. You are designing a paint that will change colour at 27 °C. You have one pigment that changes from red to colourless at that temperature and a normal yellow pigment.
   a) Describe how you could make the paint and the colour change you would see.
   b) Try to think of a useful application for your paint.

3. Carry out some research to create a booklet on attracting new recruits to the nanotechnology industry. Include some of the latest and potential applications.

4. Explain why pharmaceutical companies have a huge outlay of money before they can profit from a new product.

5. Look at the two washing powders below:

   a) Some people are allergic to certain enzymes on their skin. Which powder would be better for them?
   b) Design a fair test to see which washing powder is most effective at removing food stains.
   c) Bio-delight costs more than Sudso. What else should you take into account before deciding it will be cheaper to use Sudso?

6. Use the information on page 168 to evaluate the batch and continuous processes used to make ethanol for use as a fuel.

Further questions on page 370.

chapter 26

As you know, atoms are too small to see.
So, we have a big problem when we try to count them.
However, we can compare numbers of atoms quite easily.

Look at the R.A.M.s in the table opposite:
Can you see that oxygen is 16 times as heavy as hydrogen?
Therefore, if you have 1 gram of hydrogen and
16 grams of oxygen, you have the same number of
hydrogen and oxygen atoms. This is a lot of atoms!

Element	Relative Atomic Mass (R.A.M. or A_r)
Hydrogen	1
Carbon	12
Nitrogen	14
Oxygen	16
Fluorine	19

Chemists use atoms of carbon-12 as their standard:

One mole of a substance contains the same number of particles as atoms in 12 grams of carbon-12.

In fact, in 1 gram (1 mole) of hydrogen there are about
600 000 000 000 000 000 000 000 atoms!
No wonder we can't see them!
(This number is known as the **Avogadro constant**,
and is written 6×10^{23}).

There are huge numbers of atoms even in test-tube reactions.
So it is easier to talk about moles than the actual numbers involved.
We weigh out a substance, then say how many moles there are.

Learn this equation:

$$\text{moles of atoms} = \frac{\text{mass}}{\text{R.A.M.}}$$

*Counting atoms takes its toll.
Just weigh them out, then use
the mole!*

It's like cashing in 1 p coins at the bank. If you take
a thousand 1 p coins to your bank, does the cashier
count out each coin?
The coins are weighed out on scales. The scales convert
the mass of one hundred 1 p coins, and tell the cashier
how many pounds (£s) are on the scales.

*Would you
rather win the
National Lottery
or win a 'mole'
of 1 p coins?*

Example
How many moles of atoms are there in 2.4 g of carbon?

$$\text{Moles of atoms} = \frac{\text{mass}}{\text{R.A.M.}} = \frac{2.4}{12} = \textbf{0.2 moles}$$

Now try these yourself (the answers are on page 374).

- How many moles of atoms are there in:
  1. 2 g of hydrogen 2. 36 g of carbon 3. 160 g of oxygen 4. 1.4 g of nitrogen 5. 0.19 g of fluorine?
     (R.A.M.s   H = 1, C = 12, O = 16, N = 14, F = 19)

Changing moles to mass

You can re-arrange the equation on the last page:

mass = moles × R.A.M.

This tells us the mass of an element, if we know how many moles there are.

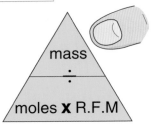

> *Example*
> What is the mass of 0.1 moles of carbon atoms?
> mass = moles × R.A.M.
> = 0.1 × 12 = **1.2 g**

If you find re-arranging equations difficult, you can use the 'magic triangle'.

To use the triangle:
Read the question. See what you need to find out.
Cover that part of the triangle with your finger.
Then you have the equation that you need!

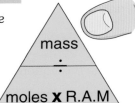

mass ÷ moles × R.A.M

Now you can try these:

- What is the mass of:
  6. 2 moles of H atoms   **7.** 5 moles of N atoms   **8.** 20 moles of O atoms
  9. 0.5 mole of F atoms   **10.** 0.01 mole of C atoms?

Amedeo Avogadro (1776–1856)

Moles of molecules

You can use the same ideas for problems about molecules.
Just use the Relative Formula Mass (R.F.M. or M_r) instead of the R.A.M.:

$$moles = \frac{mass}{R.F.M.}$$ or   mass = moles × R.F.M.

So the magic triangle becomes:

mass ÷ moles × R.F.M

> *Example*
> How many moles are there in 8 g of copper(II) oxide (formula = CuO)?
>
> **Step 1** – Work out the formula mass by adding up the R.A.M.s (see page 37):
> Cu = 64
> O = +16
> R.F.M. = 80
>
> **Step 2** – Put information from the question and the R.F.M. into the equation:
>
> $$moles = \frac{mass}{R.F.M.} \qquad moles = \frac{8}{80} = \textbf{0.1 mole}$$

- How many moles of molecules are there in:
  11. 36 g of H_2O **12.** 170 g of NH_3 **13.** 1.6 g of CH_4 **14.** 0.3 g of C_2H_6 **15.** 16 g of NH_4NO_3?

▶ Moles of gases

As you know, gases are very light. It is difficult to weigh them.
However, it is quite easy to measure the *volume* of a gas.

Do you remember your work on rates of reaction
in Chapter 16? Can you think of two ways to measure
the volume of gas given off in a reaction?

Luckily for us, we can change volumes of any gas
straight into moles. There is no need to weigh the gas.
That's because equal volumes of gas contain the same
number of particles at the same temperature and pressure.

At room temperature (20 °C) and pressure (1 atmosphere),
we can say that:
1 mole of any gas takes up a volume of 24 dm³ or 24 litres
(24 dm³ = 24 000 cm³)
Using this information we can write this equation:

$$\text{moles of gas} = \frac{\text{volume of gas (in dm}^3)}{24}$$

In our experiments in the lab, we usually measure the
volume of gas in cm³. In this case, the equation becomes:

$$\text{moles of gas} = \frac{\text{volume of gas (in cm}^3)}{24\,000}$$

The magic triangle is:

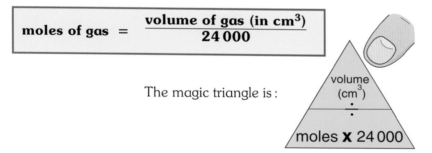

volume
(cm³)
:
:
moles **X** 24 000

Example

How many moles of hydrogen molecules are there in 24 cm³ of hydrogen gas?

$$\text{moles} = \frac{\text{volume}}{24\,000} = \frac{24}{24\,000} = \textbf{0.001 mole}$$

Now you can try these:
(Watch out for the units, dm³ or cm³!)

- How many moles of gas molecules are there in:
  1. 24 dm³ of chlorine gas   2. 6 dm³ of hydrogen gas   3. 2.4 dm³ of methane gas
  4. 120 cm³ of oxygen gas   5. 48 cm³ of nitrogen gas?

Notice that you don't need the Relative Atomic Masses or the Relative Formula Masses of the gases to work out the problems on the last page.

- Why?

Volumes of gases

You can re-arrange the equations on the last page to get:

| **volume of gas (dm³) = number of moles × 24** | or | **volume of gas (cm³) = number of moles × 24 000** |

Example

What volume does 0.1 mole of oxygen gas occupy at room temperature and pressure?

volume (in cm³) = number of moles × 24 000
= 0.1 × 24 000
= **2400 cm³**

- What volume do these gases occupy:
  6. 3 moles of hydrogen   **7.** 5 moles of chlorine   **8.** 0.1 mole of nitrogen
  9. 0.001 mole of hydrogen sulfide   **10.** 0.005 mole of sulfur dioxide?

You can get more difficult problems which include the mass of the gas.

Example

What volume does 8 g of oxygen gas occupy at room temperature and pressure?

Step 1 – Work out the number of moles of gas molecules.
Remember that the formula of oxygen gas is O_2.
Therefore, the R.F.M. of O_2 is $16 \times 2 = 32$.
So 1 mole of oxygen gas weighs 32 g.

Use the equation   moles = $\dfrac{\text{mass}}{\text{R.F.M.}}$

The number of moles of oxygen = $\dfrac{8}{32}$ = 0.25 mole

Step 2 – Now work out the volume of gas.

volume of gas (in cm³) = moles × 24 000
= 0.25 × 24 000
= **6000 cm³ of oxygen gas**

Now you can try these:

- What volume of gas is occupied by these gases at room temperature and pressure:
  (You can find the R.A.M. s on page 390.)
  11. 4 g of H_2 **12.** 8 g of CH_4 **13.** 3.55 g of Cl_2 **14.** 0.002 g of He   **15.** 8.8 g of CO_2?

▷ Moles in solution

All bottles of solutions in a lab must be labelled.
Look at a bottle in your next experiment.
You will see that the label has the name of the solution.
Most also show its **concentration**, such as 1M or 1 mol/dm^3.
The concentration is sometimes called the **molarity** of a solution.

1M or 1 mol/dm^3 means that there is 1 mole of the substance
dissolved in 1 dm^3 (or 1000 cm^3) of its solution.

Knowing this, and the concentration, we can work out
the number of moles in any solution.

Let's look at an example :

*You wil see 'M' or 'mol/dm^3' on reagent
bottles*

> *Example*
>
> How many moles of sodium chloride are there in 100 cm^3 of a
> 2 mol/dm^3 solution?
>
> 2 mol/dm^3 means that there are 2 moles in 1000 cm^3 of solution.
> 100 cm^3 is only a tenth of 1000 cm^3.
> Therefore, in 100 cm^3 we will have a tenth of 2 moles,
> which is **0.2 mole**.

This type of logical thinking can be used to solve
any moles calculations.
But here is a more tricky example :

Is this a 1 mol/dm^3 solution?

> *Example*
>
> How many moles of sodium chloride are there in 22 cm^3 of a
> 2.0 mol/dm^3 solution?
>
> Here are the logical steps :
> In 1000 cm^3 of the solution we have 2 moles.
>
> So, in 1 cm^3 of the solution we have $\dfrac{2}{1000}$ moles
>
> So in 22 cm^3 of the solution we have $\dfrac{2}{1000} \times 22$ moles
>
> $\qquad\qquad = $ **0.044 mole**

Or you can use a magic
triangle again:

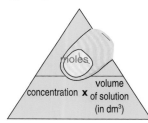

(Don't forget to change any
volumes given in cm^3 to dm^3,
by dividing by 1000)

However, if you prefer to learn equations :

> **number of moles in a solution = concentration $\times$ $\dfrac{\text{volume of solution (in cm}^3)}{1000}$**

So in the first example :
the concentration is 2 mol/dm^3, and
the volume of solution is 100 cm^3.

Therefore, the number of moles $= 2 \times \dfrac{100}{1000} = 0.2$ mole

Now you can try these:

- How many moles are there in:
  1. $2\,dm^3$ of $1\,mol/dm^3$ sulfuric acid
  2. $500\,cm^3$ of $2\,mol/dm^3$ nitric acid
  3. $250\,cm^3$ of $1\,mol/dm^3$ hydrochloric acid
  4. $100\,cm^3$ of $0.5\,mol/dm^3$ sodium hydroxide solution
  5. $50\,cm^3$ of $0.5\,mol/dm^3$ sodium chloride solution?

(Notice that you don't need the formula of the substance because you don't have to work out R.F.M.s.)

Making up solutions

Imagine that you are a lab technician.
You are asked for $250\,cm^3$ of $1\,mol/dm^3$ sodium nitrate for a chemistry lesson. How would you make up the solution?

Step 1 – How many moles are needed?
We know that $1000\,cm^3$ of a $1\,mol/dm^3$ solution contains 1 mole.
But you only need to make up a quarter of this ($250\,cm^3$).
So you will need to dissolve a quarter (0.25) of a mole of sodium nitrate.

Step 2 – Work out the mass of solid needed.
Find the R.F.M. of sodium nitrate, $NaNO_3$.
$$1 \times Na = 1 \times 23 = \quad 23$$
$$1 \times N = 1 \times 14 = \quad 14$$
$$3 \times O = 3 \times 16 = \underline{+48}$$
$$R.F.M. = \underline{85}$$
1 mole of sodium nitrate weighs $85\,g$.
Therefore, 0.25 mole of sodium nitrate weighs $0.25 \times 85 = 21.25\,g$
(Remember the magic triangle! mass = moles × R.F.M.)

So as a technician, you would weigh out **21.25 g** of sodium nitrate, then make it up to $250\,cm^3$ of solution with water.

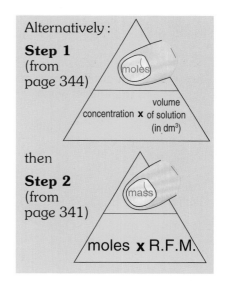

Alternatively:

Step 1
(from page 344)

volume
concentration x of solution
(in dm^3)

then

Step 2
(from page 341)

moles x R.F.M.

Now you can try these:

- What is the mass of each compound in these solutions?
  6. $500\,cm^3$ of $1\,mol/dm^3$ sodium chloride, NaCl, solution
  7. $100\,cm^3$ of $2\,mol/dm^3$ potassium hydroxide, KOH, solution
  8. $25\,cm^3$ of $1\,mol/dm^3$ sulfuric acid, H_2SO_4, solution
  9. $75\,cm^3$ of $0.5\,mol/dm^3$ lead nitrate, $Pb(NO_3)_2$, solution
  10. $13\,cm^3$ of $0.25\,mol/dm^3$ ammonium sulfate, $(NH_4)_2SO_4$, solution.

(R.A.M.s   Na = 23, Cl = 35.5, K = 39, O = 16, H = 1,
S = 32, Pb = 207, N = 14)

▷ Working out the formula

We can use moles to work out the formula of a compound.
The formula tells us the ratio of the different atoms or ions of each element in a compound.
So if we can measure the mass of each element in a compound, we can calculate its formula.
Let's look at an example:

Example
A compound of nitrogen and hydrogen was broken down into its elements.
It was found that 1.4 g of nitrogen had combined with 0.3 g of hydrogen in the compound.
What was the formula of the compound?
(R.A.M.s   N = 14, H = 1)

Step 1 – Work out the **number of moles** of each element in the compound.

Remember that: $\text{moles} = \dfrac{\text{mass}}{\text{R.A.M.}}$

$\text{moles of N} = \dfrac{1.4}{14} = 0.1$ $\text{moles of H} = \dfrac{0.3}{1} = 0.3$

Step 2 – Work out the **ratio** of the number of moles of each element, to the lowest whole numbers.

N	:	H
0.1	:	0.3
1	:	3

Therefore, we have 3 times as many H atoms as N atoms in this compound.

Its formula must be **NH_3**.

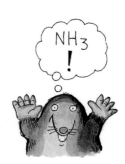

- Now see if you can work out the formula for each of the compounds made from:

  1. 12 g of carbon and 4 g of hydrogen

  2. 414 g of lead and 32 g of oxygen

  3. 1.2 g of carbon and 3.2 g of oxygen

  4. 11.2 g of iron and 4.8 g of oxygen

  5. 3.2 g of copper, 0.6 g of carbon and 2.4 g of oxygen.

  (R.A.M.s   C = 12, H = 1, Pb = 207, O = 16, Fe = 56, Cu = 64)

Using results from experiments

You can work out the formula of a compound by:
- splitting it up into its original elements, or
- making it from its elements.

In either case you need to measure the mass
of each element in the compound.

Example   Working out the formula of magnesium oxide
A student heated up some magnesium ribbon as shown:
The magnesium ribbons burns in air. It combines with
the oxygen to make magnesium oxide.

These are her results:

Mass of crucible + lid	= 25.00 g
Mass of crucible + lid + magnesium *before* heating	= 25.24 g
Mass of crucible + lid + magnesium oxide *after* heating	= 25.40 g

Step 1 – Work out the mass of each element in the compound.
Magnesium

Mass of crucible + lid + magnesium before heating	= 25.24 g
Mass of crucible + lid	= 25.00 g
Therefore, the mass of magnesium	= 0.24 g

Oxygen

Mass of crucible + lid + magnesium oxide after heating	= 25.40 g
Mass of crucible + lid + magnesium before heating	= 25.24 g
Therefore, the mass of oxygen in magnesium oxide	= 0.16 g

Step 2 – Change the masses into moles.

$$\begin{array}{cc} \text{Magnesium} & \text{Oxygen} \\ \dfrac{0.24}{24} & \dfrac{0.16}{16} \\ = 0.01 \text{ mole} & 0.01 \text{ mole} \end{array}$$

Step 3 – Work out the ratio of moles of each element.

Mg	:	O
0.01	:	0.01
1	:	1

Therefore the formula of magnesium oxide is **MgO**.

Experiment 26.1   Finding the formula of magnesium oxide
Try the experiment described above.
You will need to lift the lid slightly from time to time with a pair of tongs.
This lets oxygen gas get to the hot magnesium. But try not to let any of the
white smoke produced escape. Why not?
Work out your formula for magnesium oxide.

- How do your results compare to the student's experiment described above?
- Explain any difference.

▷ Percentage composition

You can also find the percentage mass of each element present in a compound if you know its formula. (Remember that relative atomic masses will always be given to you in an exam question.) Let's look at an example:

Example 1
What is the percentage by mass of hydrogen and oxygen in water (H_2O)?
(A_r values: $H = 1$, $O = 16$)

Step 1   Work out the relative formula mass (or molecular mass) of H_2O.

$$2 \times H = 2 \times 1 = 2$$
$$+ 1 \times O = 1 \times 16 = 16$$
$$\text{So the R.F.M. } (M_r) = 18$$

Step 2   Use the mass of each element present to find its percentage.

If we have 18 g of H_2O:
2 g will be hydrogen and 16 g will be oxygen.

So we can work out the percentage of each element using:

$$\% = \frac{\text{mass}}{M_r} \times 100$$

$$\% \text{ hydrogen} = \frac{2}{18} \times 100 = \textbf{11.1 \%}$$

$$\% \text{ oxygen} = \frac{16}{18} \times 100 = \textbf{88.9 \%}$$

I'm H_2O. My mass is made up from 11.1 % H and 88.9 % O. Can you see why my formula is **NOT** HO_8 ??

So water contains 11.1 % hydrogen and 88.9 % oxygen by mass.
This is called its **percentage composition**.

Example 2
Work out the percentage composition of the elements in iron(II) sulfate, $FeSO_4$.
(A_r values: $Fe = 56$, $S = 32$, $O = 16$)

The relative formula mass of $FeSO_4$ is:

$$1 \times Fe = 1 \times 56 = 56$$
$$1 \times S = 1 \times 32 = 32$$
$$+ 4 \times O = 4 \times 16 = 64$$
$$\overline{152}$$

$$\% \text{ Fe} = \frac{56}{152} \times 100 = \textbf{36.8 \%} \qquad \% \text{ S} = \frac{32}{152} \times 100 = \textbf{21.1 \%} \qquad \% \text{ O} = \frac{64}{152} \times 100 = \textbf{42.1 \%}$$

● Now you can find the percentage composition of these compounds:
   1. NaCl   **2.** CO   **3.** Al_2O_3 **4.** CCl_4 **5.** $Mg(NO_3)_2$
   (A_r values: $Na = 23$, $Cl = 35.5$, $C = 12$, $O = 16$, $Al = 27$, $Mg = 24$, $N = 14$)

Sometimes we have to work out the formula
of a compound given its percentage composition.
Do you think that the percentages of hydrogen
and oxygen change in different masses of water?

We know the smell of hydrogen sulfide!

Can't Jimmy work in the fume cupboard?

The proportion of each element in a compound is fixed.

Would you recognise the smell of bad eggs?
The smell is caused by a gas called hydrogen sulfide.
Let's see how we can work out its formula:

Example 3
A sample of hydrogen sulfide gas was broken down
into its elements – hydrogen and sulfur.
It was found to contain 6 % hydrogen and 94 % sulfur.
What is its empirical formula (the simplest ratio of elements present)?
(A_r values: H = 1, S = 32)

Step 1   Imagine you have 100 g of the compound.
If we have 100 g of hydrogen sulfide,
we would have 6 g of hydrogen and 94 g of sulfur.

Step 2   Find out how many moles of each element are present.

moles of H $= \dfrac{6}{1} = 6$ moles of S $= \dfrac{94}{32} = 2.9$

Step 3   Find the ratio of moles present. (Remember we use moles to 'count' atoms)

H : S

6 : 2.9

(Divide by the smallest number to get a ratio x : 1)

2 : 1

(The ratio allows for rounding off numbers and for any experimental error.)
There are twice as many hydrogen atoms as sulfur atoms
in hydrogen sulfide. So its **empirical formula is H_2S**.

**The empirical formula tells us the simplest whole number ratio
of elements in a compound.**

But how do we know that the actual formula of hydrogen sulfide is H_2S,
and not H_4S_2, H_6S_3, etc.? We need the relative formula (molecular) mass
to answer the question. In this case, M_r of hydrogen sulfide is 34.
So the actual formula (called its **molecular formula**) is H_2S i.e. (2 + 32 = 34).

- Work out the empirical formula of each compound below:
  6. 91 % phosphorus and 9 % hydrogen
  7. 75 % carbon and 25 % hydrogen
  8. 13.2 % lithium, 26.4 % nitrogen and 60.4 % oxygen.
  (A_r values: P = 31, H = 1, C = 12, Li = 7, N = 14, O = 16)

▷ Moles in equations

Imagine that you are the manager of a chemical plant.
When a customer orders some of your product,
you need to know how much raw material you need
to order yourself.

This is where chemical equations help us out.
We can use balanced equations to predict
the masses of reactants and products.
Let's look at an example:

Example
Zinc oxide is heated with carbon in a furnace.
The zinc oxide is reduced to zinc and carbon monoxide is formed.
How much zinc oxide do you need to make 130 tonnes of zinc?
(1 tonne = 1000 kg   Zn = 65, O = 16)

Step 1 – Write the balanced equation.

$$ZnO + C \longrightarrow Zn + CO$$

Step 2 – Write out the number of moles of reactants and products
from the balanced equation.

The equation above means that:
   1 mole of zinc oxide reacts with 1 mole of carbon,
   to make 1 mole of zinc and 1 mole of carbon monoxide.

Step 3 – Circle the information given and what you want to find out.

$$\boxed{ZnO} + C \longrightarrow \boxed{Zn} + CO$$

So 1 mole of zinc oxide is needed to make 1 mole of zinc.

Step 4 – Convert the moles into masses (using R.F.M.s)

$$\begin{array}{llll} ZnO & + C \longrightarrow & Zn & + CO \\ (65 + 16)\,g & & 65\,g & \\ 81\,g & \longrightarrow & 65\,g & \end{array}$$

Step 5 – Use logical steps to arrive at your final answer.

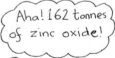

If 81 g of ZnO will give us 65g of Zn, then
81 tonnes of ZnO will give us 65 tonnes of Zn.
So how many tonnes of ZnO will give us 130 tonnes of Zn?

$\dfrac{81}{65}$ tonnes of ZnO will give us 1 tonne of Zn.

Therefore, $\dfrac{81}{65} \times 130$ tonnes of ZnO will give us 130 tonnes of Zn
            = **162 tonnes of zinc oxide**

Let's look at another example. Do you know
the test for a chloride ion? (See page 360.)
You add silver nitrate solution, then see if you get
a white precipitate of silver chloride.

Example

You start with a solution containing 0.95 g of magnesium chloride.
You add a solution of silver nitrate. If all the magnesium chloride reacts,
how much silver chloride could be made?
(Mg = 24, Cl = 35.5, Ag = 108)

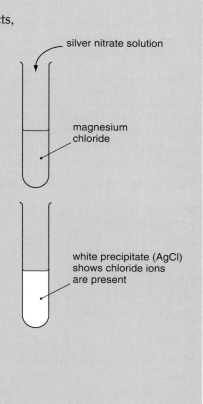

silver nitrate solution

magnesium
chloride

white precipitate (AgCl)
shows chloride ions
are present

Step 1 – Write the balanced equation.

$$2\,AgNO_3 + MgCl_2 \longrightarrow 2\,AgCl + Mg(NO_3)_2$$

Step 2 – Write out the number of moles of reactants and products
from the balanced equation.

2 moles of silver nitrate react with 1 mole of magnesium chloride,
to give 2 moles of silver chloride and 1 mole of magnesium nitrate.

Step 3 – Circle the information given and what you want to find out.

$$2\,AgNO_3 + \boxed{MgCl_2} \longrightarrow \boxed{2\,AgCl} + Mg(NO_3)_2$$

So 1 mole of magnesium chloride will give 2 moles of silver chloride.

Step 4 – Convert the moles into masses (using R.F.M.s)

magnesium chloride, $MgCl_2$ silver chloride, AgCl

$1 \times Mg = 1 \times 24 \ = \ \ \ 24$ $1 \times Ag = 108$

$2 \times Cl = 2 \times 35.5 = +\underline{71}$ $1 \times Cl \ = +\underline{35.5}$

                    95 143.5

$$\begin{array}{cccc} 2\,AgNO_3 & + & MgCl_2 & \longrightarrow & 2\,AgCl & + & Mg(NO_3)_2 \\ & & 95\,g & & (2 \times 143.5) \\ & & 95\,g & \longrightarrow & 287\,g \end{array}$$

Step 5 – Use logical steps to arrive at your final answer.

If 95 g of $MgCl_2$ give us 287 g of AgCl, then

1 g of $MgCl_2$ give us $\dfrac{287}{95}$ g of AgCl.

Therefore, 0.95 g of $MgCl_2$ give us $\dfrac{287}{95} \times 0.95$ g = **2.87 g of AgCl**

- Now you can try these problems:
  1. A student adds 4.8 g of magnesium to excess dilute hydrochloric acid.
     What mass of magnesium chloride would be made? (Mg = 24, Cl = 35.5)
  2. If you add 5.3 g of sodium carbonate to excess dilute sulfuric acid,
     what mass of sodium sulfate would be made? (Na = 23, C = 12, O = 16, S = 32)

▷ Atom economy

In the chemical industry, scientists try to convert as much of the raw materials they start with as possible into useful products. This not only increases profits, it is also good for the environment. We will use up less of the Earth's resources and produce less waste. This waste might pollute the environment.

The scientists use the term '**atom economy**' to work out the efficiency of their process. They use the formula:

$$\text{Percentage atom economy} = \frac{\text{mass of useful product}}{\text{total mass of products (or reactants)}} \times 100$$

The ideal chemical process would achieve 100% atom economy, but this isn't possible most of the time.
Many reactions also make by-products that have no use or would be uneconomic to convert to useful substances.

Let's consider the atom economy in a lime kiln:

Example

In the kiln the reaction is:

$$CaCO_3 \longrightarrow CaO(s) + CO_2(g)$$

(A_r values: Ca = 40, C = 12, O = 16)

So M_r values: $CaCO_3 = (40 + 12 + 48) = 100$
$CaO = (40 + 16) = 56$
$CO_2 = (12 + 32) = 44$

For every 100 tonnes of calcium carbonate in the kiln we can get 56 tonnes of calcium oxide (useful product).

The total mass of products always equals the total mass of reactants. (You don't create or destroy mass in a chemical reaction.)
So the percentage atom economy of the kiln is:

$$\frac{\text{mass of CaO}}{\text{mass of (CaO + CO}_2)} \times 100 = \frac{56}{100} \times 100 = \textbf{56\%}$$

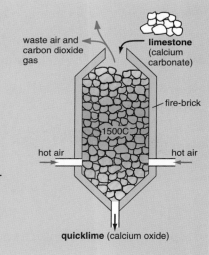

waste air and carbon dioxide gas

limestone (calcium carbonate)

fire-brick

1500C

hot air   hot air

quicklime (calcium oxide)

As well as atom economy, we also have to think about the actual yield of product in the process. So in a lime kiln, the yield of calcium oxide has to be considered:

- Not all the limestone is calcium carbonate. Limestone is made up mainly of calcium carbonate but also contains other minerals. So from 100 tonnes of limestone, we will not get 56 tonnes of calcium oxide.

- Some limestone is lost as dust in crushing.

You can see how to calculate percentage yield on the next page.
- Why will high atom economy and percentage yield help sustainable development?

We must maximise our atom economy.

MANAGER

▷ Calculating percentage yields

You have seen how to work out the mass of products you can get from a given mass of reactants. But in experiments we often get less products than we predicted. Look at the cartoon! In industrial processes we also get handling losses, unexpected reactions, and some reactions might well be reversible.

We can calculate the yield by working out how much we can get in theory, assuming all our reactants are converted into product.
That is then compared to the actual amount obtained practically:

We get losses in:
- *filtration*
- *evaporation*
- *transferring liquids*
- *heating*

$$\text{Percentage yield} = \frac{\text{actual mass obtained practically}}{\text{mass of product assuming 100\% conversion}} \times 100$$

Example

You want to convert ethanol into ethanoic acid.
You start with 92.0 g of ethanol and manage to collect 72.0 of ethanoic acid.
What is your yield?

First of all we need to work out the mass of ethanoic acid we could get assuming 100% conversion:
From the equation, 1 mole of ethanol should give 1 mole of ethanoic acid.

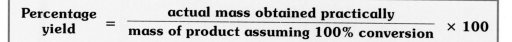

$$\underset{\text{ethanol}}{C_2H_5OH} \xrightarrow{\text{oxidising agent}} \underset{\text{ethanoic acid}}{CH_3COOH}$$

Work out the relative formula mass of ethanol and ethanoic acid.

ethanol	ethanoic acid
$(2 \times C) = 2 \times 12 = 24$	$(2 \times C) = 2 \times 12 = 24$
$(6 \times H) = 6 \times 1 = 6$	$(4 \times H) = 4 \times 1 = 4$
$+(1 \times O) = 1 \times 16 = \underline{16}$	$+(2 \times O) = 2 \times 16 = \underline{32}$
$\underline{46}$	$\underline{60}$

Calculate how much ethanoic acid could be made (assuming 100% conversion):
From the relative formula masses we see that:

46 g of ethanol could give 60 g of ethanoic acid

The question tells us that the chemist started with 92 g of ethanol.

Therefore,

92 g of ethanol could give **120 g of ethanoic acid (assuming 100% conversion)**.
(i.e. (46×2) g of ethanol could yield (60×2) g of ethanoic acid.)

Now we can work out the percentage yield:

The question tells us that the actual yield was 72 g of ethanoic acid.
Remember that:

$$\text{percentage yield} = \frac{\text{actual mass obtained practically}}{\text{mass of product assuming 100\% conversion}} \times 100$$

Therefore,

The percentage yield was $\dfrac{72}{120} \times 100 = \textbf{60\%}$

▷ Titrations

We have met the technique of titration before on page 236.
We used it to find out how much sulfuric acid
will just neutralise ammonia solution.
In any acid/alkali titration, we need a suitable indicator
to show us when the reaction is just complete.
This is called the **end point** of the reaction.
Look at the photos on the next page:

Knowing the concentration of one of the reactants,
we can use the results of titration to find out
the concentration of the other reactant.
Look at the example below:

Example

A student found that it takes 15.0 cm³ of 0.50 mol/dm³ hydrochloric acid
to neutralise 10 cm³ of sodium hydroxide solution.
What is the concentration of the sodium hydroxide solution?

Step 1   Use the balanced equation to find the ratio of moles
that react together.

$$NaOH(aq) + HCl(aq) \longrightarrow NaCl(aq) + H_2O(l)$$

This shows us that 1 mole of NaOH reacts with 1 mole of HCl.

Step 2   Find out how many moles of hydrochloric acid were used.
We use 15.0 cm³ of 0.50 mol/dm³ HCl(aq)
(On page 344 we saw how to find out the number of moles present.)
In 1000 cm³ of the solution we would have 0.50 moles

In 1 cm³ of the solution we would have $\dfrac{0.50}{1000}$ moles

So in 15 cm³ we must have $\dfrac{0.50}{1000} \times 15$

$\qquad\qquad\qquad$ moles = 0.0075 moles

Step 3   Use the information in the question and the previous two steps
to work out the concentration of the sodium hydroxide solution.

From Step 1, we know that 1 mole of NaOH reacts with 1 mole of HCl.
Therefore 0.0075 moles of NaOH will react with 0.0075 moles of HCl.
So we must have 0.0075 moles of NaOH in the 10 cm³ of solution
used in the titration. (Information from the question is used here.)
Remember that the concentration tells us the number of moles
dissolved in 1000 cm³ (1 dm³) of solution.
Working in logical steps:

In 1 cm³ of NaOH(aq) we must have $\dfrac{0.0075}{10}$ moles

So in 1000 cm³ we would have $\dfrac{0.0075}{10} \times 1000$ moles = 0.75 moles

Performing a titration

The photos below show the steps in carrying out a titration:

A pipette is used to measure the volume of alkali accurately

A suitable indicator (in this case we use methyl orange) is added. We don't use a mixed indicator, like universal indicator because we want a sharp colour change at the end point.

The acid is added from a burette. This lets us make very small additions.

We stop adding acid at the end point

Now see if you can solve this problem:

1. It takes 30 cm^3 of 1.0 mol/dm^3 hydrochloric acid to neutralise 24 cm^3 of potassium hydroxide (KOH) solution.
   What is the concentration of the potassium hydroxide solution?

Diluting solutions

We sometimes have to dilute down solutions – and not just in the chemistry lab! We dilute orange squash, some liquid medicines, and often have to 'add water' when cooking. For example, when making a sauce, we add water to make it 'thinner'.

In chemistry, we need to be more exact. We need to be able to change the concentration of solutions by adding water. For example, you might be given a bottle of a 1.0 mol/dm^3 solution and asked to dilute it to 0.5 mol/dm^3. This means that in the same volume of solution, you only want **half** the number of moles. So you add enough water to **double** the volume of the solution. If you need 50 cm^3 of a 0.5 mol/dm^3 solution, you would take 25 cm^3 of the 1.0 mol/dm^3 solution and add 25 cm^3 of water.

2. You want to prepare 100 cm^3 of a 0.1 mol/dm^3 solution from a 1.0 mol/dm^3 solution. How would you do it?

▷ Moles in electrolysis

You have learned about electrolysis in Chapter 7.
Can you remember what electrolysis means?
When a compound is broken down by electricity,
metals (or hydrogen) form at the negative electrode (cathode).
The non-metals form at the positive electrode (anode).

We can work out how much charge passes around the circuit
during *electrolysis* using this equation:

Charge	=	**current**	×	**time**
(in coulombs)		(in amps)		(in seconds)

You can measure the mass of an element formed at an electrode
in an electrolysis experiment.
You can then work out how much charge would give 1 mole
of the element.

Let's look at an example:

Example

A student set up the circuit as shown:
A current of 0.5 A passed around the circuit for
3860 seconds. The cathode had gained 0.64 g
in the experiment.
How many coulombs of charge are needed to deposit
1 mole of copper?
(R.A.M. of Cu = 64)

Use the equation: Charge = current × time
$$= 0.5 \times 3860$$
$$= 1930 \text{ coulombs}$$

So 1930 coulombs gave 0.64 g of copper.

Therefore, $1930 \times \dfrac{1}{0.64}$ coulombs will give 1 g of copper.

This means that $1930 \times \dfrac{1}{0.64} \times 64$ coulombs will give 64 g (1 mole) of copper.

193 000 coulombs will produce 1 mole of copper.

a variable resistor to keep the current low and constant

ammeter to read the current

A

copper sulfate solution

- Now you can try this example:
1. A current of 2 A is passed through molten sodium chloride
   for 9650 seconds.
   4.6 g of sodium is deposited at the cathode.
   How many coulombs of charge will give 1 mole of sodium?
   (A_r of Na = 23)

Faraday's Law

We have just seen how to work out the charge needed to deposit 1 mole of an element. This information can then be used to tell us the charge on an ion of that element.

Metal	Charge on ion	Charge needed to deposit one mole (C)
Sodium	1+	96 500
Magnesium	2+	193 000
Aluminium	3+	289 500

The table opposite shows how much charge is needed to deposit 1 mole of the elements. Can you see any pattern?

The famous scientist Michael Faraday did much of the early work on electrolysis. It was a powerful tool in the search for new elements. He found that:

Michael Faraday (1791–1867) is sometimes called 'the father of electricity'.

> **96 500 coulombs deposit 1 mole of a metal with a single charge on its ions, e.g. Na$^+$.**

We can calculate the charge on an ion from the results of an electrolysis experiment.

The charge carried by one mole of electrons (96 500 C) is called 1 Faraday. How many Faradays are needed to deposit one mole of magnesium?

Example

A student set up the circuit as shown:
She passed a current of 1 A for 16 minutes.
She collected 1.08 g of silver.
What is the charge on a silver ion?

Charge = current × time (in seconds!)
     = 1   × 16 × 60
     = 960 coulombs

960 coulombs gave 1.08 g of silver.

So $960 \times \dfrac{1}{1.08}$ coulombs will give 1 g of silver.

Therefore, $960 \times \dfrac{1}{1.08} \times 108$ coulombs will give 108 g of silver

     = 96 000 coulombs

If 96 000 coulombs give 1 mole of silver, then the charge on its ions must be **1+**.
(This allows for some experimental error.)

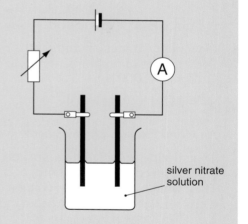

silver nitrate solution

- Now you try this example:
2. A current of 1 A was passed through molten lead bromide for 1930 seconds.
   2.07 g of lead was collected.
   What is the charge on a lead ion? (A_r of Pb = 207)

▷ Questions

1. An oxide of copper was reduced (had its oxygen removed) in a stream of hydrogen as shown:

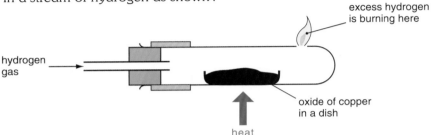

These were the results:
Mass of dish = 25.00 g
Mass of oxide of copper + dish (before reaction) = 29.0 g
Mass of copper left + dish (after reaction) = 28.2 g
Work out the formula of this oxide of copper.

(Cu = 64, O = 16)

2. Ammonium nitrate fertiliser is made by the following reaction:

$$NH_3 \quad + \quad HNO_3 \quad \longrightarrow \quad NH_4NO_3$$
ammonia + nitric acid $\longrightarrow$ ammonium nitrate

How many tonnes of ammonia is needed to make 2400 tonnes of ammonium nitrate?
(1 tonne = 1000 kg)

(N = 14, H = 1, O = 16)

3. Amy and Hassan are investigating rates of reaction.

They decide to look at the reaction between limestone (calcium carbonate) and dilute hydrochloric acid. They want to measure the volume of gas given off in a gas syringe. The syringe has a maximum volume of 100 cm³.

They plan to follow the reaction until it is complete, measuring the volume of gas at regular intervals. They will use excess acid.
What is the greatest mass of calcium carbonate that they should start with to avoid the plunger being forced out of the syringe?
(1 mole of any gas occupies 24 000 cm³ at room temperature and pressure)

(Ca = 40, C = 12, O = 16)

4. Molten potassium chloride was electrolysed.
At the end of the electrolysis, 7.8 g of potassium had been deposited at the cathode.
What volume of chlorine gas was given off at the anode?
(K = 39, 1 mole of any gas occupies 24 000 cm³)

Further questions on pages 371–372.

The alternative 'Problem Page'

Can you help these poor, unfortunate souls solve their problems?

1.
Dear Agony Aunt,
I have a terrible compound. It all started with 2.4 g of carbon and 0.8 g of hydrogen. Can you please help me find the correct formula?
(The Relative Formula Mass of the compound is 16.)

Yours hopefully,
Ray C. O.
P.S.   (C = 12, H = 1)

2.
Dear Agony Aunt,
I have been putting aside calcium carbonate for several years now.
I am naturally looking forward to the day when I can roast it all, and change it into calcium oxide and carbon dioxide.
I now have 200 tonnes of calcium carbonate. Can you tell me how much calcium oxide I can expect when I roast it?

Yours elderly,
Mrs L. Kiln
P.S.   (Ca = 40, C = 12, O = 16)

3.
Dear Agony Aunt,
I am studying for my GCSEs, but I am very worried about my concentration. I fear it is too low to ever be successful.
My doctor says that I have 0.585 g of sodium chloride in 250 cm^3 of solution. Please help. Don't try to be kind. I need to know my concentration in moles per dm^3 as soon as possible!

Yours distractedly,
Sol T. Waters
P.S.   (Na = 23, Cl = 35.5)

4.
Dear Agony Aunt,
My daughter recently added 5 g of calcium carbonate to excess dilute hydrochloric acid.
I know this sounds silly, but she forgot to measure the volume of carbon dioxide gas given off. Her lapse was at room temperature and pressure, naturally. Can you help her?

Yours motherly,
Ma Bell
P.S.   (Ca = 40, C = 12, O = 16.
And don't forget 1 mole of gas occupies 24 000 cm^3 at room temperature and pressure.)

These are tricky problems – but you can find my answers on page 374

Molly Cule
Agony Aunt

5.
Dear Agony Aunt,
Last week I found a friend of mine passing 2 amps of electricity through some copper(II) sulfate solution.
She claimed that she had only done it once; and that was only for 5 minutes. However, from the amount of copper deposited, I frankly find this difficult to believe.
Can you tell me how much copper she would get if she is telling the truth?

Yours suspiciously,
Elle Ectrolysis
P.S.   (Cu = 64)

(Answers for Chapter 26 'Moles' are available on page 374)

chapter 27

▷ Testing for anions (−)

a. Precipitation reactions

We can use precipitation reactions to test for
some negative ions.
We can see if a salt is a :

- chloride,
- iodide, or
- bromide,
- sulfate, by the tests below :

Look at the table below :
You dissolve the salt in dilute nitric acid, then do the test shown :

Type of salt (ion)	Test	Result
chloride (Cl^-)	silver nitrate solution $\longrightarrow$	a white precipitate of silver chloride
bromide (Br^-)	silver nitrate solution $\longrightarrow$	a cream precipitate of silver bromide
iodide (I^-)	silver nitrate solution $\longrightarrow$	a pale yellow precipitate of silver iodide
sulfate(VI) ($SO_4{}^{2-}$)	barium chloride solution $\longrightarrow$	a white precipitate of barium sulfate(VI)

Let's use these tests to find out about some unknown salts :

Experiment 27.1   Identifying salts

Your teacher will give you 5 unknown salts,
labelled A to E.

You can find out which salts are chlorides, bromides,
iodides or sulfates.

You will have to **dissolve each salt in
dilute nitric acid** before you do the tests.
We will use the tests from the table above.

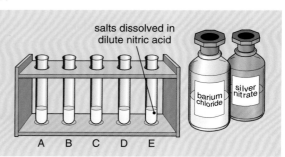

Test for chloride

$Ag^+(aq) + Cl^-(aq) \longrightarrow AgCl(s)$
chloride

Test for bromide

$Ag^+(aq) + Br^-(aq) \longrightarrow AgBr(s)$
bromide

Test for iodide

$Ag^+(aq) + I^-(aq) \longrightarrow AgI(s)$
iodide

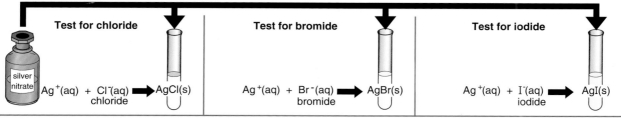

Test for sulfate

$Ba^{2+}(aq) + SO_4^{2-}(aq) \longrightarrow BaSO_4(s)$
sulfate

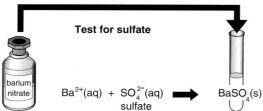

These are **qualitative** tests. They tell us what is
present but they don't tell us how much there is.
To get that information we would need to carry
out careful measurements of the mass of
precipitate made.

b. Tests for anions that give off gases

The tests on this page all release a gas. The gas produced gives you the clue to identify the negatively charged ion.

Try out the tests below:

Experiment 27.2   Testing for nitrate ions (NO_3^-)

Do this test in a fume cupboard.

Add sodium hydroxide solution to the nitrate in a boiling tube. Then add powdered aluminium and warm carefully. Hold a piece of damp red litmus paper in the mouth of the tube.
- What happens to the red litmus paper?
- Which gas causes this? (See page 235.)

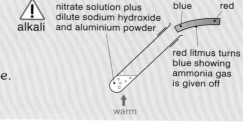

In this test the nitrate ion is reduced to form ammonia gas.

Experiment 27.3   Testing for sulfate(IV) ions (SO_3^{2-})

You might see the sulfate(IV) ion referred to as a sulfite ion.

Do this test in a fume cupboard as toxic sulfur dioxide is given off. Set up the apparatus as shown opposite:

Add dilute hydrochloric acid. Pass the gas through acidified potassium manganate(VII) solution.
- What do you see happen?

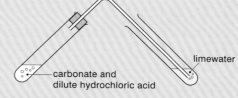

In the test for the sulfate(IV) ion, acidic sulfur dioxide gas is given off.
Here is the ionic equation:

$$SO_3^{2-}(aq) + 2H^+(aq) \longrightarrow SO_2(g) + H_2O(l)$$

This gas decolourises acidified potassium manganate(VII) solution.

Experiment 27.4   Testing for carbonate ions (CO_3^{2-})

Set up the apparatus opposite:

Add dilute hydrochloric acid. If the limewater turns milky then a carbonate ion is present.
- Which gas is given off?
- Write an ionic equation for the reaction.

NB Carbonates or hydrogencarbonates give a positive result with this test.

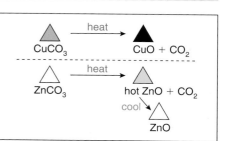

Special carbonates – You can identify copper carbonate, as it is a pale green powder that decomposes on heating to form black copper oxide powder.

Zinc carbonate is a white powder. But when heated it decomposes to zinc oxide. This is also a white powder normally, but when heated it turns bright yellow. It turns white when left to cool down.

▷ Testing for cations (+)

The tests on the previous two pages tell us
which is the negative ion in some salts.
Remember that salts also contain a metal.
The metal part of a salt is always a positive ion.

a. Flame tests

We can also identify some metals using a
flame test.

*The colours in these
fireworks are
made by metal ions.
(See page 197.)*

Experiment 27.5   Flame tests for metals
You can test the unknown salts from Experiment 27.1.

Take the colour out of a Bunsen flame by opening
the air-hole slightly.
Heat a piece of nichrome wire in the flame to clean it.
You can also dip the wire in concentrated hydrochloric
acid on a watch glass, then heat the wire to clean it.

Put the loop at the end of the wire into some water.
Then dip it into one of the unknown salts.
Hold the wire in the edge of the flame.

- Record the colour.
  Clean the wire again and test the other salts.

- Use the flame colours shown below
  and your results from Experiment 27.1
  to name salts A to E.

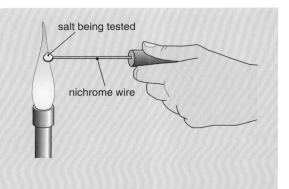

salt being tested

nichrome wire

You can find a summary of tests for ions on page 368.

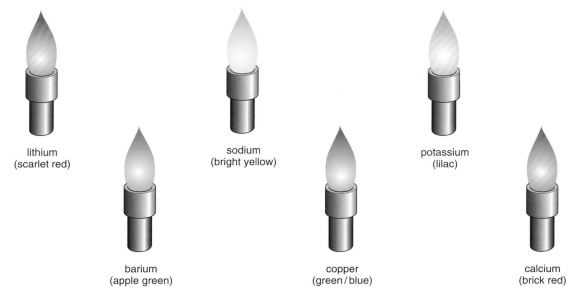

lithium
(scarlet red)

sodium
(bright yellow)

potassium
(lilac)

barium
(apple green)

copper
(green/blue)

calcium
(brick red)

b. Testing for aqueous cations

The negative ions, such as chloride ions, were identified by precipitation reactions. (See pages 360–1.)
We can also use precipitation to identify metal ions.
Look at the table below:

Metal ion	Result of adding sodium hydroxide solution
copper(II) Cu^{2+}	pale blue precipitate of $Cu(OH)_2$
iron(II) Fe^{2+}	dirty green precipitate of $Fe(OH)_2$
iron(III) Fe^{3+}	rusty brown precipitate of $Fe(OH)_3$
aluminium Al^{3+}	white precipitate of $Al(OH)_3$ which dissolves in excess sodium hydroxide
magnesium Mg^{2+}	white precipitate of $Mg(OH)_2$
calcium Ca^{2+}	white precipitate of $Ca(OH)_2$

- Which two metal ions can't be distinguished from the result of the test above?

You can tell the difference between magnesium ions and calcium ions by a flame test. Calcium ions turn the flame brick red whereas magnesium ions have no colour.

Precipitate of copper hydroxide
$Cu^{2+}(aq) + 2OH^-(aq) \rightarrow Cu(OH)_2(s)$

Precipitate of iron(II) hydroxide
$Fe^{2+}(aq) + 2OH^-(aq) \rightarrow Fe(OH)_2(s)$

Precipitate of iron(III) hydroxide
$Fe^{3+}(aq) + 3OH^-(aq) \rightarrow Fe(OH)_3(s)$

> **Experiment 27.6   Testing for metal ions**
> Collect samples of unknown solutions labelled F to K.
> Add a little sodium hydroxide (alkali) solution
> in separate test tubes.
> Use the table above to identify the metal ion
> in each solution.
>
> ⚠️ alkali

Look at the photos to see the ionic equations that we can use to describe the reactions.
Remember that ionic equations only show the ions that change in the reaction.

- What do we call the ions in the solution that **don't** change?
- Write the ionic equation for testing calcium ions.

We can also use sodium hydroxide to test for **ammonium** ions, NH_4^+.
In this case, you warm the mixture too and a sharp smelling gas is given off.
The gas turns red litmus blue.
Can you recall the only common alkaline gas from page 235?

$$NH_4Cl(aq) + NaOH(aq) \xrightarrow{heat} NaCl(aq) + H_2O(l) + NH_3(g)$$
ammonium   ammonia
  chloride

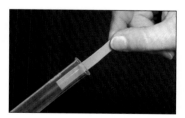
Ammonia gas turns damp red litmus paper blue

▷ Chemistry at work : Instrumental analysis

We have looked at various chemical tests for different ions. But if you are an environmental scientist checking water or air for pollution the amounts present may be tiny. This is where the use of instruments to analyse samples has become very important.

As we get to know more about electronics and computers, the instruments we now have are incredibly sensitive. They can detect amounts of pollutants that would be far too small to analyse by chemical tests in a lab.

Let's look at one instrument that identifies elements and one that identifies compounds:

Atomic absorption spectrometers

These machines are good at identifying elements present. The sample is usually ***heated in a flame***. This breaks down the sample into atoms which are then excited.

Some electrons in the atoms jump into higher energy levels (or orbitals). They absorb energy from the flame. When they fall back to lower energy levels the energy is given out as light energy. This explains the flame tests we did to identify metals on page 362.

In a spectrometer we can analyse the wavelength of the light. Each type of atom absorbs and gives out its own particular pattern of radiation.
This is called its **spectrum**.

We can use this to identify elements. We just compare peaks and troughs from the unknown sample with elements we do know. Computers hold records on each element and match up the spectra for us.
You can also see how much of the element is present in the sample.

This, and similar methods, give us accurate ways to monitor water for metals. On page 296 we saw how dangerous mercury can be. We can now trace mercury down to as little as 0.000 000 001 g. Other pollutants, such as zinc, lead and cadmium, can also be detected. Water companies can measure the concentration of calcium, magnesium, aluminium and iron in their water.

Atomic absorption spectrometers are also used in the steel industry. Remember that steel is a mixture of elements with iron. Steel makers need to control carefully the amounts of trace metals present. The quality of steel depends on this.

Sensitive instruments can detect small amounts of substance. They can be used to test for drugs in athletes and mineral ions in blood. They analyse samples in forensic labs and hospitals, as well as monitoring pollutants.

The flame atomises the sample and excites electrons in an atomic absorption spectrometer

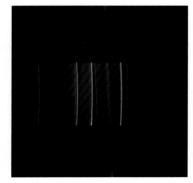

This is the characteristic spectrum of cadmium

▷ Chemistry at work : Instrumental analysis

Visible–ultraviolet spectroscopy

We have also seen how people are worried about
the pesticides that we now find in water. (See page 296.)

- Why can't we use an atomic absorption spectrometer to identify
  a pesticide in a water sample?
- What would happen to a pesticide *molecule*
  in the flame inside the spectrometer?

Visible–ultraviolet spectroscopy is a gentler method.
It doesn't use a flame to break up the molecule into atoms.
We simply shine light of different wavelengths
through the solution to be analysed. So the sample
is not destroyed in the analysis. A detector then
tells us which wavelengths have been absorbed.
These are usually *broad bands* of wavelengths.
They are not sharp peaks like the emission spectra
on the last page.

Look at the visible-ultraviolet spectrum
of a compound below:

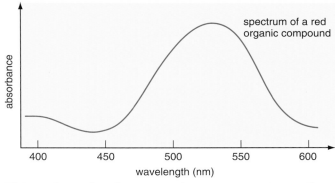

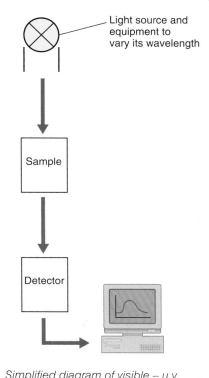

Light source and
equipment to
vary its wavelength

*Simplified diagram of visible – u.v.
spectrophotometer*

This makes it harder to match one spectra to another,
but it can still be done. We can also analyse amounts
of compound present by comparing against known amounts.

This method will only work if the substance we analyse
absorbs light in the visible or ultraviolet range.
Most of the pollutants in water are colourless.
They don't absorb in the visible range of wavelengths
(or in the ultraviolet range).

But we can react them with a compound that
changes them into a coloured molecule.
We can test for nitrate ions (from fertilisers) and for phosphate
ions (from detergents) using this method.
This helps us to guard against eutrophication.
(See page 240.)

▷ Analysing spectra

You have seen two methods of instrumental analysis on the previous two pages.
There are also many other instruments available to chemists in helping to solve crimes, diagnose illness and identify pollutants.

Most instruments produce print-outs or screen-shots of a spectrum. These mainly show wavelengths of radiation given out or absorbed.
For example, we use infra-red radiation. The different types of bonds in organic molecules absorb characteristic frequencies. So a spectrum can tell us what groups are in an unknown molecule.

Computers can now match the spectrum of an unknown compound to a huge database of stored spectra. This is called **fingerprinting**.

Samples can be passed through a machine that separates the components in a mixture before they are analysed. Gas-liquid chromatography is often used. The vaporised sample is sent through a resin. The different substances take different times to pass through.

Here is the infra-red spectrum of ethanol, C_2H_5OH:

Another technique is called **mass spectrometry**. This breaks molecules apart by bombarding them with electrons.
Then the fragments are analysed in the mass spectrum.
We can use this to identify elements or compounds.

Here is the mass spectrum of ethanol:

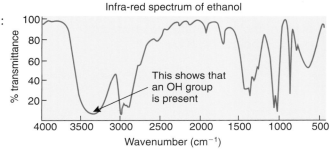

Infra-red spectrum of ethanol

This shows that an OH group is present

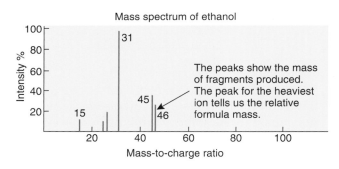

Mass spectrum of ethanol

The peaks show the mass of fragments produced. The peak for the heaviest ion tells us the relative formula mass.

Another common technique is **nuclear magnetic resonance** (n.m.r.).
This analyses small changes in the energy of nuclei in a magnetic field.

Here is the n.m.r. spectrum of ethanol:

It is used a lot to identify organic compounds.

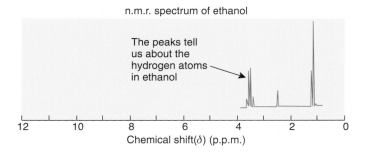

n.m.r. spectrum of ethanol

The peaks tell us about the hydrogen atoms in ethanol

▷ Quantitative analysis

The police send blood samples to forensic scientists. They analyse the concentration of ethanol in the blood of suspected drink drivers. They can find this from the size of one of the peaks in the infra-red spectrum of the sample (see previous page). This gives numerical data (**quantitative data**) that can be used in court.

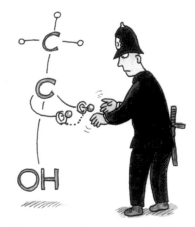

Infra-red spectra can convict drivers with too much ethanol in their blood

What's the formula?

We can also analyse an unknown organic compound chemically to find out its formula.
We burn the compound in oxygen. Then the gases made are passed through tubes that absorb water vapour and carbon dioxide formed. The gases are analysed by machines before and after they enter the absorption tubes. They tell us the mass of water and carbon dioxide produced.

Let's look at how we can process the data:

Example

A sample of a compound was found to contain carbon, hydrogen and oxygen.
Its relative formula mass was found to be 46.
On complete combustion of 9.2 g of the compound in oxygen, 17.6 g of carbon dioxide and 10.8 g of water vapour were absorbed in tubes.
What is the chemical formula of the compound?
(A_r values: C = 12, H = 1, O = 16)

This is like working out the empirical formula on page 349.
But first we have to find the mass of carbon and hydrogen from the data provided about CO_2 and H_2O.

So how much carbon is there in 17.6 g of carbon dioxide?
The relative formula mass of CO_2 is $12 + (16 \times 2) = 44$ (of which 12 is made up of carbon).
So we have $(12 \div 44) \times 17.6$ g of carbon in the unknown compound = **4.8 g of carbon**.

How much hydrogen is in 10.8 g of H_2O?
The relative formula mass of H_2O is $(2 \times 1) + 16 = 18$ (of which 2 is made up of hydrogen).
So we have $(2 \div 18) \times 10.8$ g of hydrogen in the unknown compound = **1.2 g of hydrogen**.

The rest of the compound, i.e. $9.2 - (4.8 + 1.2)$, must be oxygen.
So we have **3.2 g of oxygen.**

Now we do the usual calculation to find the simplest ratio of moles, which will give us the empirical formula (see pages 348 and 349):

C	: H	: O	
$4.8 \div 12$	: $1.2 \div 1$	: $3.2 \div 16$	
0.4	: 1.2	: 0.2	(Divide everything by 0.2 to get the simplest ratio.)
2	: 6	: 1	

So the empirical formula is **C_2H_6O.**
This is also the actual molecular formula because we are told its relative formula mass is 46,
i.e. $(2 \times 12) + (6 \times 1) + (1 \times 16) = 46$

Summary

▷ Anions

Anion	Formula	Test	Results of test
Carbonate	CO_3^{2-}	Add hydrochloric acid and test gas produced by bubbling through limewater.	Rapid effervescence and carbon dioxide produced gives white precipitate with limewater. $CO_3^{2-}(s) + 2H^+(aq) \longrightarrow H_2O(l) + CO_2(g)$
Sulfate(VI)	SO_4^{2-}	Add dilute nitric acid to solution of ion and then barium chloride solution.	Dense white precipitate of barium sulfate is produced. $Ba^{2+}(aq) + SO_4^{2-}(aq) \longrightarrow BaSO_4(s)$
Sulfate(IV) (or sulfite)	SO_3^{2-}	Add dilute hydrochloric acid.	Effervescence as acidic sulfur dioxide gas is produced. The SO_2 gas decolourises acidified manganate(VII) solution.
Chloride	Cl^-	Add dilute nitric acid followed by silver nitrate solution.	White precipitate formed. $Ag^+(aq) + Cl^-(aq) \longrightarrow AgCl(s)$
Bromide	Br^-	(as above)	Pale cream precipitate formed, i.e. $AgBr(s)$.
Iodide	I^-	(as above)	Pale yellow precipitate formed, i.e. $AgI(s)$.
Nitrate(v)	NO_3^-	Add NaOH solution to compound and aluminium powder. Heat.	Pungent smell of ammonia gas which turns damp red litmus blue.

▷ Cations

Cation	Formula	Test	Results of test
Ammonium	NH_4^+	Add NaOH solution and heat.	Gas turns damp red litmus blue.
Magnesium	Mg^{2+}	Sodium hydroxide (NaOH) solution	White precipitate formed.
Copper	Cu^{2+}		Pale blue precipitate formed.
Iron (II)	Fe^{2+}		Dirty green precipitate formed.
Iron (III)	Fe^{3+}		Rusty brown precipitate formed.
Aluminium	Al^{3+}		White precipitate formed which dissolves in excess alkali.
Sodium	Na^+	Flame test	Yellow flame.
Potassium	K^+		Lilac flame.
Lithium	Li^+		Red (scarlet) flame.
Calcium	Ca^{2+}		Brick red flame.
Barium	Ba^{2+}		Pale green (apple green) flame.
Copper	Cu^{2+}		Blue/green flame.

▷ Gases

Name of gas	Formula	Test	Results of test
Carbon dioxide	CO_2	Pass gas through limewater.	White precipitate formed in suspension (milky).
Ammonia	NH_3	Damp red litmus paper.	Red litmus turns blue.
Sulfur dioxide	SO_2	Acidified potassium manganate(VII).	Colour changes from purple to colourless.

▷ Questions

1. We can test for some anions (. . . . charged ions using . . . . reactions in which a solid forms in the solution. Other anions react to give off . . . . which we can identify e.g. red litmus turns . . . . with ammonia.
   We can use . . . . tests to identify some cations e.g. Na^+ ions. Reactions with sodium . . . . solution identify others. For example, iron(III) ions form a . . . . precipitate.

2. How can we positively identify zinc carbonate and copper carbonate?

3. An unknown white compound gives off an acidic gas that decolourises potassium manganate(VII) solution. The compound forms a white precipitate with dilute sodium hydroxide solution which dissolves in excess alkali.
   What is the unknown compound?

4. Alex has 3 unknown salts – A, B and C.
   He does a flame test on each salt. These are his results:

Salt	Colour of flame
A	yellow
B	lilac
C	green/blue
D	brick red

   a) Name the metal in each salt.
   b) Alex dissolved each salt in dilute nitric acid.
      When he added silver nitrate solution to A, a pale yellow precipitate was formed. He did the same to B, but the precipitate was white.
      C formed a white precipitate with barium nitrate solution.
      D gave off a gas with the acid, which Alex bubbled into limewater. The limewater turned milky.
      Name salts A, B, C and D.
   c) Give the test and positive result for the Fe^{2+} ion.

5. We can test for the nitrate ion by adding sodium hydroxide solution and warming with powdered aluminium. A gas is given off that turns red litmus blue.

   a) Name the gas given off.

   b) Draw a table to record the flame colours for i) sodium, ii) potassium, iii) copper, iv) calcium, v) barium.

   c) i) What do you see when you add sodium hydroxide solution to both calcium chloride and magnesium chloride?
      ii) Write a balanced symbol equation for each reaction.
      iii) What do we call this type of reaction?
      iv) Write an ionic equation to describe each reaction.
      v) How could you distinguish between calcium and magnesium in the chlorides tested?

   d) A compound gives a lilac flame test. When dissolved in nitric acid and silver nitrate is added, you get a cream precipitate.
      i) Name the compound.
      ii) Write an ionic equation for the reaction of the compound with silver nitrate solution.

6. a) Name an instrument that is useful for detecting unknown elements.
   b) Why is the instrument in a) not useful for detecting compounds?
   c) Name an instrument that is good at detecting molecules of compounds.
   d) Give some uses of these types of sensitive instrument.
   e) Why are they useful in industry?

7. An unknown hydrocarbon sample is burned in oxygen.
   It releases 17.72 g of carbon dioxide and 10.86 g of water.
   a) How many moles of carbon are there in 17.72 g of carbon dioxide?
   b) How many moles of hydrogen are there in 10.86 g of water?
   c) What is the empirical formula of the hydrocarbon? (*Answer on page 375.*)

Further questions on page 373.

▷ Making new materials

1. The chemicals made in industry can be described as bulk chemicals or fine chemicals.
   a) Explain the difference between bulk chemicals and fine chemicals. [2]
   b) Pharmaceutical drugs are fine chemicals. Making and developing a new pharmaceutical drug is usually much more expensive than a bulk chemical such as sulfuric acid.
   Describe three factors that make the drug so much more expensive. [3]

2. Some pharmaceutical chemicals are extracted from plants, others are synthesised.
   a) Describe the processes used in the extraction of chemicals from plants. [3]
   b) Explain what is meant by the term *synthesised*. [1]
   c) Suggest why chemicals extracted from plants may be more expensive than those that are synthesised. [2]

3. The development of a new pharmaceutical drug is very expensive.
   a) List three factors that contribute to the high cost of this development. [3]
   b) Briefly describe how any new pharmaceutical drug has to be tested. [3]
   c) A company intending to market a new drug carries out extensive market research. Explain why. [2]

4. Paint is a colloid.
   a) Explain what is meant by the term colloid. [2]
   b) Oil-based paints contain a solvent that evaporates as the paint dries.
      i) Describe the chemical change that also takes place when paint dries. [2]
      ii) How does this improve the qualities of the paint? [1]
   c) How do emulsion paints differ from oil-based paints? [2]

5. Thermochromic pigments can be added to acrylic paints.
   a) What is meant by the term thermochromic? [1]

b) i) Describe two uses of paint containing thermochromic pigments. [2]
   ii) For each use, explain what advantage this paint has over conventional paint. [2]

6. Many new drugs are developed from natural sources.
   Many natural substances have some drug activity but they often have harmful side effects. Scientists try to develop new drugs from these natural substances.
   The scientists follow a standard procedure for researching and developing a new medicine.

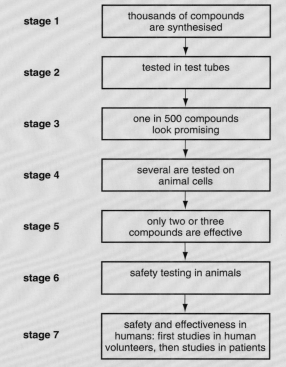

stage 1	thousands of compounds are synthesised
stage 2	tested in test tubes
stage 3	one in 500 compounds look promising
stage 4	several are tested on animal cells
stage 5	only two or three compounds are effective
stage 6	safety testing in animals
stage 7	safety and effectiveness in humans: first studies in human volunteers, then studies in patients

a) Explain why the chemicals are not tested on humans before **stage 7**. [2]
b) Explain why studies are carried out on human volunteers, before the patients in **stage 7**. [2]
c) Different people have different views about using animals in medical research.
   Suggest why some people disagree with using animals in medical research.
   One mark is for a clear, ordered answer.
   [2 + 1]   (OCR)

▷ Moles

7. Which one of the following has the greatest mass?
   (Relative atomic masses: H = 1, O = 16, Na = 23, S = 32, Cl = 35.5)
   A   1 mole Na
   B   0.5 mole NaCl
   C   0.6 mole NaOH
   D   0.2 mole Na_2SO_4 [1]   (AQA)

8. Which of the following have a concentration of 0.1 $mole/dm^3$?
   (Relative atomic masses: H = 1, N = 14, O = 16, Na = 23, S = 32)
   (1) HNO_3(aq) containing 6.3 g/litre
   (2) NaOH(aq) containing 2 g in 500 cm^3
   (3) H_2SO_4(aq) containing 0.98 g in 100 cm^3
                                [1]   (AQA)

Questions 9–11
   Use pages 390 and 391 to help you.
   From the list **A** to **D**,
   A   3
   B   6
   C   9
   D   18
   choose the number which is the:

9. Mass in grams of water produced when 1 g of hydrogen is burned completely.

$$2H_2 + O_2 \longrightarrow 2H_2O$$ [1]

10. Mass in grams of magnesium in 10 g of magnesium oxide, MgO.   [1]

11. Number of atoms of beryllium, Be, which have the same total mass as one atom of aluminium, Al.   [1]   (AQA)

12. Calculate the number of moles of water molecules, H_2O, present in 90 g of ice.
    (Relative atomic masses: H = 1, O = 16)   [1]
                                (AQA)

13. 0.5 mole of metal X combine with 1.0 mole of non-metal Y.
    What is the formula of this compound of X and Y?   [1]   (AQA)

14. Analysis of a sample of the compound sodium oxide, showed that it contained 2.3 g of sodium and 0.8 g of oxygen.
    Use this information to work out the formula of the compound sodium oxide in questions a) to e).
    a) What is the mass of each element in the sodium oxide?
    b) What is the mass of 1 mole of each element?
    c) How many moles of each element combine?
    d) What is the simplest ratio?
    e) What is the formula of sodium oxide?
                                [4]   (AQA)

15. Joseph Priestley discovered oxygen gas in 1774 by heating mercury oxide.
    A teacher repeated the experiment and found that 2.17 g of mercury oxide decomposed on heating to produce 0.16 g of oxygen.
    Calculate the formula of mercury oxide.
    (Show all your working.)
    (Relative atomic masses: Hg (mercury) = 201, O = 16)   [3]   (AQA)

16. Most of the nitric acid that is manufactured is used to make fertilisers such as ammonium nitrate.
    The equation for the reaction is:
    $$HNO_3 + NH_3 \longrightarrow NH_4NO_3$$
    a) Calculate the relative formula masses for nitric acid (HNO_3) and ammonium nitrate (NH_4NO_3).
       (H = 1, N = 14, O = 16).   [2]
    b) Calculate the mass of nitric acid needed to make 400 tonnes of ammonium nitrate.   [2]   (OCR)

17. What mass of chlorine would be needed to make 73 tonnes of hydrogen chloride?
    $$H_2 + Cl_2 \longrightarrow 2HCl$$
    The relative atomic mass (A_r) of chlorine (Cl) is 35.5.
    The relative atomic mass (A_r) of hydrogen (H) is 1.0.   [3]   (OCR)

Further questions on Synthesis and Analysis

18. Aluminium oxide is electrolysed to extract aluminium. Oxygen gas is produced at the carbon anode which is then burned away as described in the following equation:
$$C + O_2 \longrightarrow CO_2$$
Given the following relative atomic masses:
Al = 27   C = 12   O = 16.
For each 32 tonnes of oxygen used calculate:
a) the mass of carbon burned and
b) the mass of carbon dioxide produced.
Show your working. (1 tonne = 1000 kg) *[2]*

c) Calculate the mass of aluminium produced from 204 tonnes of aluminium oxide given the equation,
$$2Al_2O_3 \longrightarrow 4Al + 3O_2$$
Show your working.   *[2]*   (WJEC)

19. A student heated strongly a known mass of powdered titanium so that all the titanium combined with oxygen from the air. The following results were obtained.
Mass of container + titanium   = 35.9 g
Mass of empty container   = 31.1 g
Mass of container + contents after
heating = 39.1 g

a) What would be a suitable container for the student to use?
b) Calculate the mass of titanium and of oxygen combining in the experiment.
c) Use the table on pages 390 and 391 to write down the relative atomic mass of titanium and of oxygen.
d) Use your answers to parts b) and c) above to calculate the (empirical) formula of the titanium-oxygen compound. You must show your working.
   [8]   (EDEXCEL)

20. The symbol equation below shows the reduction of iron(III) oxide by carbon monoxide.
$$Fe_2O_3 + 3CO \longrightarrow 2Fe + 3CO_2$$
a) Calculate the formula mass of iron(III) oxide.
(Use the table on pages 390 and 391) *[2]*
b) Calculate the mass of iron that could be obtained from 32 000 tonnes of iron(III) oxide.   *[3]*   (AQA)

21. This question is about two compounds of silver.
Silver tarnishes because it reacts with hydrogen sulfide in the air to form silver sulfide.
A sample of silver sulfide contains 10.8 g of silver and 1.6 g of sulfur.
Calculate the formula of silver sulfide.
You must show all your working to gain full marks.
(Relative atomic masses: S = 32; Ag = 108)
   [4]   (AQA)

22. a) Calculate the relative formula mass (M_r) of potassium chlorate, $KClO_3$.
(Relative atomic masses: O = 16;
Cl = 35.5; K = 39)   *[2]*
b) Calculate the percentage of **oxygen** in potassium chlorate.   *[2]*   (AQA)

23. The equation for the reaction of magnesium with hydrochloric acid is:
$$Mg(s) + 2HCl(aq) \longrightarrow MgCl_2(aq) + H_2(g)$$
Use pages 390 and 391 to calculate the relative formula masses (M_r) of:
a) $MgCl_2$ *[1]*
b) H_2 *[1]*

c) Calculate the volume of hydrogen (at 1 atmosphere and 25 °C) which could be obtained from 1 g of magnesium reacting with excess hydrochloric acid.
[The relative formula mass of hydrogen measured in grams occupies a volume of 24 litres (24 dm^3) at the above conditions.]   *[2]*   (WJEC)

24. What mass of MgO would be produced by burning 12 g of magnesium metal in excess oxygen?   *[2]*
Equation for reaction:
$$2Mg + O_2 \longrightarrow 2MgO$$
(Relative atomic masses: O = 16, Mg = 24)
   (WJEC)

Answers to Further questions on 'Moles' are on page 375.

▷ Analysis

25. Column A gives the names of some anions (negative ions) and some cations (positive ions). Column B gives the tests for these ions – but they are **not in the correct order**. Match the tests to the ions : *[7]*

Column A	Column B
Sodium ion	1. Add dilute hydrochloric acid – test for carbon dioxide with limewater
Calcium ion	2. Add dilute sodium hydroxide and warm – test for ammonia with damp red litmus paper
Copper ion	3. Carry out a flame test – look for a bright yellow flame
Ammonium ion	4. Add barium chloride solution – look for a white precipitate
Chloride ion	5. Carry out a flame test – look for a blue/green flame
Sulfate ion	6. Add silver nitrate solution – look for a white precipitate
Carbonate ion	7. Carry out a flame test – look for a red flame

26. Chemical tests can be used to identify compounds. The table below shows the results of some tests carried out on three solutions, **A**, **B** and **C**. Use the information in the table to identify solutions **A**, **B** and **C**. Give the name of:
a) solution **A** *[2]*
b) solution **B** *[2]*
c) the metal ion in solution **C**. *[1]*

27. A student was asked to perform some tests on a substance labelled **A**. The results of these tests are given below.

- Substance **A** is a white solid which gives a lilac flame test.
- Substance **A** reacts with dilute acid **B** to produce gas **C** and a colourless solution **D**.
- Gas **C** turns limewater cloudy.
- Dilute nitric acid followed by silver nitrate solution was added to solution **D**. A white precipitate **E** was produced.

a) Identify:
 i) gas **C** *[1]*
 ii) compound **A** *[2]*
 iii) the negative ion present in solution **D** *[1]*
 iv) acid **B** *[1]*
 v) precipitate **E** *[1]*
b) Briefly describe how you could perform a flame test on substance **A**. *[3]* (AQA)

28. A student carries out the following reactions to identify a green solid A.

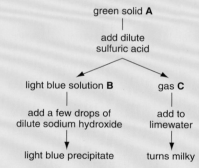

a) Name solid A, solution B and gas C. *[3]*
b) Write a balanced equation for the reaction of solution B with sodium hydroxide. *[3]* (OCR)

Solution	Flame test	Hydrochloric acid is added	Sodium hydroxide solution is added	Silver nitrate solution is added
A	Yellow	Carbon dioxide gas produced		
B	Brick-red		White precipitate insoluble in excess sodium hydroxide solution	White precipitate
C			Dark green precipitate	

Answers to 'Moles' Chapter 26

Pages 340 and 341
1. 2 moles   2. 3 moles   3. 10 moles
4. 0.1 mole   5. 0.01 mole
6. 2 g   7. 70 g   8. 320 g   9. 9.5 g
10. 0.12 g
11. 2 moles   12. 10 moles   13. 0.1 mole
14. 0.01 mole   15. 0.2 mole

Pages 342 and 343
1. 1 mole   2. 0.25 mole   3. 0.1 mole
4. 0.005 mole   5. 0.002 mole
6. $72 \, dm^3$ ($72\,000 \, cm^3$)
7. $120 \, dm^3$ ($120\,000 \, cm^3$)
8. $2.4 \, dm^3$ ($2400 \, cm^3$)
9. $0.024 \, dm^3$ ($24 cm^3$)
10. $0.12 \, dm^3$ ($120 \, cm^3$)
11. $48 \, dm^3$ ($48\,000 \, cm^3$)
12. $12 \, dm^3$ ($12\,000 \, cm^3$)
13. $1.2 \, dm^3$ ($1200 \, cm^3$)
14. $0.012 \, dm^3$ ($12 \, cm^3$)
15. $4.8 \, dm^3$ ($4800 \, cm^3$)

Page 345
1. 2 moles   2. 1 mole   3. 0.25 mole
4. 0.05 mole   5. 0.025 mole
6. 29.25 g   7. 11.2 g   8. 2.45 g
9. 12.41 g   10. 0.429 g

Page 346
1. CH_4 2. PbO   3. CO_2 4. Fe_2O_3
5. $CuCO_3$

Page 348 and 349
1. Na = 39.3 %, Cl = 60.7 %
2. C = 42.9 %, O = 57.1 %
3. Al = 52.9 %, O = 47.1 %
4. C = 7.8 %, Cl = 92.2 %
5. Mg = 16.2 %, N = 18.9 %, O = 64.9 %
6. PH_3 7. CH_4 8. $LiNO_2$

Page 351
1. $Mg + 2HCl \longrightarrow MgCl_2 + H_2$
   Answer = **19 g**
2. $Na_2CO_3 + H_2SO_4 \longrightarrow Na_2SO_4 + CO_2 + H_2O$
   Answer = **7.1 g**

Page 355
1. $1.25 \, mol/dm^3$ (or 1.25 M) potassium hydroxide solution
2. Take $10 \, cm^3$ of the $1.0 \, mol/dm^3$ solution and add $90 \, cm^3$ of water to make up $100 \, cm^3$ of the $0.1 \, mol/dm^3$ solution, i.e. we dilute the solution by 10.

Pages 356 and 357
1. Charge (in coulombs) $= 2 \times 9650 = 19\,300 \, C$
   4.6 g of Na are given by $19\,300 \, C$
   1 g of Na is given by $\dfrac{19\,300}{4.6}$

   23 g (1 mole) of Na are given by

   $$\dfrac{19\,300}{4.6} \times 23 \, C = \mathbf{96\,500 \, C}$$

2. Charge $= 1 \times 1930 \, C$
   2.07 g of Pb are given by $1930 \, C$
   1 g of Pb is given by $\dfrac{1930}{2.07} \, C$

   207 g (1 mole) of Pb are given by

   $$\dfrac{1930}{2.07} \times 207 \, C = 193\,000 \, C$$

   Therefore the charge on the lead ion is **2 +**.

Page 358
1. CuO
2. 510 tonnes of ammonia.
3. 0.416 g of calcium carbonate
4. $2.4 \, dm^3$ ($2400 \, cm^3$) of chlorine gas

Page 359

Aunt Molly Replies

1. CH_4 – Don't worry, Ray.
   This gas is perfectly natural!
2. 112 tonnes of calcium oxide
   Quite a nest-egg Mrs Kiln!
3. 0.04 moles per dm^3 (0.04M)
   I'm afraid your concentration is weak, Sol.
   But at least the solution is clear!
4. $1.2 \, dm^3$ or $1200 \, dm^3$
   What a forgetful ding-a-ling!
5. 0.2 g (almost!)
   That's not a lot of copper, but it has obviously given you a shock!

Answers to question 7 in 'Synthesis' Chapter 27

Page 369
7. a) 0.4
   b) 1.2
   c) CH_3

Answers to Further questions on Moles
(in Further Questions on Synthesis and Analysis)

Page 371

7. **B**

8. (1), (2) and (3)

9. **C**

10. **B**

11. **A**

12. 5 moles

13. XY_2

14. a) 2.3 g of Na, 0.8 g of O
    b) Na = 23 g, O = 16 g
    c) 0.1 mole of Na with 0.05 mole of O
    d) 2 : 1
    e) Na_2O

15. HgO

16. a) $HNO_3 = 63$; $NH_4NO_3 = 80$
    b) 315 tonnes

17. 71 tonnes

Page 372

18. a) 12 tonnes
    b) 44 tonnes
    c) 108 tonnes

19. a) a crucible
    b) Ti = 4.8 g, O = 3.2 g
    c) Ti = 48, O = 16
    d) TiO_2

20. a) 160
    b) 22 400 tonnes

21. Ag_2S

22. a) 122.5
    b) 39%

23. a) 95
    b) 2
    c) 1 litre (or 1 dm^3)

24. 20 g

Key Skills

As you study Science or Chemistry, you will need
to use some general skills along the way.
These general learning skills are very important
whatever subjects you take or job you go on to do.

The Government has recognised just how important
the skills are by introducing a new qualification.
It is called the **Key Skills Qualification**.
There are 6 key skills:

Key skills are important in all jobs !

- **Communication**
- **Application of number**
- **Information Technology (IT)**
- **Working with others**
- **Problem solving**
- **Improving your own learning**

The first 3 of these key skills will be assessed by exams
and by evidence put together in a portfolio.
You can see what you have to do to get this qualification
in the criteria below.
You will probably be aiming for Level 2 at GCSE.
If you go for Level 2, you will cover the Level 1 criteria as well.

Communication

In this key skill you will be expected to:

- hold discussions
- give presentations
- read and summarise information
- write documents

You will do these as you go through your course,
and producing your coursework will help.
Look at the criteria below:

What you must do ...	Evidence
Contribute to a discussion.	Make clear, relevant contributions; Listen and respond to what others say; Help to move the discussion forward.
Give a short talk, using an image.	Speak clearly; Structure your talk; Use an image to make your main points clear.
Read and summarise information from two extended documents (which include at least one image).	Select and read relevant material; Identify accurately main points and lines of reasoning; Summarise information to suit your purpose.
Write two different types of document (one piece of writing should be an extended document and include at least one image).	Present information in an appropriate form; Use a structure and style of writing to suit your task; Make sure your text is legible and that spelling, punctuation and grammar are accurate, so that your meaning is clear.

Application of number

In this key skill you will be expected to:

- obtain and interpret information
- carry out calculations
- interpret and present the results of calculations

What you must do ...	Evidence
Interpret information from two different sources (including material containing a graph).	Choose how to obtain the information, selecting appropriate methods to get the results you need; Obtain the relevant information.
Carry out calculations to do with: a) amount and sizes b) scales and proportions c) handling statistics d) using formulas.	Carry out your calculations, clearly showing your methods and level of accuracy; Check your methods and correct any errors, and make sure that your results make sense.
Interpret the results of your calculations and present your findings. You must use at least one diagram, one chart (table) and one graph.	Select the best ways to present your findings; Present your findings clearly and describe your methods; Explain how the results of your calculations answer your enquiry.

Information Technology

In this key skill you will be expected to:

- use the Internet and CD ROMs to collect information
- use IT to produce documents to best effect

What you must do ...	Evidence
Search for and select information for two different purposes.	Identify the information you need and where to get it; Carry out effective searches; Select information that is relevant to your enquiry.
Explore and develop information, and derive new information, for two different purposes.	Enter and bring together information using formats, such as tables, that help development; Explore information (for example, by changing information in a spread sheet model); Develop information and derive new information (for example by modelling a process on computer).
Present combined information for two different purposes. This work must include at least one example of text, one example of images and one example of numbers.	Select and use appropriate layouts for presenting combined information in a consistent way (for example by use of margins, headings, borders, font size, etc.); Develop the presentation to suit your purpose and types of information; Make sure your work is accurate, clear and saved appropriately.

Doing your Coursework

Your coursework or practical assessment is important.
It is an important part of becoming good at Science. And it
counts as a significant part of your GCSE or IGCSE grade.
Doing good coursework can be one of the best ways to
boost your final grade.

'Practical skills are important'

The different examination boards vary in the ways in which
they assess your coursework. There is a short summary
lower down this page. Your teacher will give you full details of
what you need to do for your practical assessment. You can
also research the details yourself on the exam board's web-site,
see below.

Different examination boards allocate different percentages
of marks for coursework or practical assessment.
They may also call it by different names.

For example the following 5 examination boards use the
name "controlled assessment". This means that your
teacher will observe you while you do some practical work.
- AQA GCSE examination board: controlled assessment = 25%
- Edexcel GCSE examination board: controlled assessment = 25%
- OCR GCSE examination board: controlled assessment = 25%
- WJEC GCSE examination board: controlled assessment = 25%
- CCEA GCSE examination board: controlled assessment = 25%

Other (IGCSE) exam boards use different methods:
- Edexcel IGCSE examination board has an 'assessment of practical
  skills' within the written papers (see page 380) = 20%
- Cambridge Cam IGCSE examination board has 3 alternative ways:
  'Coursework' (assessed by your teacher); or a 'Practical Test'; or
  an 'Alternative to Practical' written paper (see page 382).
  You do only one of these alternatives = 20%

Your teacher will give you more details, or you can find
more details by visiting the correct web-site in the list below.
Some of these sites have revision materials to help you.

www.aqa.org.uk
www.edexcel.com
www.ocr.org.uk
www.wjec.co.uk
www.rewardinglearning.org.uk (CCEA)
www.cie.org.uk

► Helping you investigate !

Use the questions below to help you at each stage in your investigations.

PLANNING

Before you start to collect evidence, think about:

- What is the best way to tackle this particular problem?
- How can you make your tests as *fair* as possible?
- How can you make your tests *safe*?
- Can you *predict* what will happen?
- Can you explain your prediction if you make one?
- How many observations or measurements will you make?
- What *range of values* will give suitable results?
- Will you need to repeat results to make them more *reliable*?
- Which apparatus will you choose to get *accurate* results?
- Will *datalogging* equipment help you get greater accuracy?
- Can you use *other sources of information* to help you plan your tests?
- Should you *try out some tests* to check your ideas before writing down your final plan?
- How will you *record* your results clearly and accurately?

ANALYSING AND CONCLUDING

When you have collected your results, think about:

- Can your results be shown on a *bar-chart or a line-graph*? (Remember to use a line-of-best-fit, and to point out any unexpected (anomalous) results – check these if you have time.)
- Can you see a *pattern* in your results?
- Do your results support your prediction (if you have made one)? Explain why.
- Can you *explain* your results using the work you have done in Science?

EVALUATING YOUR METHOD AND DATA

When you have finished your conclusion, think about:

- Can you comment on any *improvements* you could make to:
  - the way you did your tests
  - the accuracy of your readings
  - the reliability of your results (would someone else get the same results if they were to repeat your tests again?).
- Are there any *results which do not fit in* with the general pattern?
- Are your results good enough to draw a *firm conclusion*? Do you have enough evidence to be sure? Explain why. How could you gain further evidence?

Oxidising
These substances provide oxygen which allows other materials to burn more fiercely.

Harmful
These substances are similar to toxic substances but less dangerous.

Highly flammable
These substances easily catch fire.

Corrosive
These substances attack and destroy living tissues, including eyes and skin.

Toxic
These substances can cause death. They may have their effects when swallowed or breathed in or absorbed through the skin.

Irritant
These substances are not corrosive but can cause reddening or blistering of the skin.

▷ Practical work

1. Damp litmus paper is used to test for some gases.

Gas	Damp blue litmus paper	Damp red litmus paper
ammonia	stays blue	turns blue
carbon dioxide	turns red	stays red
chlorine	turns white	turns white
hydrogen	stays blue	stays red
sulfur dioxide	turns red	stays red

A student is given five gas jars, labelled **P**, **Q**, **R**, **S** and **T**, each containing one of the gases in the table above. Each gas was tested with damp litmus paper.

The student was told to use the information in the table above to write a conclusion.
The results and conclusions are shown below.

Gas	Result	Conclusion
P	blue litmus turns red red litmus stays red	P must be carbon dioxide
Q	blue litmus turns white	Q has to be chlorine
R	blue litmus turns red red litmus stays red	R is sulphur dioxide
S	blue litmus stays blue red litmus turns blue	S can only be ammonia
T	blue litmus stays blue red litmus stays red	T must be hydrogen

a) Identify **two** gases for which the conclusions are **definitely** correct. [2]

b) Identify **two** gases for which the conclusions are **possibly** correct. [2]

(Total 4 marks)
(Edex IGCSE, 2008)

2. Hydrochloric acid reacts with solid calcium carbonate.
$$2HCl(aq) + CaCO_3(s) \rightarrow CaCl_2(aq) + H_2O(l) + CO_2(g)$$
Some students investigate the effect on the rate of the reaction of changing the temperature of the hydrochloric acid.

The method is:
- use a measuring cylinder to pour 50 cm^3 of dilute hydrochloric acid into a conical flask
- heat the acid to the required temperature
- place the flask on a balance
- add 10 g (an excess) of calcium carbonate chips to the flask
- time how long it takes for the mass to decrease by 1.00 g.

The experiment is repeated at different temperatures.

The table shows the students' results.

Temperature of acid (°C)	Time to lose 1.00 g (s)
22	93
35	68
46	65
57	40
65	33
78	26

a) i) On graph paper, draw a graph of these results. The x-axis should be labelled 'Temperature in °C' and span 10–80 °C and the y-axis should be labelled 'Time to lose 1.00 g in seconds' and span 20–100 seconds. [3]

ii) One of the points is anomalous. Circle this point on your graph. [1]

iii) The students did not make an error in reading the stopwatch. Suggest a possible cause of this anomalous result. [1]

b) i) Use your graph to find the time taken to lose 1.00 g at 30 °C and at 52 °C. [2]

ii) The rate of the reaction can be found using the equation:

$$\text{rate of reaction} = \frac{\text{mass lost}}{\text{time taken to lose this mass}}$$

Use this equation and your results from b)i) to find the rate of reaction at 30 °C and at 52 °C in g/s. [2]

iii) How does the rate of reaction change when the temperature increases? [1]

iv) Give an explanation for this change in terms of particles and collisions. [3]

c) One student suggests that the results would

be more accurate if they insulate the conical flask before adding the calcium carbonate. Explain how insulating the conical flask would make the results more accurate. [2]

d) The students did not obtain any results at temperatures below room temperature, 22 °C. Describe how the method could be changed to obtain results below room temperature. [1]

(Total 16 marks)
(Edex IGCSE, 2007)

3. Air contains about 20% oxygen. When a fuel burns in air it reacts with the oxygen.

A student investigates the length of time a candle burns when it is covered by an upturned beaker. The diagram shows the apparatus she uses.

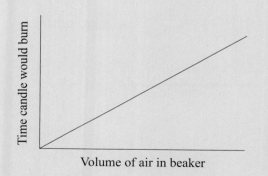

She repeats the experiment using different sizes of beaker.

a) Before she started the experiment the student sketched a graph to show how she thought the length of time the candle would burn would depend on the volume of air in the beaker.

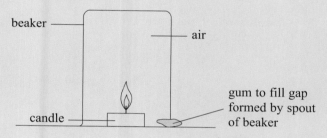

i) Describe the relationship shown in her sketch graph. [2]

ii) Suggest why she thought the graph was this shape. [2]

iii) Why is it important to seal the spout of the beaker with gum? [1]

The results the student obtains are shown in the table.

Beaker	Volume of air in beaker (cm³)	Time for which candle burned (s)			Mean time (s)
		Run 1	Run 2	Run 3	
A	240	14	16	18	16
B	460	27	28	29	28
C	700	59	61	66	62
D	1020	68	69	73	70
E	1250	95	96	91	

b) Suggest a method the student could use to measure the volume of the beaker used in each run accurately. [1]

c) Which beaker (**A**, **B**, **C**, **D** or **E**) has results which are the **most** reliable? Explain your answer. [2]

d) Calculate the mean time for beaker **E**. [1]

e) i) Draw a graph of the mean time for which the candle burned against volume of air in the beaker. [4]

ii) One of the points on your graph is anomalous. Circle this point. [1]

iii) Suggest and explain what may have happened in the experiment to produce this anomalous point. [2]

iv) The student was not sure whether or not the graph line went through (0,0). What further practical work should she do to help her decide? [1]

f) Another student repeats the experiment using pure oxygen in place of air. She finds the candle burns for about five times longer than when air is used.

i) Explain why the candle burns about five times longer in pure oxygen than in air. [1]

ii) Use your graph to help you calculate how long a candle would burn in a beaker containing 600 cm³ of oxygen. You must show your working. [2]

(Total 20 marks)
(Edex IGCSE, 2007)

Further questions on practical work

4. Hydrogen peroxide breaks down to form oxygen.

The volume of oxygen given off can be measured using the apparatus below.

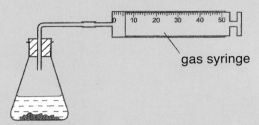

gas syringe

Solids **W** and **X** both catalyse the breakdown of hydrogen peroxide. The syringe diagrams show the volume of oxygen formed every 20 seconds using these catalysts at 25 °C.

time/s	using catalyst W	using catalyst X
0		
20		
40		
60		
80		
100		

a) Use the gas syringe diagrams to copy and complete the table below.

time/s	volume of oxygen/cm^3	
	using catalyst W	using catalyst X
0		
20		
40		
60		
80		
100		

[3]

b) Plot a graph to show each set of results. Use time/s as the x-axis. Clearly label the curves. [6]

c) Which solid is the better catalyst in this reaction? Give a reason for your choice. [2]

d) Why is the final volume of oxygen the same in each experiment? [1]

e) Sketch a line on your graph to show the shape of the graph you would expect if the reaction with catalyst **X** was repeated at 40 °C. [2]

(Total 14 marks)
(Cam IGCSE)

5. Diesel is a liquid fuel obtained from crude oil. Biodiesel is a fuel made from oil obtained from the seeds of plants such as sunflowers.

Using the apparatus below, plan an experiment to investigate which of these two fuels produces more energy.

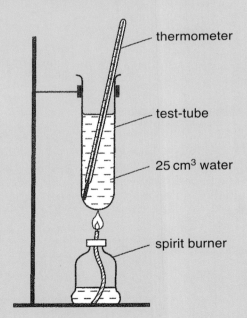

thermometer

test-tube

25 cm^3 water

spirit burner

[6]
(Total 6 marks)
(Cam IGCSE)

Suggestions for a revision programme

1. Read the summary at the end of the chapter to gain some idea of the contents.
Then read through the chapter looking at particular points in more detail, before reading the summary again.

2. Covering up the summary, check yourself against the fill-in-the-missing-word sentences at the end of the chapter.

   Remember to **re-read** the summary and to **review** each chapter after the correct revision intervals of 10 minutes, then 1 day, then 1 week (as explained on page 385). Continue in this way with all the other chapters in the book.

3. While reading through the summaries and chapters like this, it is useful to collect together all the statements in red and yellow boxes that you need to know. At this stage, you will also find the checklists and revision quizzes from the Support Pack useful.

4. While you are going right through the book, attempt the questions on the 'Further Questions' pages for every chapter. These are GCSE questions plus new questions written for the latest specifications. Your teacher will be able to tell you which are the most important ones for your specification.
You can check against your own particular specification on **www.chemistryforyou.co.uk**

5. Read the section on 'Examination technique' on page 386, and check the dates of your exams. Have you enough time to complete your revision before then ?

6. A few weeks before the examination, ask your teacher for copies of the examination papers from previous years. These 'past papers' will help you to see :
   – the particular style and timing of your examination
   – the way the questions are asked, and the amount of detail needed
   – which topics and questions are asked most often and which suit you personally.

When doing these past papers, try to get used to doing the questions **in the specified time**.

It may be possible for your teacher to read out to you the reports of examiners who have marked these papers in previous years.

The Support Pack contains some extra past paper questions with guidance given on how to answer them. It also contains the answers to all the past paper 'Further questions' included in this book.

Revision techniques

Why should you revise?

You cannot expect to remember all the Chemistry that you have studied unless you revise. It is important to review all your course, so that you can answer the examination questions.

Where should you revise?

In a quiet room (perhaps a bedroom), with a table and a clock. The room should be comfortably warm and brightly lighted. A reading lamp on the table helps you to concentrate on your work and reduces eye-strain.

When should you revise?

Start your revision early each evening, before your brain gets tired.

How should you revise?

If you sit down to revise without thinking of a definite finishing time, you will find that your learning efficiency falls lower and lower and lower.

If you sit down to revise, saying to yourself that you will definitely stop work after 2 hours, then your learning efficiency falls at the beginning but *rises towards the end* as your brain realises it is coming to the end of the session (see the first graph).

We can use this U-shaped curve to help us work more efficiently by splitting a 2 hour session into 4 shorter sessions, each of about 25 minutes with a short, *planned* break between them.
The breaks *must* be planned beforehand so that the graph rises near the end of each short session.
The coloured area on the graph shows how much you gain:

For example, if you start your revision at 6.00 p.m., you should look at your clock or watch and say to yourself, 'I will work until 6.25 p.m. and then stop – not earlier and not later.'
At 6.25 p.m. you should leave the table for a relaxation break of 10 minutes (or less), returning by 6.35 p.m. when you should say to yourself, 'I will work until 7.00 p.m. and then stop – not earlier and not later.'

Continuing in this way is more efficient *and* causes less strain on you.

You get through more work *and* you feel less tired.

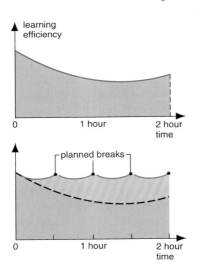

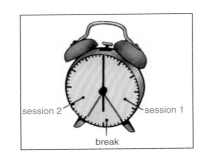

How often should you revise?

The diagram shows a graph of the amount of information that your memory can recall at different times after you have finished a revision session :

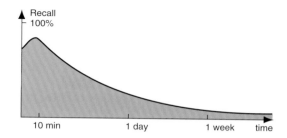

Surprisingly, the graph rises at the beginning. This is because your brain is still sorting out the information that you have been learning.
The graph soon falls rapidly so that after 1 day you may remember only about a quarter of what you had learned.

There are two ways of improving your recall and raising this graph.

● 1. If you briefly **revise the same work again after 10 minutes** (at the high point of the graph) then the graph falls much more slowly.
This fits in with your 10-minute break between revision sessions.
Using the example on the opposite page, when you return to your table at 6.35 p.m., the first thing you should do is **review**, briefly, the work you learned before 6.25 p.m.

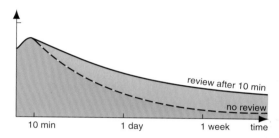

The graph can be lifted again by briefly reviewing the work **after 1 day** and then again **after 1 week**. That is, on Tuesday night you should look through the work you learned on Monday night and the work you learned on the previous Tuesday night, so that it is fixed quite firmly in your long-term memory.

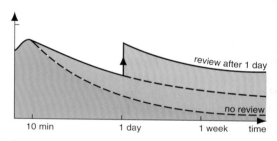

● 2. Another method of improving your memory is by taking care to try to **understand** all parts of your work. This makes all the graphs higher.
If you learn your work in a parrot-fashion (as you have to do with telephone numbers), all these graphs will be lower. On the occasions when you have to learn facts by heart, try to picture them as exaggerated, colourful images in your mind.

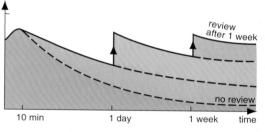

Remember: **the most important points about revision are that it must occur often and be repeated at the right intervals.**

		Mon	Tues	Wed	Thur	Fri
Chemistry		chap 3	chap 4	p25-9		
Physics	Science	p 24-9	chap 5	p67		
Biology		p36-42				
History		chap 3				

Examination technique

In the weeks before the examinations :

Attempt as many 'past papers' as you can so that you get used to the style of the questions and the timing of them.

Note which topics occur most often and revise them thoroughly, using the techniques explained on previous pages.

Just before the examinations :

Collect together the equipment you will need :
- Two pens, in case one dries up.
- At least one sharpened pencil for drawing diagrams.
- A rubber and a ruler for diagrams.
  Diagrams usually look best if they are drawn in pencil and labelled in ink.
  Coloured pencils are usually **not** necessary (but may sometimes make part of your diagram clearer).
- A watch for pacing yourself during the examination. The clock in the examination room may be difficult to see.
- **A calculator (with good batteries).**

It will help if you have previously collected all the information about the length and style of the examination papers (for **all** your subjects) as shown below :

Date, time and room	Subject, paper number and tier	Length (hours)	Types of question : – structured ? – single word answers ? – longer answers ? – essays ?	Sections ?	Details of choice (if any)	Approximate time per page (minutes)
4th June 9.30 Hall	Science Paper 2 (Chemistry) Higher Tier	$1\frac{1}{2}$	Structured questions (with single-word answers and longer answers)	1	no choice	4–6 min.

In the examination room :

Read the front of the examination paper carefully. It gives you important information. How is your examination paper different from the one shown opposite?

Some hints on answering questions are given in the box below.

NATIONAL EXAMINING BOARD

Science : Chemistry
Higher Tier
4th June
9.30 a.m.

Time : 2 hours

Answer **all** the questions.

In calculations, show clearly how you work out your answer.

Calculators may be used.

Mark allocations are shown in the right-hand margin.

In what ways is your examination paper different from this one?

Answering 'structured' questions :

- Read the information at the start of each question carefully. Make sure you understand what the question is about, and what you are expected to do.

- Pace yourself with a watch so you don't run out of time. If you have spare time at the end, use it wisely.

- *How much detail do you need to give?*
  The question gives you clues :
  – Give short answers to questions which start :
    '**State**...' or '**List**...' or '**Name**...'.
  – Give longer answers if you are asked to '**Explain**...' or '**Describe**...' or asked '**Why does**..?'.

- Don't explain something just because you know how to! You only earn marks for doing exactly what the question asks.

- Look for the marks awarded for each part of the question. It is usually given in brackets, e.g. [2]. This tells you how many points the examiner is looking for in your answer.

- The number of lines of space is also a guide to how much you are expected to write.

- Always show the steps in your working out of calculations. This way, you can gain marks for the way you tackle the problem, even if your final answer is wrong.

- Try to write something for *every* part of each question.

- Follow the instructions given in the question. If it asks for one answer, give only one answer. Sometimes you are given a list of alternatives to choose from. If you include more answers than asked for, any wrong answers will cancel out your right ones!

You can find some structured questions in the Support Pack – together with answers and useful hints on examination technique.

If your exam includes 'multiple-choice' questions :

- Read the instructions very carefully.

- If there is a separate answer sheet, mark it exactly as you are instructed. Take care to mark your answer opposite the correct question number.

- Even if the answer looks obvious, you should look at all the alternatives before making a decision.

- If you do not know the correct answer and have to guess, then you can improve your chances by first eliminating as many wrong answers as possible.

- Ensure you give an answer to every question.

Careers using Chemistry

Have you thought what you want to do when you leave school?
Chemistry opens the door to many careers.
It is sometimes called 'the central Science', because it lies between Biology and Physics.
So if you would like to study these further, Chemistry is a great help.
But hopefully you will want to learn more about Chemistry because you enjoy it!
Here are some careers where further studies in Chemistry will be an advantage, if not a requirement:

What are you going to do when you leave school?

?

agricultural scientist
animal technician
art restorer
bacteriologist
biochemist
brewer
chemical engineer
chemist
civil engineer
Civil Service science officer
conservationist
cosmetics scientist
dental technician
dentist
doctor
electrical engineer
environmental health officer
food scientist
forensic scientist
geologist
horticulturalist
information scientist
journalist (science)
laboratory technician
lecturer in science
managers in industry
marine scientist
marketing
materials scientist
medical technician
metallurgist
patent lawyer
pathologist

personnel manager
pharmacist
photographer
pollution controller
production manager
quality controller
research scientist
safety officer
sales person
scientific archaeologist
sport scientist
teacher of science
technical support staff
technical writer
veterinary surgeon/assistant
waste disposal manager/scientist
water technologist
wine taster
zoologist

You need a good sense of taste and smell (and chemistry!) to explain about wines

There is a lot of chemistry in developing a photograph. (See page 254.)

a food scientist

a research scientist using an electron microscope

a biochemist

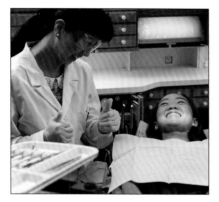

a dentist

an art restorer

an environmental scientist collecting water to test

a cosmetics scientist developing a perfume

a pharmacist

a brewer

Table 1   Physical properties of some elements

Element	Symbol	Atomic number	Relative atomic mass	State at 25°C	Melting point (°C)	Boiling point (°C)	Density (g/cm³) (gases at 25°C)
Aluminium	Al	13	27	s	660	2350	2.70
Argon	Ar	18	40	g	−189	−186	0.001 66
Arsenic	As	33	75	s	613 (sublimes)		5.78
Barium	Ba	56	137	s	710	1640	3.59
Beryllium	Be	4	9	s	1285	2470	1.85
Boron	B	5	11	s	2030	3700	2.47
Bromine	Br	35	80	l	−7	59	3.12
Caesium	Cs	55	133	s	29	669	1.88
Calcium	Ca	20	40	s	840	1490	1.53
Carbon (diamond)	C	6	12	s	3550	4827	3.53
Carbon (graphite)	C	6	12	s	3720 (sublimes)		2.25
Chlorine	Cl	17	35.5	g	−101	−34	0.002 99
Chromium	Cr	24	52	s	1860	2600	7.19
Cobalt	Co	27	59	s	1494	2900	8.80
Copper	Cu	29	64	s	1084	2580	8.93
Fluorine	F	9	19	g	−220	−188	0.001 58
Gallium	Ga	31	70	s	30	2070	5.91
Germanium	Ge	32	73	s	959	2850	5.32
Gold	Au	79	197	s	1064	2850	19.28
Helium	He	2	4	g	−270	−269	0.000 17
Hydrogen	H	1	1	g	−259	−253	0.000 08
Iodine	I	53	127	s	114	184	4.95
Iron	Fe	26	56	s	1540	2760	7.87
Krypton	Kr	36	84	g	−157	−153	0.003 46
Lead	Pb	82	207	s	327	1760	11.34
Lithium	Li	3	7	s	180	1360	0.53
Magnesium	Mg	12	24	s	650	1100	1.74
Manganese	Mn	25	55	s	1250	2120	7.47
Mercury	Hg	80	201	l	−39	357	13.55
Neon	Ne	10	20	g	−249	−246	0.000 84
Nickel	Ni	28	59	s	1455	2150	8.91
Nitrogen	N	7	14	g	−210	−196	0.001 17
Oxygen	O	8	16	g	−219	−183	0.001 33
Phosphorus (white)	P	15	31	s	44	280	1.82

Table 1 continued

Element	Symbol	Atomic number	Relative atomic mass	State at 25°C	Melting point (°C)	Boiling point (°C)	Density (g/cm³) (gases at 25°C)
Platinum	Pt	78	195	s	1772	3720	21.45
Potassium	K	19	39	s	63	777	0.86
Rubidium	Rb	37	85	s	39	705	1.53
Scandium	Sc	21	45	s	1540	2800	2.99
Selenium	Se	34	79	s	220	685	4.81
Silicon	Si	14	28	s	1410	2620	2.33
Silver	Ag	47	108	s	962	2160	10.50
Sodium	Na	11	23	s	98	900	0.97
Sulfur	S	16	32	s	115	445	1.96
Strontium	Sr	38	88	s	769	1384	2.6
Tin	Sn	50	119	s	232	2720	7.28
Titanium	Ti	22	48	s	1670	3300	4.51
Uranium	U	92	238	s	1135	4000	19.05
Vanadium	V	23	51	s	1920	3400	6.09
Xenon	Xe	54	131	g	−112	−108	0.005 5
Zinc	Zn	30	65	s	420	913	7.14

Table 2 Charges on some ions

Positive ions			Negative ions		
Charge	Name of ion	Formula	Charge	Name of ion	Formula
1+	ammonium copper(I) hydrogen lithium potassium silver sodium	NH_4^+ Cu^+ H^+ Li^+ K^+ Ag^+ Na^+	1−	bromide chloride hydroxide fluoride iodide nitrate	Br^- Cl^- OH^- F^- I^- NO_3^-
2+	barium calcium copper(II) iron(II) lead(II) magnesium nickel(II) strontium zinc	Ba^{2+} Ca^{2+} Cu^{2+} Fe^{2+} Pb^{2+} Mg^{2+} Ni^{2+} Sr^{2+} Zn^{2+}	2−	carbonate oxide sulfate(VI) sulfate(IV) sulfide	CO_3^{2-} O^{2-} SO_4^{2-} SO_3^{2-} S^{2-}
3+	aluminium iron(III)	Al^{3+} Fe^{3+}	3−	nitride phosphate	N^{3-} PO_4^{3-}

Periodic table of the elements

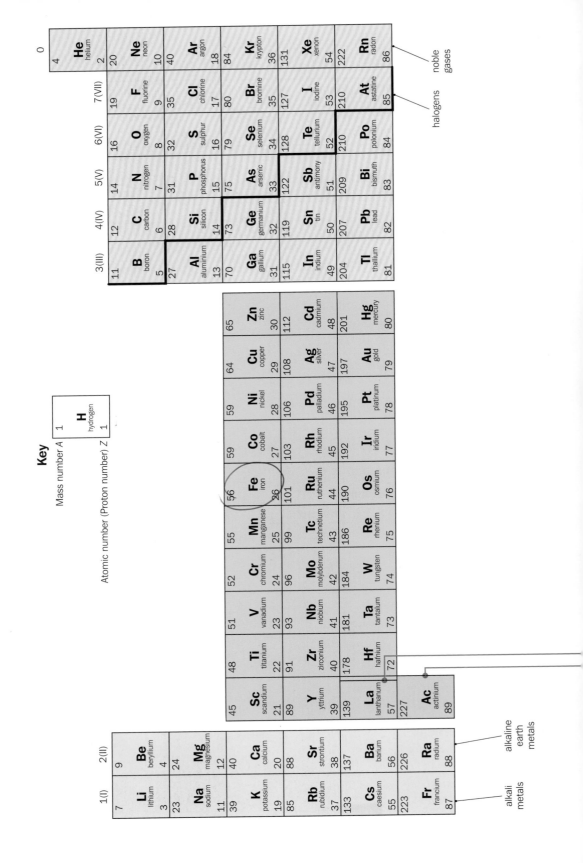

140	141	144	147	150	152	157	159	163	165	167	169	173	175
Ce cerium 58	**Pr** praseodymium 59	**Nd** neodymium 60	**Pm** promethium 61	**Sm** samarium 62	**Eu** europium 63	**Gd** gadolinium 64	**Tb** terbium 65	**Dy** dysprosium 66	**Ho** holmium 67	**Er** erbium 68	**Tm** thulium 69	**Yb** ytterbium 70	**Lu** lutetium 71
232	231	238	237	242	243	247	245	251	254	253	256	254	257
Th thorium 90	**Pa** protactinium 91	**U** uranium 92	**Np** neptunium 93	**Pu** plutonium 94	**Am** americium 95	**Cm** curium 96	**Bk** berkelium 97	**Cf** californium 98	**Es** einsteinium 99	**Fm** fermium 100	**Md** mendelevium 101	**No** nobelium 102	**Lr** lawrencium 103

(The mass numbers shown are those of the most common isotope)

'Ode to Dmitri'

The Periodic Table was a scientific break-through.
For chemistry made sense, it was easier too.
A Russian named Dmitri was first to spot the pattern,
But some elements were wrong in the spaces they sat in.
"I know," thought Dmitri, "I'll just leave some gaps."
And a stroke of genius had just come to pass.
Some of the elements were as yet undiscovered
So he made predictions from the properties of others.

A few years later when germanium was found,
Scientists agreed his ideas were sound.
Even now we use the Table on which we never dine,
Still based on that discovery in 1869.

Dmitri Mendeleev was the youngest of 17 children!
He started working on his Periodic Table while writing a text book. A copy of the Table was held over his coffin at his funeral. Element 101 was later named in honour of him (Mendelevium).

Index

Acknowledgements

I would like to thank the science staff and students (particularly Rebecca Henry, Tamara Ranatunga and Richard Secker) at Bramhall High School for their help in developing material for this book. I must also thank Geoff Bailey, John Bailey, Gareth Bell, Peter Borrows, Phil Bunyan, Sonia Clark, John Donneky, Roger Frost, Alan Goodwin, Janet Hawkins, Hilary Herrick, Sam Holyman, David Horrocks, Sandra Humphries, Alan Jackson, Sarah Jones, Steve Lund, John Mallott, Claire Penfold, Nina Ryan, Gordon Sutton, Malcolm Tomlin, Susannah Wills, Mike Wooster, John Hepburn, Janet Oswell, Tom Spicer, Adrian Wheaton and Gareth Williams.
Their contributions have been invaluable. Finally, my thanks go to Keith Johnson for his advice and insight into the way we learn and what makes a good textbook.

Acknowledgement is made to the following examining bodies for permission to reprint questions from their examination papers:

AQA	Assessment and Qualifications Alliance
EDEXCEL	Edexcel Foundation
OCR	Oxford, Cambridge and RSA Examinations
WJEC	Welsh Joint Education Committee
NI	Northern Ireland Council for the Curriculum, Examinations and Assessment
Cam IGCSE	University of Cambridge Local Examinations Syndicate (Cambridge IGCSE Chemistry Paper 6 May/June06 Q6; Paper 6 Oct/Nov07 Q7)
Edex IGCSE	Edexcel IGCSE Examinations

Photograph acknowledgements

Actionplus : 184B Richard Francis; Adams Picture Library: 109M, 305MR; AFP: 156B; Alamy: 287B Photofusion Picture Library, 315T Dinodia Images, 334B Alan Oliver; Alban: 297T; ALCAN: 90B; Ancient Art & Architecture Collection: 271; AP/Press Association Images: 49M, 98B, 179B; Arthur Bell & Sons: 169T; Associated Press: 285B Kasumi Kasahara; BananaStock (NT): 5T, 45; BASF UK: 232T, 232M; Bass Brewers: 286M, 328B; BOC: 301T, 305ML; Bruce Coleman: 128 C M Pampaloni; Cheshire County Council: 103; Copper Development Association: 96T Ampthill Metal Company; Corbis: 174B David Lees, 315BR Akhtar Soomro/EPA, 331B & 335T Ariel Skelley, 260B Patrick Robert/Sygma; Corbis (NT): 169B V94; Corel (NT): 57B C456, 58BR, 59MT, 114L C437, 142 C681, 150b C231, 150c BST02, 230 C603, 258T C541, 291 C39, 389ML C344; Corus: 99; Crown Copyright: 59B; Delices Napolean (a French company offering food and snacks in self-heating and cooling packagings): 197T; Digital Vision (NT): 5B DV6, 55T DV5, 55B DV15, 58MR DV12, 59T DVBP, 126 DVLC Gerry Ellis/Michael Durham, 150a DV11, 159T DV11, 183B DVJA, 200B DV6, 270 DVC James Lauritz, 285T DV7, 303ML DV6; Ecoscene: 288 Nick Hawkes, 289T; Eye Ubiquitous: 315BL J Dakers; Flexon: 272M; Fotolia: 41, 47, 49, 54TL, 54TR, 72L, 76L, 77, 78T, 79T, 79BL, 79BR, 90M, 91B, 109B, 112, 113TL, 113TR, 113B, 114TR, 114M, 115B, 116T, 117B, 133, 140, 141T, 148M, 150e, 150f, 150h, 157, 159B, 164T, 164B, 167T, 167B, 172T, 174M, 177B, 180T, 181T, 181MT, 183T, 185B, 194T, 203BL, 206, 214, 216M, 226T, 226B, 254T, 256, 258B, 259, 260T, 272T, 273T, 273M, 273B, 285M, 286B, 294, 303T, 306B, 328T, 329, 332, 337B, 388R, 389TR, 389BC; Geoscience Features Photolibrary: 97, 240, 311T, 311M, 311B, 317; Getty Images: 287T; Getty Stone: 129T A Husmo; Getty Telegraph Colour Library: 72M D James, 89 P Tweedie, 106B P Boulat/Cosmos, 389TC M Fielding; IBM: 336T; ICI: 102, 150d, 168M, 239T; Illustrated London News (NT): 58BL; Image (NT): 48L; Impact Photos: 389BR C Bluntzer; iStockphoto: 6, 25, 44, 46, 76R, 78ML, 78MR, 91M, 98T, 106T, 110, 124B, 125, 141B, 144, 148T, 154, 158, 166B, 169MT, 171, 172M, 174T, 177T, 181B, 184T, 185T, 203T, 203BR, 210, 239B, 262, 289B, 296T, 303MR, 305TR, 318B, 320, 331T, 337T, 362, 388L; J Allen Cash: 78B, 216T; John Bailey: 117T; Johnson Matthey: 48R, 148B; Koolpack.co.uk: 197MT; KP: 301B; London Fire Service: 305B; Magnum Pictures: 296B Chris Steele-Perkins; Martyn Chillmaid: 8, 37, 54B, 86, 94, 109, 115T, 124T, 127, 129B, 147T, 162T, 168T, 170T, 170M, 170B, 176T, 176M, 178, 179T, 180B, 181MB, 193, 197MB, 212, 216B, 220, 236, 248, 252, 254B, 261T, 292T, 293, 305TL, 331M, 333, 344, 355a-d, 363a-d; Mary Rose Trust: 160; NASA: 306T; Nick Cobbing: 14; Perkins Elmer: 364M; PhotoDisc (NT): 90T PD31, 114BR PD76, 136M PD76, 139 PD17, 150g PD71, 155 PD51, 161 PD31, 389TL PD40; Photofusion Picture Library: 169MB; Photographers Library: 197B; Photri International: 234; Popperfoto: 156T; Public Domain: 57T; Robert Harding Picture Library: 260MT; Sanosil: 297B; SCIENCE PHOTOLIBRARY: 7T, 116M, 147B, 176B, 183M, 224, 228T, 228B, 232B, 253, 269T, 286T, 330, 335B, 341, 357, 393; 7B ANDREW LAMBERT PHOTOGRAPHY, 334T CHRIS PRIEST & MARK CLARKE, 116B & 336B COLIN CUTHBERT, 389MR COLIN CUTHBERT/NEWCASTLE UNIVERSITY, 314 DAVID PARKER, 318T DAVID WEINTRAUB, 364B DEPT. OF PHYSICS, IMPERIAL COLLEGE, 366 DR JURGEN SCRIBA, 192 ERICH SCHREMPP, 263TMB GEOFF TOMPKINSON, 132 GUSTOIMAGES, 301M HANK MORGAN, 172B HARVEY PINCIS, 272B HEINI SCHNEEBELI, 364T JAMES HOLMES/CELLTECH, 319B JAMES KING-HOLMES, 319T JEREMY BISHOP, 166T JIM VARNEY, 58T, 58MC JOHN GREIM, 338 JOHN WATNEY, 269B MANFRED KAGE, 72B MARK A SCHNEIDER, 389MC MAURO FERMARIELLO, 58ML MEHAU KULYK, 107 NASA, 131 PASCAL GOETGHELUCK, 15 PHILIPPE PLAILLY/EURELIOS, 389BL PHILIPPE PSAILA, 136T RUSS MUNN, 136B SCOTT CAMAZINE, 20 & 292B SHEILA TERRY, 261B SINCLAIR STAMMERS, 116B image provided by Stone Circle Countertops; 260MB SUSAN LEAVINES, 200T VOLKER STEGER, 59MB WILL & DENI MCINTYRE; topfoto.co.uk: 49T uppa.co.uk, 168B John Maier Jr/The Image Works; 228M; Win Health Ltd: 162B; Zeneca Bioproducts: 162M.